THE HORSEMEN OF ATHENS

PUBLISHED FOR THE
CENTER FOR HELLENIC STUDIES

GLENN RICHARD BUGH

The Horsemen of Athens

PRINCETON UNIVERSITY PRESS
PRINCETON, NEW JERSEY

Published by Princeton University Press, 41 William Street,
Princeton, New Jersey 08540
In the United Kingdom: Princeton University Press,
Guildford, Surrey

Library of Congress Cataloging-in-Publication Data

Bugh, Glenn Richard, 1948-
The horsemen of Athens / Glenn Richard Bugh.
p. cm.
Bibliography: p.
Includes index.
ISBN 0-691-05530-0 (alk. paper) :
1. Cavalry—Greece—Athens—History. 2. Athens (Greece)—
History, Military. I. Title.
U33.B84 1988
357'.1'09385—dc19 88-3210

Publication of this book has been aided by the
Whitney Darrow Fund of Princeton University Press

This book has been composed in Linotron Sabon

Clothbound editions of Princeton University Press books
are printed on acid-free paper, and binding materials are
chosen for strength and durability. Paperbacks, although
satisfactory for personal collections, are not usually
suitable for library rebinding

Printed in the United States of America by
Princeton University Press,
Princeton, New Jersey

CONTENTS

LIST OF ILLUSTRATIONS

(following page 262)

ACKNOWLEDGMENTS

EVERY BOOK is a collective effort in spirit and substance; mine is no exception. I owe special thanks to my mentors and friends, Achilles Avraamides, who inspired my love for ancient history, and Ken Holum, who guided this study through its dissertation stage. During my year at the American School of Classical Studies at Athens, 1976–77, I received much good advice and support from Professors M. Wallace, who has to accept the credit (or blame) for suggesting a new study of the Athenian horsemen, W. K. Pritchett, A. Raubitschek, C. Edmonson, and E. Vanderpool, who graciously shared some of his personal correspondence dealing with the *grammateis* of the Athenian cavalry. And I am grateful for the warm hospitality shown me by Dr. D. Peppas Delmouzou, Director of the Epigraphical Museum in Athens, and her able staff during visits in the spring of 1977 and the summers of 1981 and 1985, and in this country by Professor Ch. Habicht, Director of the Epigraphical Library at the Institute for Advanced Study at Princeton. In the past three years, Professors S. V. Tracy, J. H. Kroll, and M. H. Hansen have generously clarified a number of technical points.

The research for this book was funded by the Department of History and the Small Project Grant program at Virginia Polytechnic Institute and State University; a summer stipend from the American Philosophical Society, 1981, for work on Lemnos and in Attika; and a Junior Fellowship at the Center for Hellenic Studies, Washington, D.C., 1983–84, where I compiled the Hellenistic material and wrote much of the book. I am deeply indebted to its two directors, first Professor Bernard Knox for providing such an ideal working environment, and then Professor Zeph Stewart for his acute editorial suggestions. Of course, I accept full responsibility for any remaining flaws.

For the process of "computerizing" the manuscript and then patiently answering my endless x-editing questions, I am indebted to our secretaries, Patty Mills, Debbie Rhea, Teresa Phipps, and Laura Hall. For help on the index, my thanks to Russ Friedman.

No one has been more supportive over these long years than my wife, Suzanne. To her I lovingly dedicate this book.

Blacksburg, Virginia
May, 1987

INTRODUCTION

First of all, let me state what this book is not. It is not a study of the various equestrian competitions dominated by the upper classes of Athens; fortuitously, a very recent book by D. G. Kyle (*Athletics in Ancient Athens*, 1987) covers this ground, at least for the Classical period. And it is not really a study of battlefield tactics or campaign strategies, although some discussion appears where appropriate, nor is it a study of military equipment; very fine books by Anderson, Pritchett, and others have dealt with these matters at great length. My interests lie elsewhere, in people, institutions, and society.

This book is a history of the Athenian cavalry from the Archaic period through the Hellenistic age. Within it is a story about the rise and fall of the Athenian cavalry. Imbedded into the narrative is a detailed administrative and prosopographical analysis of a military and social organization whose members came predominantly from the upper classes of Athens, whose outlook, consequently, was fundamentally aristocratic, and whose loyalty to the Athenian democracy could and did fluctuate with the changing fortunes of Athens in the Greek world of the Classical and Hellenistic periods. This is a study of the changing images of the cavalry within Athenian society. What will become apparent is that Athenian attitudes towards the cavalry and cavalry service were not static in these periods. In the larger sense, this book addresses the issue of the uneasy relationship between aristocratic and democratic ideologies. Certainly, to speak of the Athenian cavalry is not merely to define a military institution, but a social class with political expectations. To understand the dynamics of the Athenian cavalry and its personnel is to grasp a little more firmly the unique character of the Athenian democracy.

A new study of the Athenian cavalry is clearly warranted. The last major study devoted exclusively to the subject appeared in 1886 (M. A. Martin, *Les cavaliers athéniens*). Moreover, this book appeared before Kenyon's 1891 publication of Aristotle's *Athenaion Politeia*, which provides critical information on the administration of the Athenian cavalry in the fourth century B.C. This work has elicited a good deal of debate over the years. For example, Aristotle's authorship, the chronological development of the Athenian democracy in the early chapters, and even the static analysis of Athenian institutions in the fourth century have been criticized, and at times simply labeled as propaganda or historically inaccurate. In the following chapters I myself will have occasion to argue that certain sections are chronologically ambiguous, most notably for the Athenian cavalry at 49.1–2, but the work, whether by Aristotle or his school, remains the starting place for all serious discussion of the administration of the Athenian cavalry. Each piece of information must be evaluated on its own merits, on its own historical reasonableness, and the onus of proof that it cannot have happened as described rests with the critic. This same principle applies to other slippery sources, particularly the fourth-century Attic orators whose aims are to persuade, not necessarily to tell the truth, and to the lexicographers of the Roman and Byzantine periods.

The small monograph by M. W. Helbig (*Les hippeis athéniens*, 1902) was really only a response to Martin's conclusions concerning the date of the establishment of a genuine Athenian cavalry corps. Furthermore, Helbig's interpretations of the evidence from vase paintings have now been convincingly discredited. More recently, two very useful works have appeared: J. K. Anderson, *Ancient Greek Horsemanship*, 1961, and P.A.L. Greenhalgh, *Early Greek Warfare: Horsemen and Chariots in the Homeric and Archaic Ages*, 1971, but they address themselves generally to the issue of cavalry and horsemanship, with Athens being but a subset of the overall topic.

The need for a new study of the Athenian cavalry is even more clearly demonstrable by the discoveries from the excavations in Athens. Since the 1930s, the American excavations of the Athenian Agora have yielded important epigraphical evidence directly relating to the Athenian cavalry. And new material has appeared within the last three years. The finds from the Agora are especially instructive in the areas of cavalry personnel and administration. In 1971 a deposit of thin lead tablets recording the names of the horsemen and technical information about their mounts was discovered. This cavalry archive duplicates in part a deposit unearthed in 1965 by the German archaeologists in the area of the ancient cemetery of Athens. These tablets, published by K. Braun, *AM* 85 (1970) and J. H. Kroll, *Hesperia* 46 (1977), dating to the fourth and third centuries B.C., have opened a world of prosopographical possibilities. This new data now makes it possible to "flesh out" at least some of the Athenian horsemen in the Classical and Hellenistic periods.

I have endeavored to incorporate all the new evidence into the larger narrative study. In the course of my research, I have examined all the relevant epigraphical material to which I had access during several sojourns in Greece. The end product is a fresh look at the horsemen of ancient Athens.

SOME EXPLANATION on the use of the Greek and the abbreviations. Every writer on Classical topics strives for consistency of format, but it is inevitable that idiosyncratic variations and the strong tradition of Latinizing and/or Anglicizing Greek words makes this goal unachievable. My hope is that the inconsistency will not distract or confuse the reader.

In the main I have followed the sigla for journal titles found in *L'année philologique*, with a few exceptions such as *TAPA* for *TAPhA* (*Transactions of the American Philological Association*), *AJP* for *AJPh* (*American Journal of Philology*), *HSCP* for *HSPh* (*Harvard Studies in Classical Philology*), and more radically, *AM* for *MDAI(A)* (*Mitteilungen des deutschen archäologischen Instituts. Athenische Abteilung*). Similarly, for Classical authors and their works, I have most often turned to the standard Latinized abbreviations found in Liddell-Scott-Jones, *Greek-English Lexicon*, but recommendations from the press and my own preferences have resulted in variants, e.g., Aristophanes' comedies with English titles; Xenophon and his works, for which I have adhered more closely to the Greek titles than to their Latin counterparts; and the fourth-century Attic orators whose titles I have expanded into English in the text, but for which I have retained the *Lexicon* usage in their abbreviated footnote and text-inserted forms, with the exceptions of Xen. for X. (Xenophon) and Dem. for D. (Demosthenes).

The reader will quickly discern that I have Anglicized certain well-known Classical writers such as the historians Herodotus and Thucydides, and the philosophers Socrates, Plato, and Aristotle, but for most others I have preserved their Greek forms. I have transliterated the Greek throughout, even at those infrequent times it occurs in modern book and journal titles, e.g., Helbig, *Hippeis* and Threpsiades-Vanderpool, "Pros tois Hermais." Some features of the transliterations: circumflexes appear over those "e's" and "o's" whose quantity is long in Greek except for the familiar words *strategos* and *demos* and for proper nouns. Finally, although the Greek alphabet has no letter "c," I could not bring myself to transliterate the letter "chi" as "kh," rather than its Latinized form, "ch," as in *promachoi*.

For those readers unfamiliar with prosopographical terminology, an Athenian male citizen could be fully identified in the Classical and Hellenistic periods by his given name (*nomen*), the name of his father (patronymic), and the name of his deme (one of about 140 parishes in Athens and Attica). All too commonly, unfortunately, one or two of these identifying elements has not been preserved. In addition, a horseman's tribal affiliation may be known but not his deme; this occurs most frequently in the epigraphical evidence for the Hellenistic festivals.

ABBREVIATIONS

Books and Articles

Alföldi, "Reiterei"	A. Alföldi, "Die Herrschaft der Reiterei in Griechenland und Rom nach dem Sturz der Könige," *Gestalt und Geschichte, Festschrift K. Schefold* in *Antike Kunst,* Beiheft 4, Bern, 1967, 13–47
Anderson, *AGH*	J. K. Anderson, *Ancient Greek Horsemanship*, Berkeley, 1961
AP	Aristotle, *Athenaion Politeia*, ed. F. G. Kenyon (Oxford Classical Texts, 1920)
APF	J. K. Davies, *Athenian Propertied Families 600–300* B.C., Oxford, 1971
Beloch, *GG*[2]	K. J. Beloch, *Griechische Geschichte,* vols. I–IV, 2d ed., Berlin and Leipzig, 1912–27
Braun, "Funde"	K. Braun, "Der Dipylon-Brunnen B1: Die Funde," *AM* 85 (1970): 197–269
Busolt-Swoboda, *GS*	G. Busolt and H. Swoboda, *Griechische Staatskunde,* 2 vols., Munich, 1920–26
CAH	*Cambridge Ancient History*
Connor, *New Politicians*	W. R. Connor, *The New Politicians of Fifth-Century Athens*, Princeton, 1971
Dar.-Sag.	Ch. Daremberg and E. Saglio, eds., *Dictionnaire des antiquités grecques et romaines d'après les textes et les monuments*, Paris, 1877–1919
Ferguson, *HA*	W. S. Ferguson, *Hellenistic Athens,* London, 1911
FD	M.-G. Colin, *Fouilles de Delphes*, vol. III, fasc. 2: *Epigraphie, Inscriptions du Trésor des Athéniens*, Paris, 1907–13
FGrH	F. Jacoby, *Die Fragmente der griechischen Historiker,* I–, Berlin and Leiden, 1922–
Gomme et al., *HCT*	A. W. Gomme, A. Andrewes, K. J. Dover, *A Historical Commentary on Thucydides,* 5 vols., Oxford, 1945–81
Greenhalgh, *EGW*	P.A.L. Greenhalgh, *Early Greek Warfare: Horsemen and Chariots in the Homeric and Archaic Ages,* Cambridge, 1973
Habicht, *Studien*	Ch. Habicht, *Studien zur Geschichte Athens in hellenistischer Zeit*, Göttingen, 1982 (= *Hypomnemata* 73)
Habicht, *Untersuchungen*	Ch. Habicht, *Untersuchungen zur politischen Geschichte Athens im 3. Jahrhundert v. Chr.*, Munich, 1979 (= *Vestigia* 30)

Heinen, *Untersuchungen*	H. Heinen, *Untersuchungen zur hellenistischen Geschichte des 3. Jahrhunderts v. Chr.*, Wiesbaden, 1972; *Historia Einzelschriften*, 20
Helbig, *Hippeis*	W. Helbig, *Les hippeis athéniens*, Paris, 1902
Hignett, *HAC*	C. Hignett, *A History of the Athenian Constitution to the End of the Fifth Century* B.C., Oxford, 1952
Hölscher, *Historienbilder*	T. Hölscher, *Griechische Historienbilder des 5. und 4. Jahrhunderts v. Chr.*, Würzburg, 1973
ID	*Inscriptions de Délos*, Paris, 1926–72
IG	*Inscriptiones Graecae*, Berlin, a multivolume work covering the Greek world, certain volumes of which, notably vols. I–III for Attika, are in their second edition (1913–), with a third edition of vol. I now appearing in fascicles. These volumes and editions are noted by the appropriate Roman numeral and superscript, e.g., II^2, I^3.
Keil, *AA*	B. Keil, *Anonymus Argentinensis: Fragmente zur Geschichte des perikleischen Athens aus einem Strasburger Papyrus*, Strasburg, 1902
Kroll, "Archive"	J. H. Kroll, "An Archive of the Athenian Cavalry," *Hesperia* 46 (1977) 83–140
Kyle, *AAA*	D. G. Kyle, *Athletics in Ancient Athens*, Leiden, 1987 (= *Mnemosyne Suppl.* 95)
MacKendrick, *AthArist*	P. MacKendrick, *The Athenian Aristocracy 399 to 31* B.C., Cambridge, Mass., 1969 (Martin Classical Lectures, vol. XXIII, Oberlin)
Martin, *Cavaliers*	M. A. Martin, *Les cavaliers athéniens*, Paris, 1886
ML	R. Meiggs and D. Lewis, *A Selection of Greek Historical Inscriptions to the End of the Fifth Century* B.C., Oxford, 1969
NPA	J. Sundwall, *Nachträge zur Prosopographia Attica*, Helsinki, 1910
PA	J. Kirchner, *Prosopographia Attica*, 2 vols., Berlin, 1901–3
Pritchett, *GSW*	W. K. Pritchett, *The Greek State at War*, 4 vols., Berkeley, 1974–1985 (vol. 1 is a reprint of *Ancient Greek Military Practices*, Berkeley, 1971–University of California Classical Studies, no. 7)
Pritchett, *Studies*	W. K. Pritchett, *Studies in Ancient Greek Topography*, 4 vols., Berkeley, 1965–82
Raubitschek, *DAA*	A. Raubitschek, *Dedications from the Athenian Acropolis*, Cambridge, Mass., 1949
RE	A. Pauly, G. Wissowa, and D. Kroll, eds., *Real-Encyclopädie der klassischen Altertumswissenschaft*, Stuttgart, 1893–
Rhodes, *CAP*	P. J. Rhodes, *A Commentary on the Aristotelian Athenaion Politeia*, Oxford, 1981

SEG	*Supplementum Epigraphicum Graecum*
SIG^3	W. Dittenberger, *Sylloge Inscriptionum Graecarum*, 4 vols., 3d. ed., Leipzig, 1915–24
Snodgrass, *AAG*	A. M. Snodgrass, *Arms and Armour of the Greeks*, Ithaca, N.Y., 1967
Threpsiades-Vanderpool	J. Threpsiades and E. Vanderpool, "Pros tois Hermais," *Deltion* 18 (1963): 99–114
Tod, *GHI*	M. N. Tod, *A Selection of Greek Historical Inscriptions*, 2 vols., Oxford, 1933–48
Tracy, *First Fruits*	S. V. Tracy, *IG* II^2 *2336. Contributors of First Fruits for the Pythais*, Meisenheim am Glan, 1982 (*Beiträge zur klassischen Philologie*, 139)

Journals

AAA	Archaiologika Analekta ex Athenon
AC	L'Antiquité classique
AD	Archaiologikon Deltion
AE	Archaiologike Ephemeris
AHR	American Historical Review
AJA	American Journal of Archaeology
AJAH	American Journal of Ancient History
AJP	American Journal of Philology
AM	Mitteilungen des deutschen archäologischen Instituts. Athenische Abteilung
AncSoc	Ancient Society
JDAI	Jahrbuch des deutschen archäologischen Instituts
BCH	Bulletin de correspondance hellénique
C&M	Classica et Mediaevalia
CQ	Classical Quarterly
CR	Classical Review
CW	The Classical World
GRBS	Greek, Roman and Byzantine Studies
HSCP	Harvard Studies in Classical Philology
JHS	Journal of Hellenic Studies
JRS	Journal of Roman Studies
Num.Chron.	Numismatic Chronicle
REG	Revue des études grecques
RFIC	Rivista de filologia e di istruzione classica
SO	Symbolae Osloenses
TAPA	Transactions of the American Philological Association
YCS	Yale Classical Studies
ZPE	Zeitschrift für Papyrologie und Epigraphik

THE HORSEMEN OF ATHENS

CHAPTER ONE

Aristocratic Horsemen of Archaic Athens

> Marathon, 490 B.C. ". . . the Athenians were few in number and coming at a run without support of cavalry or bowmen."
> (Hdt. 6.112)

NO SINGLE piece of evidence can prove, incontestably, the existence of an Athenian cavalry before the mid-fifth century B.C. That there were *hippeis* cannot be denied, but that these "horsemen" belonged to a cavalry corps still eludes unanimous consent. A. Martin, in his monumental study,[1] attributed the creation of an Athenian cavalry corps to Solon in 594 B.C. W. Helbig replied[2] that, on the basis of the literary and ceramic evidence, an Athenian cavalry did not exist until after the Persian Wars. He argued that vase paintings depicting men on horseback in a military context represented one of three possibilities: (1) foreign cavalrymen, e.g., the famous horse-breeding Thessalians; (2) mounted hoplites, i.e., hoplites who rode their horses to the battlefield, but like the Homeric heroes with their chariots, dismounted and fought on foot; or (3) mounted squires accompanying the mounted hoplites.

Since 1904 scholars have routinely placed themselves in one camp or the other in this debate,[3] prompting A. Snodgrass to confess:

> Our ignorance of much of the military field should not be underestimated: there is still room for learned discussion, and complete disagreement over many fundamental questions, such as whether or not the Athenian 'cavalryman' of the 7th, 6th, and early 5th centuries actually rode his horse into battle.[4]

[1] *Cavaliers.*

[2] *Hippeis.*

[3] Concentrating on the ceramic evidence, Helbig himself followed the lead suggested by Ed. Meyer, *Forschungen zur alten Geschichte*, 2 vols. (Halle, 1892–99; repr. Hildesheim, 1962) 2: 184, n. 1. He has since gained the powerful support of Busolt-Swoboda, *GS* 2, p. 978, n. 3 and p. 1128; Snodgrass, *AAG*, p. 85; and J.A.O. Larsen, *Greek Federal States* (Oxford, 1968), 106–7, to name but a few. Lammert, on the other hand, s.v. *Hippeis*, *RE*, vol. 8, pt. 2 (1913), col. 1696, simply commented: "seine [i.e., Helbig's] Beweise sind indessen nicht durchschlagend."

[4] *AAG*, p. 77.

Recently, P.A.L. Greenhalgh, reexamining all the relevant ceramic material,[5] has proposed a compromise: the mounted hoplite and the cavalryman were but alternative forms of service in the sixth-century Athenian army, thus both existed at the same time.

What are we to make of all this? A survey of the relevant evidence will, at least, allow us to choose with more confidence. Certainty will probably never be attainable. For my part, I am persuaded that an Athenian cavalry did in fact exist in the Archaic period.[6]

It remains to prove it. Because of the difficult nature of the evidence, I have chosen to divide the chapter into four distinct, but ultimately interrelated, sections, each of which addresses the question of the existence and nature of the Athenian cavalry: (a) all the relevant literary testimonia, excluding the problematic Solonian census classes which will be discussed at length in the third section; (b) the ceramic evidence of the sixth and early fifth centuries B.C.; (c) Solon's "cavalry" class and horse traditions in Attika; and (d) various political, military, and societal considerations which militated against a *major* Athenian cavalry force in this period.

Testimonia

The most explicit statement supporting the position that Athens had a true cavalry in the Archaic period comes from the lexicographer Pollux.[7] Under the lemma *naucraria*,[8] he writes that each naucrary provided two horsemen (*duo hippeas*) and a ship (*naun*), from which it gets its name. Since there were forty-eight naucraries in Archaic Ath-

[5] *EGW*, pp. 111–36. Greenhalgh was unable to make use of T.B.L. Webster, *Potter and Patron in Classical Athens* (London, 1973), especially his chapter entitled "Horsemen and Chariots," a catalog of the pertinent ceramic evidence. Webster comments that although the type of warfare in which the heavily armed man dismounted to fight and left his horse with his squire was out of date by the early sixth century, it perhaps continued as a "method of training after the invention of the cavalry phalanx. Certainly the tradition persists in vases" (p. 181).

[6] In recent years, there has been a proliferation of general surveys on Archaic Greece. Most of them cover much the same ground, but one has been particularly informative, A. Snodgrass, *Archaic Greece: The Age of Experiment* (London, 1980).

[7] *Onomastikon* 8.108.

[8] On naucraries, see Hignett, *HAC*, p. 68; more recent bibliography and discussion can be found in Rhodes, *CAP*, pp. 151–53; *CAH*[3] III.3 (1982): 368–69; and now V. Gabrielson, "The Naukrariai and the Athenian Navy," *C&M* 36 (1985): 21–51; C. J. Haas, "Athenian Naval Power before Themistocles," *Historia* 34 (1985): 29–46; and S. D. Lambert, *Historia* 35 (1986): 105–12. For translation of relevant documents, see Ch. Fornara, *Archaic Times to the End of the Peloponnesian War*, 2d ed. (Cambridge, 1983), no. 22.

ens, the cavalry force would have numbered ninety-six. But because the information linking two horsemen to each naucrary is cited only here, we find ourselves in the unenviable position of accepting the testimony of Pollux without benefit of supporting evidence. Busolt and Swoboda considered the information "of very dubious worth."[9] Nevertheless, the testimony of Pollux should not be rejected out of hand.

An obvious question follows: what type of force? Mounted hoplites or true cavalry? Alföldi argued for cavalry,[10] Greenhalgh for mounted hoplites;[11] Anderson remained noncommittal.[12]

In my own mind, there is no reason to believe that Pollux thought of these horsemen as anything but true cavalry, whatever the source of his information may have intended. In any event, Anderson (p. 130) is probably correct to suggest that these ninety-six *hippeis* constituted a border or coastal patrol. To counter pirate raids or rapid sorties across the frontier border, mobility was essential.

The Atthidographer Kleidemos[13] says that Kleisthenes replaced the four tribes by ten[14] and divided them into fifty parts which were called naucraries. Martin, writing before the discovery of the *Athenaion Politeia* attributed to Aristotle, supposed that the Athenian cavalry was correspondingly increased to 100 men in the Kleisthenic reforms.[15] The *Athenaion Politeia* (21.5), however, clearly states that the naucraries were replaced by demes, whose number is attested in the literature at 100 (Hdt. 5.69) or at 174 (Strabo 9.1.16 = 396 C).[16] In fact, we hear no more of the naucraries after Kleisthenes.

[9] *GS* 2, p. 824, n. 1, following Keil, *AA*, p. 140; this judgment is shared by Gabrielson, *C&M* 36 (1985): 44–47.

[10] "Reiterei," p. 13.

[11] *EGW*, pp. 76–77.

[12] *AGH*, p. 130.

[13] *FGrH* 323 F 8.

[14] The reference in Hdt. 5.69 to tribal commanders (*phylarchs*) now numbering ten instead of four has been used by Ch. Fornara, *The Athenian Board of Generals from 501 to 404* B.C. (Wiesbaden, 1971), 1, to prove the existence of tribal cavalry commanders, as was the case in the later Athenian cavalry. But there is no reason to assume that they were anything but civil magistrates at the time of Kleisthenes. An analogous use might be the "phylarchs" appointed by the five Athenian ephors to oversee the tribes after the disaster at Aigospotami in 405 B.C. (Lys. 12.43–44); there is no indication that they are to be identified with the tribal cavalry commanders for that year.

[15] *Cavaliers*, pp. 94–95, 124. Martin even suggested that in the fifth and fourth centuries B.C., when the Athenian cavalry numbered 1,000, each of these 50 naucraries continued to furnish 20 horsemen, with each tribe furnishing 100 (p. 98).

[16] See now J. Traill, *Political Organization of Attica* (Princeton, 1975; *Hesperia Sup-*

In the end, we must acknowledge that there is no evidence to support the claim for an Athenian cavalry numbering 100 connected with post-Kleisthenic naucraries. Even if we accept Pollux on the naucraric-cavalry connection, we cannot say how Kleisthenes might have integrated the cavalry quota into the deme organization.[17]

Herodotus (1.63) offers a tantalizing piece of information connected with the victory of the Athenian aristocrat Peisistratos at Pallene in 546 B.C.[18] In this battle Peisistratos succeeded in defeating the Athenian forces with the help of mercenaries, notably the *hippeis* of Eretria (*AP* 15.2–3), and thereby secured his return from exile. While the Athenian army was in retreat, Peisistratos ordered his sons, "to mount (their) horses," overtake the Athenians, and persuade them to return quietly to their homes. How and Wells[19] repeat Helbig's claim[20] that this passage shows that the sons of Peisistratos were serving as mounted hoplites, not as true cavalrymen. This is far from certain. One can easily imagine a situation where the sons have dismounted to congratulate and converse with their father—himself perhaps fighting as a hoplite—then remount in obedience to his command. Thus they may have fought as true cavalrymen. Furthermore, the fact that the Athenian force appears to be on foot in their retreat does not mean that there might not have been some Athenian cavalrymen present. In the end all that this passage confirms is that the sons of Peisistratos rode horses in the concluding stages of the battle at Pallene. We must not facilely suppose that all members of the wealthiest class in sixth-century Athens raised horses, but Peisistratos at least probably did, even if he is not explicitly identified as a *hippotrophos*, i.e., a man who raises horses.[21] He clearly had a penchant for "horsey" names,[22] wit-

---

plement XIV) who reduces the deme count to ca. 139; see now D. Whitehead, *The Demes of Attica 508/7–ca. 250 BC: A Political and Social Study* (Princeton, 1986), 19–21.

[17] Recently, a good deal of work has been devoted to the military organization of Athens in the Archaic period: H. van Effenterre, "Clisthène et les mesures de mobilisation," *REG* 84 (1976): 1–17; P. Siewert, *Die Trittyen Attikas und die Heeresreform des Kleisthenes* (Munich, 1982); and F. J. Frost, "The Athenian Military before Cleisthenes," *Historia* 33 (1984): 283–94. Each emphasizes the critical impact of Kleisthenes.

[18] See also *AP* 17.4; Polyainos 1.21.1; and J. H. Schreiner, "The Exile and Return of Peisistratus," *SO* 56 (1981): 13–17.

[19] *A Commentary on Herodotus*, 2 vols. (Oxford, 1928) 1:85.

[20] *Hippeis*, p. 191.

[21] For the family of Peisistratos, see *APF* 11793, IV–V. There Davies argues, however, that "there is no evidence that the family ever engaged in horse-breeding" (p. 454). See now Kyle, *AAA*, pp. 112–13, 158–59, Appendix B, A55.

[22] Who can forget the comic scene in Aristophanes' *Clouds* 63–64 where Strepsiades laments his wife's efforts to give their son a sufficiently "horsey" name, e.g., "Xanthippos," "Charippos," or "Kallippides." He finally conceded to "Pheidippides."

ness his sons Hippias, Hipparchos, and Thessalos, the last, a reference to the land famed for its horse traditions. Later, Hippias, succeeding his father as tyrant, will enlist the Thessalian cavalry as allies against the Spartans (Hdt. 5.63). And there is some evidence that Peisistratos married off one of his daughters to a certain Thrasyboulos, famed as a *hippotrophos*.[23]

In the early decades of the fifth century B.C. Persia twice invaded Greece. The detailed narrative of these wars by the historian Herodotus should have settled the question of whether or not Athens possessed a cavalry, but, unfortunately, that is not the case. In the second invasion (480 B.C.) Herodotus reports that the mighty Persian army under King Xerxes is descending upon southern Greece. The Athenians debate whether to stay and fight or to flee to safety before the Persians arrive in Attica. Into this debate Plutarch adds an anecdote involving Kimon, scion of the illustrious clan of the Philaidai. Reportedly, Kimon, probably about thirty years old at the time,[24] marched triumphantly with his companions through the Kerameikos, ascended the Acropolis, and dedicated his bridle to Athena, claiming that in the present crisis Athens needed sea-fighting men (*naumachôn andrôn*), not ones with equestrian expertise, *hippikês alkês* (*Kim.* 5.2–3). This anecdote assumes that Kimon served as a cavalryman, because the juxtaposition of horsemanship and naval service clearly implies a military contrast. Only a horseman is limited to land warfare, while a hoplite can fight as a marine (*epibatês*) on board a ship. According to Plutarch's account, Kimon thereupon borrowed a shield from the temple and marched down to the ships.

Kimon's interest in horses should not surprise us. He belonged to a family with a brilliant reputation for *hippotrophia* in the sixth century B.C. Kimon's great-uncle, Miltiades III, won the four-horse chariot race at Olympia[25] and his grandfather, Kimon I, won the same event at three consecutive Olympiads (536–528 B.C.).[26] This tradition continues in Kimon's own son, Lakedaimonios, who was a hipparch around the middle of the fifth century—incidentally, our first indispu-

[23] See *APF* 11793, p. 449.

[24] See *APF* 8429, p. 302.

[25] Hdt. 6.36.1; Paus. 6.10.8; see *APF* 8429, p. 299 and Kyle, *AAA*, pp. 111–12, 158 for discussion.

[26] Hdt. 6.103.1–2; see *APF* 8429, p. 300. In addition, H. R. Immerwahr, "Stesagoras II," *TAPA* 103 (1972): 181–86, has tentatively suggested that the chariot team depicted on a black-figured pyxis dated to 540–530 B.C. and inscribed with the name Stesagoras (identified as Stesagoras II, son of Kimon I) may be this winning team.

table reference to a true cavalry in Athens.[27] Epilykos, son of Nikostratos, who served as hipparch in 325 B.C., may also have belonged to the Philaidai.[28]

What then can we conclude from the Kimon incident? If the anecdote reflects a genuine historical event—and there is no compelling reason to dispute it—then it seems inescapable that Kimon had expected to serve his country as a cavalryman and was so doing in 480 B.C. But it is disturbing that supporting evidence for this conclusion is not forthcoming from Herodotus. In fact Herodotus will state directly, or imply by silence, that at no time during the Persian Wars (490–479) did the Athenians field a cavalry. Let us take a look at the evidence.

During the Persian invasion of 490, the exiled tyrant Hippias led the Persian army to the plain of Marathon.[29] The two armies faced each other for a number of days and then, under the generalship of Miltiades IV, the father of Kimon, the Athenians charged the enemy lines, almost a mile distant. This "Marathon" run by the Athenian hoplites has been recorded by Herodotus 6.112:

> The Persians, seeing them advancing at a run, prepared to receive them, though they thought the Athenians mad and bent on their own destruction, inasmuch as the Athenians were few in number and coming at a run without support of cavalry or bowmen.

How else can one interpret the Greek words *oute hippou* than that the Athenians had no cavalry force at Marathon?[30] This could simply

[27] *IG* I² 400; see Raubitschek, *DAA*, p. 146, no. 135 and *APF* 8429, pp. 305–6. It is interesting to note that one of his brothers was named Thettalos.

[28] M. Crosby, "Greek Inscriptions," *Hesperia* 6 (1937): 462, no. 10 and discussion, *APF* 8429, p. 312.

[29] For a good discussion of the campaign, see N.G.L. Hammond, *Studies in Greek History* (Oxford, 1973), 170–250, an expanded version of his article "The Campaign and Battle of Marathon" in *JHS* 88 (1968): 13–57. For topographical considerations, see Pritchett, *Studies* 1, pp. 83ff; E. Vanderpool, "The Deme of Marathon and the Herakleion," *AJA* 70 (1966): 319; and J.A.G. Van der Veer, "The Battle of Marathon: A Topographical Survey," *Mnemosyne* 35 (1982): 290–321.

[30] Attempts have been made to posit an Athenian cavalry fighting at Marathon. An inscription unearthed in the Athenian agora preserving a fragmentary epigram related to the Persian Wars, published by J. H. Oliver, in "Selected Greek Inscriptions," *Hesperia* 2 (1933): 480–494, no. 11 (see *SEG* x.404), was restored to include the phrase "on swift-footed colts" by Paul Maas, "Zu den Perserepigrammen," *Hermes* 70 (1935): 235–36. In the same journal (pp. 236–38), L. Wickert defended this restoration, arguing that because of the small size and tactical unimportance of the cavalry at Marathon, the memory of their presence disappeared completely within a few decades. There is little wrong with this argument, but the restoration has now been almost universally rejected in favor of "swift-faring ships," referring to the battle of Salamis, not that of Marathon;

mean that an Athenian cavalry did not exist at that time,[31] but two other possibilities bear serious consideration.

First of all, the Athenian cavalry may have been stationed elsewhere, perhaps in the plain of Phaleron near Athens in anticipation of a seaborne attack from that direction. There may be some support for this suggestion in the *Suda* s.v. *chôris hippeis*, a gloss which states that Datis and the Persian cavalry (perhaps only a part?) had withdrawn from Marathon before the decisive battle. Since the Persian fleet sailed to Phaleron immediately after the battle in hopes of attacking Athens before the victorious Athenian hoplites returned from Marathon (Hdt. 6.116), one might suppose that Datis and the horse-transports had preceded them there. According to this scenario, the Athenian cavalry would have been forced to hurry home to meet them *before* the battle of Marathon and therefore were not present. But there is no evidence for this.

Moreover, Herodotus 6.112 plainly implies that Persian bowmen and horsemen were present.[32] If not, how do we account for the incredulity of the Persians? Besides, Pausanias, writing in the second century A.D., speaks of neighing horses at Marathon (1.32.4) and stables of Artaphernes' horses (1.32.7). If we disregard the heroizing elements, Pausanias' account does confirm a strong tradition.[33] For what it is worth, Aelius Aristides' *Panathenaikos* 106–8 (Behr), also second century A.D., claims that the Athenians even captured some of the Persian horses. Furthermore, the exiled Athenian tyrant, Hippias, had guided the Persian army from Eretria on Euboia to the plain of Marathon not

---

for discussion of relevant bibliography, see ML 26, pp. 54–57 and now D. L. Page, *Further Greek Epigrams* (Cambridge, 1981), no. xx, pp. 219–25; for translation, Fornara, *Archaic Times*, no. 51.

[31] So Anderson, *AGH*, p. 130. He suggests that "after the expulsion of the tyrants, cavalry were evidently felt to be an unnecessary luxury."

[32] Recent interpretations of *chôris hippeis* seem to favor the view that they were present at Marathon, but that they were "quartered apart" *(chôris)* from the rest of the troops and that the surprise nature of the Athenian charge rendered them unprepared and ineffective: G. Shrimpton, "The Persian Cavalry at Marathon," *Phoenix* 34 (1980): 20–37 and H. Y. McCulloch, "Herodotus, Marathon, and Athens," *SO* 57 (1982): 35–55.

[33] The tradition of Persian cavalry at Marathon may also be supported by the depiction of Persians fighting by their ships, one grasping his horse, on the Roman sarcophagus at Brescia, see E. Vanderpool, "A Monument to the Battle of Marathon." *Hesperia* 35 (1961): 105, n. 26, pl. 35, and by the Persian cavalrymen shown fighting on the south frieze of the Nike temple in Athens, see E. G. Pemberton, "The East and West Friezes of the Temple of Athena Nike," *AJA* 76 (1972): 303 and E. Harrison, "The South Frieze of the Nike Temple and the Marathon Painting in the Painted Stoa," *AJA* 76 (1972): 353–78, pls. 73–78.

only because it was the nearest convenient disembarkation point on the coast of Attika, but because it was the most suitable region in Attica for cavalry maneuvers (*epitêdeotaton chôrion* tês Attikês enhippeusai: Hdt. 6.102). The plain of Phaleron, south of Athens, on the other hand, could hardly be so described. In 510 B.C. Hippias, then tyrant of Athens, had to remove obstructions in the plain—thereby making it "more suitable to cavalry" (*hippasimon*)—to enable his Thessalian cavalry allies to carry out their assault against Anchimolios and the Lakedaimonian camp (Hdt. 5.63).[34] The advantage for the Persian cavalry, therefore, lay at Marathon, not at Phaleron.

Secondly, the Athenians may have deliberately chosen not to field their small cavalry force because of the superiority of the Persian cavalry.[35] This suggestion deserves serious consideration. A cavalry numbering only ninety-six men would hardly have offered much resistance against a better trained and more numerous Persian cavalry. These Athenian cavalrymen would have been deployed more efficiently in the hoplite ranks.[36] There is absolutely no doubt that all those who could provide themselves with hoplite armor were expected to fight as hoplites if the occasion demanded. More to the point, every cavalryman was by the very presumption of wealth that that service entailed, a hoplite. When he left cavalry service, he automatically became a hoplite until the end of his normal military obligations (around sixty). Thus the richest members of Athenian society were in a sense always hoplites. Aristotle notes that hoplite service is appropriate to the "well-to-do" (*euporoi*) not to the "poor" (*aporoi*, *Pol.* 1321a12–14).[37]

This dual military role can be seen in the careers of a few well-known Athenians. Kimon may have been a *hippeus* in 480 B.C., but during his ostracism his companions fought bravely at Tanagra (458/7 B.C.) around his hoplite armor (Plut. *Kim.* 17.5; *Per.* 19.7). And Al-

[34] The Thessalian cavalry, numbering 1,000, was commanded by their king, Kineas. The Spartans subsequently sent Kleomenes with a second force which defeated the Thessalian cavalry: Hdt. 5.64; cf. *AP* 19.5.

[35] See Martin, *Cavaliers*, pp. 121–22; H. Delbrück, *History of the Art of War* (Westport, Conn., and London, 1975), 59–60, n. 4; Alföldi, "Reiterei," p. 13; and for Greece in general, Snodgrass, *AAG*, p. 89.

[36] Compare Xen. *Cyr.* 4.5.49 where Xenophon has Cyrus recommend that his Persian cavalry dismount and fight on foot if they are more serviceable as infantry—a tactic not uncommon in our American Civil War; on this last, see S. Z. Starr, *The Union Cavalry in the Civil War*, 3 vols. (Baton Rouge, La., 1979–85), 3:589–94 (summary).

[37] *AP* 26.1 comments that the many expeditions of the mid-fifth century B.C. had led to heavy casualties among respectable members of the people (*demos*) and the well-to-do (*euporoi*). Of course, Aristotle's use of *euporoi* included members of what we term today the middle class.

kibiades, young lion of the famous Athenian clan of the Alkmeonidai, fought as a hoplite at Poteidaia in 431 (Pl. *Ap.* 28 E–Plut. *Alk.* 7.6, *Smp.* 221 A). His son and namesake was registered as a hoplite in 395, but his illegal attempts to pass himself off as a mounted archer presuppose at least a passable degree of equestrian training, though apparently not cavalry service ([Lys] 15.5–8). Lysias 14.10 also remarks that some legitimate cavalrymen had been ordered to serve as hoplites in this campaign. One, at least, volunteered to do exactly that. Mantitheos, a young aristocratic cavalryman, had been selected to serve by his cavalry officer. Mantitheos successfully petitioned his commander to strike his name from the cavalry register (*katalogos*) in order to serve as a hoplite; he apparently was still a hoplite in the campaigns of the following year (Lys. 16.13-17). And during the troubled year of 404/3, the Athenian cavalry patrolled as horsemen by day, while at night they guarded the city walls as hoplites (Xen. *Hell.* 2.4.24).

That the Athenians elected not to deploy their small cavalry force can likewise explain its apparent absence during the second Persian campaign in 480 B.C. The numbers of Persian and allied cavalry, admittedly exaggerated,[38] do confirm an enormous superiority in that arm. The superiority lay not only in sheer numbers. Aeschylus[39] proclaims the international reputation of the Persians in the use of the horse and bow: "masters of the bow and horsemen, fearful to behold, terrible in battle, steadfast in spirit."[40] It is no wonder that the Greeks were elated when on one celebrated occasion they beat back the attacks of the Persian horse trying to recover the body of their fallen commander, Masistios, and even forced them to retreat (Hdt. 9.25).

It may, then, have been by deliberate design that the Athenians deployed no force of horsemen to meet the mighty host of Asia. Throughout the Plataian campaign of 479, when the Persian cavalry repeatedly harassed the Greek allies, not one mention is made of an opposing cavalry.[41] The mounted messenger (*spheôn hippea*) whom the Athe-

[38] Hdt. 7.184 records 80,000 horsemen; to these must be added the cavalry contributions of the Greeks who had defected, e.g., Macedonians, Thessalians, and Boiotians (Hdt. 9.31–32).

[39] *Pers.* 26–28. Cf. Aristides *Panathenaikos* 274 (Behr).

[40] A. E. Wardman, "Tactics and the Tradition of the Persian Wars," *Historia* 8 (1959): 49–60, has discussed the importance of the various types of arms used by the combatants in the Persian Wars on the basis of the *Persai* and the *History* of Herodotus. Cf. Snodgrass, *AAG*, p. 57: Persian Wars symbolized as spear (hoplite) against the bow.

[41] Pritchett, *GSW* 2, p. 224 suggests that the 300 *logades* mentioned at Hdt. 9.21.3 were cavalrymen. This finds support only in the disputed reference to Masistios being killed by an Athenian horseman in Paus. 1.27.1 and is contradicted by the account in Herodotus. For a list of Greek forces, see Hdt. 9.28–30.

nians dispatched to receive instructions from Pausanias (Hdt. 9.54) is identified as a herald (*kêryx*) in 9.55–56. The same function as messenger can be claimed for the horseman sent from Pausanias to the Athenians in Herodotus 9.60 although this can certainly not be used as proof that the Spartans had a cavalry at this date—in fact, Thucydides 4.55.2 tells us that the Spartans did not organize a cavalry until 424 B.C. In view of the distances sometimes separating even allied armies, it would have been foolish not to have had some rapid means of communication. That these messengers, at least in the case of Athens, were drawn from horsemen normally expected to furnish cavalry service is quite possible. Thus Martin assumed that the cavalry, being too few to give the Persian cavalry a good fight, served as *ordonnances* (aides-de-camp), and thus as messengers.[42] Alföldi suggested that the *hippeus* in 9.54 must have been a member of the aristocratic cavalry corps (*ein Ritter*) because of the importance of the mission.[43] This does not necessarily follow, but the idea of cavalrymen being used as bearers of important information is reasonable.[44]

There is also the slight possibility that even if the Athenians had wanted to field a cavalry, they may not have had the horses to do it. When Mardonios, the Persian general, and his army approached the territory of Attika, the Athenians were forced to flee to Salamis, Troizen, and Aigina. Herodotus records that the Athenian fleet returned

[42] *Cavaliers*, pp. 121–22.

[43] "Reiterei," p. 30.

[44] Cf. Xen. *Hell.* 4.5.7 where a cavalryman brings the news to Agesilaos of the disaster of the Lakedaimonian *mora* at Lechaion ca. 390 B.C. Moreover, I have argued elsewhere ("The Athenian Cavalry in the Sicilian Expedition. Some Notes on Thucydides 6.43," in *Studia Pompeiana et Classica in Honor of Wilhelmina F. Jashemski*, ed. by R. I. Curtis, vol. 2, New Rochelle, N.Y., 1986) that the thirty Athenian cavalrymen sent on the first expedition to Sicily in 415 B.C. (Thuc. 6.43) were later dispatched as heralds to win over various Sicilian cities (Thuc. 6.48). Now, why the Athenians did not send a horseman instead of a runner to Athens to announce the news of the great victory at Marathon only serves to cast doubt on the historicity of the entire episode (not reported in Herodotus); for recent discussion, see F. J. Frost, "The Dubious Origins of the 'Marathon,' " *AJAH* 4 (1979): 159–63 with reply by E. Badian, pp. 163–66. It is curious, however, that the Pheidippides (or Philippides) sent by Athens to Sparta to solicit aid in the forthcoming battle at Marathon is specifically identified as a runner (*hêmerodromos*) by Herodotus (6.105–6). I suspect that over such a distance (approx. 150 miles) a professional runner would have had an easier time of it than one who had to worry about welfare and provisions not only for himself but for his horse. The Greeks had no postal relay system like that of the Persians. Also, a man would have been reluctant to injure his own horse on such a mission—the state did not provide horses for its cavalry at this date. For the skill and usefulness of *hêmerodromoi*, see Livy 31.24.4. The reader might be interested to know that ancient tradition is being kept alive in modern Greece: an annual ultra-marathon between Athens and Sparta has been recently introduced.

from Artemision to Athens to allow each Athenian to convey his family and household belongings to safety (8.40–41; 9.6). Perhaps the Athenians were forced also to abandon many of their horses out of haste and the difficulty of transporting them by sea.[45] This suggestion may find some support in the Kimonian anecdote.

An interesting aspect of the Plataian campaign is that we discover that Athens now possesses an archery force, apparently organized between 490 and 480 B.C.[46] A group of archers accompanied the 300 picked men of Athens who were dispatched to relieve the Megarians harassed by Masistios and the Persian horse (Hdt. 9.22). Masistios' horse is felled by an Athenian arrow.[47] And Aeschylus mentions Athenian bowmen annihilating a force of Persians on the island of Psyttaleia.[48]

In sum, the evidence from the Persian Wars is not decisive enough to reject the notion that the Athenians possessed a cavalry in the early decades of the fifth century B.C. The failure to detect it in the wars was a consequence of its modest size; the affair of Kimon may in fact serve to affirm this point. In the face of the redoubtable Persian cavalry, Athens wisely chose not to sacrifice its cavalrymen. Those who did not join the hoplite ranks were used as messengers.

More problematic is the reference in Plato's dialogue *Meno*. Socrates says that Themistokles had taught his son Kleophantos to be a good horseman and that this training included standing upright "on the horses" (*epi tôn hippôn*) and even hurling a javelin in that position "from the horses" (*apo tôn hippôn*) as well as many other marvelous feats (93 D). To judge from the keen interest shown in his son's proper equestrian education, it seems probable that Themistokles himself had been a *hippeus* in his youth. If we place the birth date of Themistokles at ca. 524 B.C.,[49] then he could have been riding as early as the last decade of the sixth century.

What interests us is whether the use of *hippea* by Plato[50] assumes the

[45] Thuc. 2.56.2 reports that horse-transports made out of old ships were used for the first time in the summer of 430 B.C. Unfortunately, this passage has been variously interpreted, see Gomme et al., *HCT*, p. 163; my Chapter Three.

[46] See A. Plassart, "Les archers d'Athès," *REG* 26 (1913): 151–213, esp. pp. 195–96.

[47] Cf. Aristides *Panathen.* 183 (Behr) and Paus. 1.27.1. For other references to Athenian bowmen in the Persian Wars: Hdt. 9.60; Plut. *Them.* 14; *FGrH* 688 Ktesias F 13a, sect.26.

[48] *Pers.* 460–62; cf. Hdt. 8.95: Ch. Fornara, "The Hoplite Achievement at Psyttaleia," *JHS* 86 (1966): 51–54, argues that Herodotus' account of this action cannot be trusted.

[49] *APF* 6669, pp. 214–15.

[50] And repeated by Plut. *Them.* 32.1 and Athenaios 12.45–533 D.

duties of a cavalryman. Equestrian prowess, however showy, was a perfectly natural preoccupation of an aristocrat interested in racing and hunting.[51] Besides, the use of the plural *hippôn* in conjunction with the "many other marvelous feats" recalls certain acrobatics described in Homer (*Il.* 15.679–83) where a man leaped on and off several horses to the amazement of those watching. Thus, what Plato describes in and of itself need not have anything to do with cavalry service.

Nevertheless, a young man so proficient in horsemanship would have been prime material for cavalry service. This may have been the case with Perikles' sons Paralos and Xanthippos, whom Socrates describes as "the foremost horsemen in Athens" (*Meno* 94 B). Since we can be quite certain that they were *hippeis* at a time when Athens did in fact possess a cavalry, i.e., the mid-fifth century B.C., it is quite possible that they served as cavalrymen. Considering the context of the passage and the proximity of the two examples (93 D and 94 B), we may suppose that Plato, at least, thought that Kleophantos had also served as a cavalryman—and his father before him. The same argument applies to Perikles as well.[52] Unfortunately, there is no corroborating evidence that any of the individuals mentioned above served in a cavalry force.

Ceramic Evidence

What seems to be secure, however, is that Athenian vases of the sixth and early fifth centuries B.C. do show true cavalrymen. It would be superfluous to detail this ceramic evidence since that task has been admirably performed by Webster[53] and Greenhalgh.[54] Greenhalgh's examination shows that representations of true cavalry are depicted as early as the 580s (on an Attic vase from the Acropolis, p. 111) and "steadily increase throughout the rest of the century . . . in particular the representations of cavalry in battle."[55] Since most of them were presumably painted for well-to-do Athenian patrons, we can assume

[51] So Anderson, *AGH*, p. 130, reads this passage. For evidence of fancy riding as early as the seventh century, see Alföldi, "Reiterei," p. 22.

[52] It is worth noting that Perikles' father, Xanthippos, had married into the powerful Alkmeonidai whose equestrian victories at the Panhellenic Games in the sixth century B.C. are well attested: *APF* 9688, p. 371 and Kyle, *AAA*, pp. 114, 157–58.

[53] *Potter & Patron*, pp. 179–95.

[54] *EGW*, pp. 96–145, esp. 111–36.

[55] *EGW*, p. 118, e.g., figs. 57 (ca. 550 B.C.), 58 (ca. 565 B.C.), 64 (ca. 550 B.C.), 65 (ca. 550–525 B.C.), 67 (ca. 525-500 B.C.), 68 (ca. 500 B.C.), 69 (late sixth century B.C.), and 70 (late sixth century B.C.).

that they depict scenes familiar to their patrons' military activities, even if heroic or exaggerated. One should not conclude, however, that the increasing number of vases depicting cavalry action reflects an increasing importance of the Athenian cavalry in actual warfare. Anderson cautions: "it is true that cavalrymen of various types frequently appear on Attic vases of the sixth century B.C., but the historical evidence does not suggest that they were important."[56]

Greenhalgh does seem to have demonstrated one major point, that both true cavalrymen and mounted hoplites existed contemporaneously, at least in vase paintings. In so doing he has effectively refuted both Helbig's thesis that there were no true cavalrymen in the sixth century and also Alföldi's belief that there were no horsemen at that time who were not true cavalry. Why is it so hard to accept the idea that some rich men rode their horses to battle and dismounted to join the hoplite ranks—a carryover from the Homeric tradition, albeit with chariots—while other rich men or youths rode horses to battle and fought as cavalrymen?[57]

Within the vast body of ceramic material dealing with horses and horsemen at least two vase paintings have been interpreted as depicting an administrative procedure associated with the Athenian cavalry, the so-called *dokimasia* or scrutiny. This *dokimasia* has been detailed by Aristotle in the *Athenaion Politeia* (49.1–2). There are in fact two *dokimasiai* described in this passage: (1) a scrutiny by the Council of Five Hundred of the horses of the existing cavalrymen to determine if they are being well cared for (49.1), and (2) a scrutiny by the same council of the new recruits to determine if they may be exempt from service because of bodily incapacity or lack of sufficient wealth (49.2). The second procedure in no way implies that the recruits already owned horses for cavalry service or presented them at this time.

The first cup, Berlin F 2296,[58] painted by the so-called Dokimasia Painter and dated to ca. 470 B.C.,[59] shows a defile of young men

[56] *AGH*, p. 130.

[57] So Greenhalgh, *EGW*, p. 151. For depictions of "squires" in service to young nobles, see G. Richter, *Archaic Gravestones of Attica* (London, 1961), nos. 20, 64, 70, and 71.

[58] First described in detail by G. Körte, "Dokimasie der attischen Reiterei," *Archaeologische Zeitung* 38 (1880): 177–81, pl. 15; see Adolf Greifenhagen, ed., *Corpus Vasorum Antiquorum*, (Berlin, 1962) vol. 2, pl. 75.

[59] J. D. Beazley, *Attic Red-figure Vase-Painters*, 2d ed. (Oxford, 1963), 412. 1, and H. Cahn, "Dokimasia," *Revue archéologique*, ser. 7 (1973): 10. Cahn (p. 10, no. 3) notes in astonishment the date of 500 B.C., albeit with a question mark following, attributed to it by P. J. Rhodes, *The Athenian Boule* (Oxford, 1972), 175. J. Boardman, *Athenian Red-figure Vases: The Archaic Period* (London: 1971), 137, claims that the Dokimasia

dressed in *chlamydes* (cloaks) and *petasoi* (wide-brimmed hats) and carrying two spears each (Figure 1). They escort their horses towards two individuals, one standing and one seated. The seated man appears to be jotting down something on a tablet resting in his lap. Körte interpreted these two men as a special examining committee of the council.[60] He suggested that the vase presented pictorially a *dokimasia* of cavalry recruits. Note that he wrote these words before the discovery of the *Athenaion Politeia.*

Strictly speaking, the Berlin cup ought to show the *dokimasia* of the cavalry mounts of existing cavalrymen (*AP* 49.1), not that of the recruits (*AP* 49.2). Another possibility exists: the vase depicts the form of *dokimasia* prior to the introduction of the system which Aristotle describes in the fourth century. That is to say, the young men may in fact be cavalry recruits presenting their horses (and themselves) for inspection—horses which they already owned.

The second red-figure cup, recently reconstructed from a number of fragments, can be found in the collection of Herbert Cahn (Bâle 133) and has been discussed by the owner (Figure 2).[61] This cup, belonging to the group of the Thalia Painter, is dated to the last decade of the sixth century B.C.[62] Like the Berlin cup, it shows a defile of young men, several of whom are wearing the Thracian hat (the *alopekis*), and others with the *petasoi* fastened behind their heads and clothed in *himatia* and *chitoniskoi.* They each carry two spears and appear to be leading their horses towards a standing man who is also wearing a *chitoniskos* along with a sash (*écharpe*) draped behind him extending to the backs of his knees. His apparel is quite different from that of the seated scribe in the Berlin cup, who wears a long *chiton* draped about his lower torso and legs. As a matter of fact, the scribe is dressed like the young horsemen. He wears not only the *chitoniskos*, but also the high boots with four leather strips hanging from the tops ("chaussé de souliers hauts à revers à quatre festons," Cahn, "Dokimasia," p. 5). Whereas the scribe in Berlin F 2296 has the appearance of a civil magistrate, the scribe in Bâle 133 may be a horseman. Perhaps he is the

Painter is still working in the 460s. Helbig, *Hippeis*, p. 79, was happy to accept a date post-479 B.C. for this cup because it accommodated his theory that no Athenian cavalry existed before the Persian Wars. With the publication of a second cup depicting the same theme and dated to ca. 510–500 B.C., Helbig's thesis was undermined (see below footnote 61).

[60] Körte, "Dokimasia," p. 178.

[61] Cahn, "Dokimasia," pp. 3–32.

[62] Ibid., pp. 8, 12. See Beazley, *Attic Red-figure Vase-Painters*, p. 1708 (Addenda III to p. 1626) and Beazley, *Paralipomena* (Oxford, 1971), 332.

commander of the troop. The scribe in Bâle 133 is jotting something down on a five-tablet book, a *pentaptychon*, which he holds in his left hand. In the space between the scribe and the first horse, Cahn read archaic Attic letters spelling in retrograde the name of the horse, *Melanthis*, "Black Flower" (p. 6). Elsewhere traces of letters prove that the other horses were also identified by name.

Cahn interpreted the scene as a representation of the *dokimasia* of the Athenian cavalry and identified the scribe as one of the *katalogeis* (enrollers) mentioned at *AP* 49.2. Like Körte (p. 181), he was struck by the highly unusual theme of the scenes: with only one possible exception, these two vases illustrate the only nonreligious official act of the state shown in Attic vase paintings.[63]

Their importance is not limited to this. Two vases depicting the same theme and bridging a forty-year period of great moment in Athenian history (ca. 510[64] to ca. 470 B.C.), casts serious doubt on Helbig's thesis that there was no Athenian cavalry until after the Persian Wars. Cahn concluded that these vases prove that "Athenian cavalrymen were chosen by a *dokimasia* in order to be enrolled (*pour s'enrôler*) in a regular troop (*une troupe régulière*) from before 500 B.C." ("Domikasia," p. 12).

We cannot accept this statement without qualification. First of all, Cahn's vase could also depict a type of *dokimasia*, e.g., a scrutiny or inspection, of existing cavalrymen, undertaken by military officials *at the time of a campaign*. Secondly, Cahn's identification of the scribe as a *katalogeus* is not as secure as he believes. Judging from this man's costume, he is more likely to be a member of the cavalry, presumably an officer. The *katalogeis*, in the fourth century at least, were civilian officials elected by popular vote. Why should they be attired in such a fashion? In fact, I have argued elsewhere that the board of *katalogeis* may have been a recent institution at the time when the *Athenaion Politeia* was composed (ca. 330s).[65] Finally, it cannot be the case that they represent "a regular troop," at least not in the way that Aristotle

[63] Cahn, "Dokimasia," p. 22. Cahn discusses other vase paintings depicting youths with horses which he thinks may be related to the *dokimasia* of the cavalry (pp. 15–22). What all these others lack, however, is an official with tablet in hand, so certain identification is impossible.

[64] This is not the place to resurrect the problem of the existence of the Solonian Council of Four Hundred (*AP* 8.4) even if the vase belongs to the period prior to the creation of the Kleisthenic Council of Five Hundred in 508/7. If the Council of Four Hundred did not exist, the scrutiny could have been carried out by military commanders or specially appointed officials.

[65] "Introduction of the *katalogeis* of the Athenian Cavalry," *TAPA* 112 (1982): 23–32; see Chapter Five.

and Xenophon have described the Athenian cavalry in the fourth century, that is to say, a cavalry for which the state spent forty talents each year to maintain on full-time call throughout the entire year.[66]

From a different perspective, it is even possible to argue that the *dokimasia* depicted in the two vase paintings discussed above do not involve an administrative procedure connected with cavalry service, but simply the preliminaries to a procession or festival in which men or youths on horses participate.

On the famous Parthenon frieze (440s), (admittedly a glorification of the Athenian cavalry), on the West side, Slab XII, two figures (22 and 23) stand facing each other with a horse between them (Figure 3). The figure on the left, a young man, nude save for a cloak, extends his right hand and index finger towards the horse's mane—which is differently coiffured from the other horses in the cavalcade—and his left hand is hidden behind his right forearm. The figure on the right, in front of his horse, appears dramatically posed, his right arm bent and stretched upward in a gesture of seeming protestation. It has been suggested that the man on the left "might be writing something down on a tablet, as in the *dokimasia* (inspection of young knights and their horses). Could he be booking the knight, whose gesture might be one of protest, for having his horse's mane wrongly dressed?"[67]

F. Brommer has disputed Robertson's sighting of a tablet, claiming that "von einem Täfelchen ist nichts zu sehen und der rechte Zeigefinger ist ausgestreckt."[68] Brommer is quite right. A tablet cannot be discerned, though it may be obscured by the right forearm. True, the index finger cannot be said to be in a writing position, but it is possible that the sculptor wanted to emphasize the exact cause of the problem to the horse's owner by an unmistakable pointing of his finger. The "marshal" may have just crossed him and his horse from the list of participants.

If the so-called "marshal" were engaged in this activity, perhaps we could interpret the *dokimasia* cups as an inspection for a religious procession involving youthful, aristocratic horsemen, but not necessarily cavalrymen. Is it just coincidence that the young horseman being examined in Bâle 133 has his left arm stretched out with his hand resting on his horse's mane? Along these lines, Bruno Keil, who did not believe in a state-organized cavalry before the mid-fifth century B.C., would have interpreted the horsemen on the vase paintings: "als Pa-

[66] Xen. *Hipparch* 1.19.

[67] M. Robertson and Alison Frantz, *The Parthenon Frieze* (New York, 1975), pl. 9 with notes.

[68] F. Brommer, *Der Parthenonfries*, 2 vols. (Mainz am Rhein, 1977) 2: 19.

rade-und Luxus-truppe, eine Vertretung des Ritterstandes, mochte ein Reitercontingent bei feierlichen Aufzügen und Wettrennen sich zeigen."[69] In any event, the "inspection" depicted on the Parthenon frieze has no direct connection with the *dokimasiai* described in *AP* 49.1–2.

Another point. The young men portrayed on the two vases carry spears, two to be exact. Is this emblematic of military service? We know that spears were used in certain athletic events, not only in the Panathenaia, but in other festivals as well.[70] On this point, Webster points out that it is not always easy to distinguish between cavalrymen, hunters, and jockeys. An early amphora of Panathenaic shape showing a rider with a spear near a tripod might be an example of the last.[71] Despite this ambiguity, Webster believes that "when a man has a chlamys and two spears, whether he leads two horses or one, I am inclined to regard him as a cavalryman leaving home for training or action" (p. 182). By this criterion the horsemen on the *dokimasia* cups should be identified as cavalrymen.

The preceding discussion has not been intended to reject the two vase paintings as evidence for Athenian cavalry prior to the Periklean age, only to indicate that the interpretation of their scenes is not invariable and that the "examinations" described in *AP* 49.1–2 may not be directly applicable. And of course to suggest that the scenes have to do with festival activity, not military service, does not preclude the young horsemen also being cavalrymen. On balance, I am willing to accept the vase paintings as proof of some sort of *dokimasia* associated with cavalry service.

A final note. H. Metzger and D. van Berchem have suggested that, since there was an annual procession in the fourth century B.C. of the Athenian cavalry in honor of Zeus (Plut. *Phok.* 37) and the eagle is mythologically associated with Zeus, the eagles which appear in certain black-figure vases flying above mounted hoplites, i.e., horsemen with armor and hoplite shields, should indicate symbolically "l'appartenance du cavalier à la classe des Hippeis."[72] Unfortunately, the

[69] *AA*, p. 141.

[70] For numerous examples of riders with spears, possibly associated with athletic events, see Webster, *Potter & Patron*, p. 182, nn. 1, 2; for inscriptions, the Panathenaia (*IG* II² 2311.II.68; fourth century B.C.); Theseia (*IG* II² 958.II.95; second century B.C.)—see H. W. Parke, *Festivals of the Athenians* (Ithaca, N.Y., 1977) 36, and the Pythais (*FD* III.2, nos. 37, 44; second century B.C.). For a list of Athenian festivals in which equestrian events were included, see M. B. Walbank, *Regulations for an Athenian Festival, Hesperia Supplement* XIX (Princeton, 1982), 176, n. 11.

[71] Webster, *Potter & Patron*, p. 181 and n. 7; description in Beazley, *Attic Black-Figure Vase-Painters* (Oxford, 1956), p. 86, no. 7.

[72] H. Metzger and D. van Berchem, "Hippeis," *Gestalt and Geschichte, Festschrift K.*

authors facilely accept Helbig's thesis and make the rather casual association of military service ("cavalier" = mounted hoplite) with the census class called "Hippeis" (*AP* 7.4). Although I am not persuaded by their central thesis, the issue they raise, the relationship between this Solonian census class and the Athenian cavalry, demands closer scrutiny.

Solon's Hippeis and Hippotrophia in Attika

Probably the most controversial, and potentially most important, evidence concerning the Athenian cavalry in the Archaic period involves the constitutions of the early Athenian lawgivers, Drakon and Solon. The evidence is located in the *Athenaion Politeia* at 4.2–4 and 7.3–4. What is at stake is the exact meaning of the second highest of the four census classes, that known as *hê hippas*, the members of that class as *hippeis* (horsemen).

Referring to the Drakontian constitution, *AP* 4.2–4 reports:

> Political power had been handed over to those who provided their own armour. They chose the nine Archons and the Treasurers from those men who had an unencumbered property qualification of not less than ten minae; the lesser magistrates were chosen from those who armed themselves, while the *strategoi* and the cavalry commanders had to show unencumbered property to the value of at least a hundred minae and legitimate children, by citizen wives, not less than ten years old. The Prytaneis had to receive sureties for them, as also for the *strategoi* and cavalry commanders of the previous year until after their *euthuna*, the sureties being four citizens from the same class as the *strategoi* and cavalry commanders. There was a Boule of 401 members, selected by lot from the citizen body. All those over thirty years old cast lots for this and the other offices, and nobody could hold the same office twice until all those eligible had held it; then the allotment started again from the beginning. If a member of the Boule failed to attend a sitting of the Boule or Ekklesia, he was fined three drachmae if he was a *pentakosiomedimnos*, two if he was a *hippeus* and one if he was a *zeugites*.[73]

Schefold in *Antike Kunst*, Beiheft 4 (Bern, 1967), 155–58, pls. 55–57. In addition, Prof. E. Pemberton informs me that eagles and other birds are regular features on black-figure vases.

[73] *AP* 4.2–4; translated by J. M. Moore, *Aristotle and Xenophon on Democracy* (Berkeley, 1975), 149.

Drakon's work as a lawgiver is usually dated to around 620 B.C. In this passage we see a military organization that already includes generals and hipparchs (cavalry commanders) with the obvious implication of an existing cavalry corps. Furthermore, we see evidence for the existence of census classes, one of which includes individuals called "horsemen." If we believed in the historicity of this passage, the question of a cavalry in the Archaic period would be laid to rest. But the whole passage is so suspect that it has been rightly labeled "an invention of some oligarchical pamphleteer who wrote in the last quarter of the fifth century."[74] For example, the provision that citizenship and eligibility for political office depend on one's ability to provide arms (*tois hopla parechomenois*) suspiciously recalls the requirements for membership in the Five Thousand in 411 B.C.: *hopla parechontai* (Thuc. 9.97.1) and *tois ek tôn hoplôn*.[75] The several references to the hipparchs and to the census classes, not to mention the existence of coinage, are therefore probably anachronistic. Helbig was incorrect to use this passage to prove the antiquity of the office of hipparch,[76] though he would not have claimed that the hipparch commanded a true cavalry. In the end *AP* 4.2–4 cannot be safely used to posit an Athenian cavalry in the Archaic period.

With respect to the Solonian constitution (594 B.C.)[77] *AP* 7.3–4 supplies the following information:

> [Solon] divided the people into four property classes according to wealth, as had been done before; the four classes were: *pentakosiomedimnoi, hippeis, zeugitai* and *thetes*. He distributed the other magistracies to be held by the *pentakosiomedimnoi, hippeis*, and *zeugitai*, allotting the nine Archons, the Treasurers, the *poletai*, the Eleven, and *kolakretai* to various classes in accordance with their property qualification. The *Thetes* received only the right to sit in the *Ekklesia* and the *dikasteria*. The property qualification for the *pentakosiomedimnos* was a minimum yearly return from his own property of 500 measures, dry or liquid. The *hippeis* had a minimum of 300, and some say that the class was also restricted to those able to maintain a horse; they deduce this from early dedications, for

[74] Hignett, *HAC*, p. 5 and n. 5 (with bibliography); and more recently, Rhodes, *CAP*, pp. 108–118; Moore, *Arist. & Xenoph.*, pp. 212–14, 260–62.

[75] *AP* 33.1–2; compare *Pol.* 1279a37-b4; 1297b1-34.

[76] *Hippeis*, p. 55.

[77] See R. Sealey, "Zum Datum der solonischen Gesetzgebung," *Historia* 28 (1979): 238–41.

there is a statue of Diphilos on the Acropolis with the following inscription:

> Anthemion, the son of Diphilos,
> made this dedication to the gods,
> having risen from the *thetes* to the
> class of the *hippeis*.

A horse stands by, showing the connection between the *hippeis* and being able to maintain a horse. None the less, it is more plausible that this class should have been defined by measures of produce like the *pentakosiomedimnoi*. The minimum qualification for the *zeugitai* was 200 measures, wet and dry combined, while the remainder of the population formed the *thetes* and were not entitled to hold office. This is why even now, when they are about to cast lots for a magistracy and a man is asked what his class is, nobody would say that he was one of the *thetes*.[78]

This is not the time to dwell on the troubling phrase, "as had been done before." I am content for the moment to accept Solon as the author of this system[79] and to suspect that this phrase was inserted to maintain an internal consistency with the Drakontian constitution mentioned shortly before (4.2–4).[80] This does not mean that the terms Solon chose did not exist in some form prior to his reforms.

[78] Translated by J. M. Moore, *Arist. & Xenoph.*, pp. 151–52.

[79] As apparently Ar. *Pol.* 1274a16–23; Plut. *Solon* 18.1–2; and Harpokration s.v. *hippas, pentakosiomedimnon*. The distribution of the Attic population into census classes, to my mind, reflects the establishment of a new system. It has been suggested that the census classes, at least that of the *pentakosiomedimnoi*, have a greater antiquity than Drakon, a suggestion based on the discovery of an unusual ceramic chest and lid in a Geometric grave in the Athenian Agora, see E. L. Smithson, "The Tomb of a Rich Athenian Lady, ca. 850 B.C.," *Hesperia* 37 (1968): 83, 94–97, pls. 24–27. The lid is adorned with what is interpreted as five model-size granaries which Smithson supposes symbolically represent five measures of grain, that is, 500 measures, therefore indicating the deceased woman's relationship as a wife or daughter, to a *pentakosiomedimos* (pp. 83, 96). An intriguing idea, but it has been seriously questioned by C. Roebuck, "Three Classes (?) in Early Attica," *Hesperia* 43 (1974): 489, and Rhodes, *CAP*, pp. 137–38, n. 10. Moore, *Arist. & Xenoph.*, p. 218, argued that the four classes had existed before Solon, but that Solon replaced those termed *eupatridai* ("well-sired ones") with *pentakosiomedimnoi*. Rhodes, *CAP*, p. 137, and R. Sealey, *A History of Greek City-States 700 B.C.–338 B.C.* (Berkeley, 1976), 119, think that the class of *pentakosiomedimnoi* was added by Solon to the three other classes, already in existence, by simply dividing the *hippeis* into the "rich" and "very rich."

[80] Rhodes, *CAP*, p. 137, properly points out that at other places in the text the inconsistencies remain, but still, no one would claim that the *Athenaion Politeia* is a finished, polished work as we have received it, so we should be chary in expecting *thorough-going* consistency.

It is obvious that by Aristotle's day the exact meaning of the *hippeis* class in the Solonian system was in doubt. It is clear where Aristotle stood on this issue, even if his use of "more plausible" (*eulogôteron*) demonstrates that the evidence at his disposal was not decisive. But the very fact that the variant tradition had to rely upon ancillary proofs shows that in none of the extant poems or preserved laws of Solon did the phrase "those able to maintain a horse" (*tous hippotrophein dynamenous*) appear in connection with the second highest census class. Aristotle understood the system as one based on agriculture, pure and simple. Just as the term *pentakosiomedimnoi* literally translates as "500 measuremen" (a *medimnos* equals approximately a bushel and a half), Aristotle concluded that the rest of the system was similarly determined.

But "some say" (*enioi phasi*) that the name of the class originated from the ability to maintain a horse. These writers supported their claim on the basis of (a) the implication from the name, *hippeus* or *tên hippada telôn*,[81] and (b) the fact that a certain individual, Anthemion, son of Diphilos, commemorated his elevation from the class of *thetes* to the *hippeis* class with a statue of a horse.[82]

The problem of interpretation is compounded by the fact that *hippotrophia* can mean simply raising horses for personal use, i.e., hunting, athletic competitions, etc. or raising/maintaining a horse (or horses) specifically for military use.[83] As a *hippotrophos*, one may do both or one to the exclusion of the other. Obviously, a man who ordinarily raised horses for personal use would have been an obvious choice for cavalry service. But one does not necessarily follow upon the other. In Lykourgos' speech, *Against Leokrates*, the orator's bitter accusation of *hippotrophia* against Leokrates makes sense only if Ly-

[81] Ste. Croix has shown that the translation in Liddell-Scott-Jones *Greek-English Lexicon* of *tên hippada telôn* as "paying the Knights' tax" is inaccurate—it really means "belonging to the Knight's class" and did not involve a tax: "Demosthenes' *timêma* and the Athenian *eisphora* in the 4th Century," *C&M* 14 (1953): 42–45.

[82] Pollux 8.131 essentially records the same information, but fails to comment on exactly what this is supposed to demonstrate. Rhodes, *CAP*, pp. 144–45, reports a new interpretation of this epigram offered by L. H. Jeffrey in an unpublished note to the effect that the two lines do not belong together; that Anthemion was really a "hippeus," even perhaps one of Kimon's companions in his famous march to the Acropolis in 480 B.C., who made a dedication to the gods (first line); and that the second line, added by a contemporary wit, meant just the opposite of what is its usual translation, that is, he changed "hippic" status to "thetic" to fight on the ships. Rhodes rightly regards this theory as unlikely.

[83] For succinct discussion and examples illustrating this ambiguity, see W. Wyse, *The Speeches of Isaeus* (Cambridge, 1904), 471–75. Note particularly: Plut. *Them.* 5–2; Xen. *Hipparch.* 1.11–12; Is. 5.43; and Ar. *Clouds* (Pheidippides).

kourgos is criticizing him for maintaining horses for ostentatious personal display rather than for the benefit of the state, i.e., cavalry (139; cf. Xen. *Hipparch.* 8.7). And Plato recommends, as if the opposite were currently the norm, in his *Laws* (834 D) that equestrian events should be relevant to military preparedness, adding that cavalry officers should supervise these activities.

If we may digress, the situation in Sparta is even more revealing. The Spartans established a true cavalry force in 424 B.C. (Thuc. 4.55.2), but the men who actually rode the horses, i.e., *hoi hippeis*, were not the same individuals as those who, because of their wealth, raised them, i.e., *hoi hippotrophoi* (Xen. *Hell.* 6.4.11). Furthermore, Pausanias informs us that after the Persian invasions, that is, well before 424 B.C., the Spartans became the "most honor-covetous" (*philotimotata*) of all the Greeks in the raising of horses (*pros hippôn trophas*), i.e., for the Panhellenic Games. He even lists the most prominent *hippotrophoi.*[84] The most celebrated *hippotrophos* was Kyniska, the sister of the Spartan king Agesilaos, who entered racing teams at the Olympic Games in the early fourth century B.C.[85]

The implication is clear—one could be a *hippotrophos* without having any connection with cavalry service. Perhaps, then, the 300 *medimnoi* requirement was considered the minimum amount of annual production necessary to maintain a horse—for whatever purpose. Likewise, Anthemion's equestrian statue need not have implied cavalry service—perhaps it was simply symbolic of the hippic class.[86] One thing, at least, does seem evident to me. If a military obligation is implied in the expression *tous hippotrophein dynamenous*,[87] then those who provided the explanation were thinking of true cavalry and not mounted hoplites. The Anthemion epigram and statue belong to the period after the sack of the Acropolis by the Persians in 480 B.C.,[88] at a date sometime in the fifth century (or fourth) when men termed *hippeis* were unmistakenly cavalrymen, not mounted hoplites.

Later writers, all of whom appear to have borrowed their information directly or indirectly from Aristotle, were reluctant to commit

[84] 6.2.1–2; see A.H.M. Jones, *Sparta* (Cambridge, Mass., 1968), 36–37. Isokrates (6.55) remarks that in the 360s the Spartans were more preoccupied with their racing horses than in preventing the Theban general Epameinondas from invading the Peloponnesos.

[85] Paus. 3.8.1 (= 3.15.1, 6.1.6); Xen. *Ages.* 20.1.

[86] As suggested by Moore, *Arist. & Xenoph.*, p. 219.

[87] As seems to be the case in the scholion to Aristophanes *Knights* 627 (repeated by the *Suda* s.v. *hippeis*).

[88] Rhodes, *CAP*, p. 144.

themselves to one or the other explanation for the Solonian census class. Pollux, at times a valuable source for Athenian institutions, repeated the two options without confidence: "those who belong to the hippic class seem to be named after their ability to raise horses, and they produced three hundred measures" (8.103). The use of "seem" (*dokousin*) confirms his uncertainty. The *Suda* (s.v. *hippas*) is so confused that it reports that Solon instituted five classes, the second highest being *hippeis*, but a fifth being *hippas!* Plutarch (*Solon* 18.1–2) simply lists both possibilities, separated by the conjunction "or." He spared himself the trouble of deciding. Other late sources mention the class only in agricultural terms.[89]

The question remains, did Solon choose the name for the second highest census class from military considerations, i.e., cavalry or mounted hoplites, or was it simply an agricultural classification? Almost a century ago the military explanation was forcefully advanced by C. Cichorius[90] on the basis of a linguistic analysis of the three terms, *hippeis*, *zeugitai*, and *thetes*. According to this analysis, the *zeugitai* equated to the *hoplitai*, defined by the minimum requirement for purchase of hoplite armor and infantry service (thus excluding the thetes), while the *hippeis* equated to cavalrymen, defined by the minimum requirement for raising horses and for cavalry service (to include, of course, the *pentakosiomedimnoi*). This explanation has gained wide support[91] and appears at this writing to be winning the day.[92]

But problems remain. First of all, the name *hippeis*. This term and its hippic variants, as we have seen with *hippotrophia*, need not mean cavalrymen at all. Homer refers to the charioteers and warriors in

[89] E.g., I. Bekker, ed., *Anecdota Graeca*, 3 vols. (1814–21), 1:267, line 13 and Hesychios s.v.v. *hippada* and *hippas*.

[90] "Zu den Namen der attischen Steuerklassen," in *Griechische Studien H. Lipsius dargebracht zum sechzigsten Geburtstag* (Leipzig, 1894), 135–40.

[91] See e.g., Beloch, *GG*² 1.1.303 with n. 1; J. E. Sandys, *Aristotle's Constitution of Athens* (London, 1912), 28; A. Andrewes, *Greek Tyrants* (New York, 1963; Harper Torch Books), 87 and repeated in his article in *CAH*² III.3, p. 385; L. H. Jeffrey, *Archaic Greece* (London & Tonbridge, 1976), 93, 107, n. 6; V. Ehrenberg, *From Solon to Socrates*, 2d ed. (London, 1973), 65–67; R. Sealey, *A History of the Greek City-States 700–338* B.C. (Berkeley, 1976), 119–20; Anderson, *AGH*, pp. 128 and 212, n. 2; D. Whitehead, "The Archaic Athenian Zeugitai," *CQ* 31 (1981): 282–86; and Rhodes, *CAP*, p. 138.

[92] Moore argues that the economic consideration was primary (though he concedes that the class "was almost certainly in existence before Solon, and at that time membership probably did depend on a man being able to maintain a horse or horses at his own expense": p. 219), as does J. A. Vartsos, "Class Divisions in Fifth-Century Athens," *Platon* 30 (1978): 226–44, esp. 228, n. 6 and 232–33.

chariots as *hippees* in contrast to infantry,[93] yet the poet makes it abundantly clear that the Homeric heroes usually got off their chariots and fought on foot. The chariot, then, was primarily used as a means of transportation to and from the battlefield, not as a weapon of war.[94] Secondly, who would argue that the so-called "charioteers and companions" that made up the 300 Theban hoplites present at the battle of Delion in 424 B.C. (Diod. 12.70.1) really still used chariots at any stage of the battle? Then there are the famous Spartan *hippeis*, a corps of 300 select Spartans who acted as a bodyguard for the kings and fought as hoplites.[95] These are presumably the same *hippeis* whom the geographer Strabo explains as "not maintaining horses" (*tous de mê hippotrophein:* 10.4.18–482 C). After 424 B.C., then, the Spartans possessed two distinct kinds of *hippeis*, the select 300 hoplites and the true cavalry.[96]

The definition of the term *hippeis*, therefore, can be elusive—a point of which Aristotle, author of countless city-state constitutions, was probably not unaware.

It is commonly stated, as I have noted above, that only those who belonged to the top three Solonian census classes were liable for military service as hoplites, the thetes being excluded.[97] If this is true, why then did Solon not call his third class *hoplitai* instead of *zeugitai*, to make it clear that the term implied "the ability to maintain (or afford) hoplite armor."[98] If the identification of *zeugitai* with hoplites were

[93] Cf. *Iliad* 2.810; 12.66; 15.270; 23.130–33, 262, 273, 287, etc.

[94] Greenhalgh, *EGW*, p. 146, has suggested that Homer used the heroic chariots to mask the real military practice of his day—horses and mounted hoplites. Thus chariot use may belong to the Mycenaean period, but not to the Homeric world of the Geometric period.

[95] Hdt. 1.67; 8.124; Thuc. 5.72.4; see Hodkinson, "Social Order and the Conflict of Values in Classical Sparta," *Chiron* 13 (1983): 239–81, esp. 247 and H. Michell, *Sparta* (Cambridge, 1964), 249.

[96] There may have been a time when Sparta had had cavalrymen, but the evidence is slight. That they, like other states, had mounted hoplites ca. 625 B.C. is more defensible. For the artistic evidence for both of these possibilities, see Greenhalgh, *EGW*, pp. 94–95. Dion. Hal. 2.13.4 mentions that they served both on horse and on foot, but we probably should not make too much of this late remark; see Michell, *Sparta*, p. 349, n. 2.

[97] See Harpokration s.v. *thetes;* also G. Gilbert, *The Constitutional Antiquities of Sparta and Athens,* trans. E. J. Brooks and T. Nicklin (New York, 1895; repr. Amsterdam, 1968), 315, and A.H.M. Jones, *Athenian Democracy* (Oxford, 1957), 142, n. 50.

[98] D. Whitehead, "The Archaic Athenian Zeugitai," *CQ* 31 (1981): 282–86, has in fact argued that on the basis of philological usage and a reference in Plut. *Pelopidas* 23.4, the term *zeugitai* meant "those yoked-together," i.e., hoplites fighting side by side in close infantry formation. But Plutarch is writing in the second century A.D. and his choice of words can hardly be considered a reliable example of contemporary usage for the Archaic and Classical periods. Moreover, Asklepiodotos, a first century B.C. writer

that patent, we should expect Aristotle's opponents to have applied a corresponding military equivalent for them as they have done for the *hippeis*. To suppose that they did and that Aristotle chose not to record it, perhaps out of fear of weakening his argument, strikes me as hypercritical.

Even if we accept the interpretation that 300 *medimnoi* was the minimum needed to maintain a horse adequately, can we really believe that the difference in cost between maintaining a horse and owning hoplite armor was only 100 *medimnoi*? No doubt, hoplite armor was moderately expensive at the time of Solon and did require periodic upkeep and repair, but how can that compare with the cost of buying, stabling, grooming, and feeding of even one horse?[99] If it is truly linked to *hippotrophia*, I would have expected the rate for the *hippeis* class to have been more obviously differentiated from that for the *zeugitai*.

There are other considerations as well. By the sixth century hoplite warfare had become the predominant form of combat in southern Greece.[100] With the exception of Boiotia, Chalkis, and Eretria, no state south of Thessaly possessed a respectable cavalry.[101] Athens was no exception to this military reality. The reasons why Athens, in particular, did not develop a cavalry of note ultimately have to do with topography, military technology, land use, and the dynamics of Athenian society.

Attika was never a land famed, or even acknowledged, for its *hippotrophia* in any sense of the word. Some literary sources seem to imply otherwise, but they belong to the world of mythology not history. For example, Sophokles in his *Oedipus at Colonus* (668–70, 707–15, *OCT*) has his chorus, standing next to the Hill of Horses (*Hippios Kolonos*) recall Poseidon's gift to the Athenians:

---

on military tactics, uses the word *zugein* (cf. its substantive, *zugon*, 2.5) to mean soldiers positioned side by side, not *zeugein* (2.6). The rest of Whitehead's argument is compelling, but not conclusive; see Vartsos, "Class Divisions," p. 228, n. 6.

[99] We know very little about the cost of armor (see W. K. Pritchett, "The Attic Stelae," *Hesperia* 25 [1956]: 306–8), but even the finest panoply would have fallen short of the 1,200 drachmas suggested as the amount of the state loan for the purchase of a warhorse in the Classical and Hellenistic periods (on this last, see Kroll, "Archive," p. 89—more on this in subsequent chapters). Perhaps relevant is the late sixth century B.C. inscription detailing regulations for the cleruchs of Salamis: the cleruchs are required to provide their own *hopla* to the value of 30 drachmas (*IG* I[3] 1, lines 8–10; ML 14, pp. 25–27). The figures mentioned in Aristophanes' *Peace* (421 B.C.) serve the purpose of comic exaggeration, i.e., 1,000 drachmas for a corselet—a splendid one (lines 1224–25) and 100 drachmas for a helmet (lines 1250–51).

[100] See Snodgrass, *AAG*, Chapter 3, for a good survey.

[101] See Larsen, *Greek Federal States*, p. 107, although I believe, with Greenhalgh, that he is incorrect in concluding that these southern Greek states had *no* cavalry at all.

Here, stranger,
here in the land where the horses are a glory . . .
And I have another praise to sound in song
a mighty gift bestowed our mother-city—
the splendor of a majestic, ancient god
  the pride and power of our earth—
the glory of horses, glory of young horses
the glorious rippling sinews of the sea!
O Poseidon, you have throned her in this power,
    lord god of the sea-lanes, you were first
to forge the bit and bridle, first to curb
the fiery rage of stallions in these roads—[102]

Poseidon was the god credited with having given the horse and horsemanship to mankind (Paus. 7.21.8–9); not surprisingly he became the patron god of horsemen.[103] Homer describes Menestheus, the leader of the Athenians against Troy, as a "marshaller of horses" (*Il.* 2.554). Aelius Aristides quotes this line in his *Panathenaikos* (377, Behr) and mentions also the Athenian cavalry battling the Amazons (83), Athena bequeathing to the Athenians racing chariots and war-horses (*hippôn polemisteriôn*) and teaching Erichthonios, early king of Athens, "all about horsemanship" (*tên teleian hippikên*, 43).

On the other hand, it is worth noting that Athens did not produce a chariot winner at the Olympic Games until Alkmeon in 592 B.C. (Hdt. 6.125.5; Isok. 16.25), whereas, by contrast, Pagondas of Thebes (a city replete with cavalry traditions) won the first four-horse chariot race in 680 B.C. (Paus. 5.8.7). Moreover, the list, albeit incomplete, of Athenian equestrian victors at Panhellenic Games in the sixth century is almost completely dominated by a very select number of familiar noble families, e.g., the Alkmeonidai, the Philaidai, and the Peisistratidai.[104] Also, it might not be irrelevant to note that the Panathenaic athletic programme, including equestrian events, was not organized until 566/5 B.C.[105]

[102] R. Fagles, trans., *The Three Theban Plays* (New York, 1982), 309–10. This praise may reflect the contemporary setting of the Peloponnesian War in which the cavalry did play an important part.

[103] See Homer *Il.* 23.583–84; Ar. *Knights* 551–64; *Clouds* 83; L. R. Palmer, "Poseidon and the World of Water," *Nestor* 11 (1984): 1804, suggests that Poseidon's "horsiness" comes from the Indo-European origin of the Greeks, to wit, they "brought with them . . . a Stallion-god who stamped water out of the ground with his hoof."

[104] See J. K. Davies, *Wealth and the Power of Wealth in Classical Athens* (New York, 1981), 167.

[105] See Parke, *Festivals of the Athenians*, pp. 34–36 and Kyle, *AAA*, pp. 33–39.

This information is really not so startling. Attika was a land of poor soil[106] and small farms committed to barley, wheat, and olives,[107] not large estates possessed of pasturage for serious horse-breeding. Aristotle observed generally:

> in places where the country happens to be suitable for horsemanship (*hippasimon*), there natural conditions favor the establishment of an oligarchy that will be powerful (for the security of the inhabitants depends on the strength of this element, and keeping studs of horses (*hippotrophia*) is the pursuit of those who own extensive estates (*makras ousias*).[108]

He had already noted in an earlier section that "as for horse-breeding (*hippotrophia*), it is not easy to do this unless you are rich" (*Pol.* 1289b35–36). J. K. Davies in his *Athenian Propertied Families* rightly uses *hippotrophia* as one criterion for membership in the Athenian liturgical class (pp. xxv–xvi). When Aristotle argues that the political form of oligarchy and the military preeminence of cavalry go hand in hand, he specifically mentions those states famed for their cavalry,[109] e.g., Eretria, Chalkis, Magnesia on the Maiander (*Pol.* 1289b39–41) and Thessaly (*Pol.* 1269a36). Athens is not mentioned, even though Aristotle knew very well that it had once been ruled by an oligarchy of aristocratic families. These families may have lorded it over the populace, but they did not do it from horses.

Other ancient writers were acutely aware of the importance of topography. In the *Odyssey* Telemachos tells Menelaos not to offer him horses as a gift:

> I will not take the horses to Ithaka, but will leave them here, for your own delight, since you are lord of a spreading plain, there is plenty of clover here, there is galingale, and there is wheat and millet here and white barley, wide grown. There are no wide courses in Ithaka, there is no meadow; a place to feed goats; but lovelier than a place

[106] Thuc. 1.2. (*leptogeôn*); Plato *Critias* 111 B–C; and see M. Cary, *The Geographical Background of Greek and Roman History* (Oxford, 1949), 75–76: "only one-quarter of Attica is estimated to be cultivable."

[107] V. D. Hanson, *Warfare and Agriculture in Classical Greece* (Pisa, 1983) 37 and n. 2, notes that the largest farm attested in Attika was probably no larger than 100 acres, and that appears exceptionally large (i.e., the product of rhetorical exaggeration).

[108] *Pol.* 1321a12–13 (Loeb trans.).

[109] Although Greenhalgh (*EGW*, p. 147) has argued that even in the cases of Eretria and Chalkis, they may have been mounted hoplites, not true cavalry until the sixth century B.C. Perhaps, but Aristotle surely thought of them as cavalry in his analysis. Greenhalgh accepts only Thessaly as a cavalry power throughout this period (pp. 149–50).

> to feed horses; for there is no one of the islands that has meadows for driving horses; they are all sea slopes; and Ithaka more than all the others (4.600–608).[110]

Simonides, the lyric poet, writes "the rearing of horses (*hippotrophia*) is not for Zakynthos, but for those places with fields that bear wheat."[111] In the *Laws*, Plato, through his speaker, Kleinias, contrasts "uneven" (*anomalos*) Crete with "level" (*pedias*) Thessaly (652 D). Later in the same work he concludes that "horses are not much needed in Crete, at least in large numbers, so less attention is paid to the rearing and racing of them" (834 D; Loeb).

The topographical evidence for Attika is even more revealing. We recall that in 490 B.C., Hippias directed the Persian forces to the plain of Marathon because it was the best cavalry terrain in all of Attika (Hdt. 6.102). We know also why the plain of Phaleron was not considered as suitable by the Persians—in 510 B.C. this same Hippias had been forced to make it *hippasimon* for his Thessalian cavalry allies. This probably meant tearing down the "field walls of rubble, earthen banks, or planted hedges" that surrounded the Attic farms in the plain.[112] Boiotia, by contrast, had extensive level and uninterrupted terrain—ideal for cavalry. In 480/79, Mardonios transferred his army from Attika to Boiotia because the terrain of Attika was not suitable for cavalry (Hdt. 9.13). And in 86 B.C., the Roman general Sulla repeated this strategy by withdrawing his forces from Attika to Boiotia to meet the armies of Mithridates of Pontos. Why? Attika was "rough" (*tracheian*) and "ill-suited for horses" (*dyshippon*).[113] During the Peloponnesian War Thucydides reports that in skirmishes against Spartan-held Dekeleia and in the patrolling of Attika some horses of the Athenian cavalry kept going lame because of the "hard ground" (*en gêi apokrotôi*) and the increasing hardships of service (7.27.5).

The failure of the Greeks to develop stirrups and horseshoes[114]

[110] R. Lattimore, trans., *The Odyssey of Homer* (New York, 1965), 80–81.

[111] *Lyra Graeca* II, frag. 44 (Loeb).

[112] See Hanson, *Warfare and Agriculture*, p. 37; cf. Thuc. 6.64 and 66 on the importance of obstructions to hamper free movement of the Syracusan cavalry during the Sicilian Expedition, 415–413 B.C.

[113] Plut. *Sulla* 15.2–3; for general strategic considerations, Onasander 31 observes that when an enemy is superior in cavalry, a good general should choose a location that is rough (*trachea*) and hemmed in (*stena*) near mountains least suitable for cavalry maneuvers (*hippasima*); cf. also Aineias Taktikos 6.6, 8.4, 26.4; Xen. *Hell.* 7.2.12; and Diod. 18.15.4 (Thessalian cavalry against Macedonian phalanx in Lamian War, 323–22 B.C.).

[114] See Snodgrass, *AAG*, p. 85 and Anderson, *AGH*, pp. 90–91. Cf. also Hanson, *Warfare and Agriculture* p. 92, n. 13 and p. 103, n. 1.

makes the question of terrain that much more relevant. The Athenian Xenophon, noted authority on *ta hippika*, has much to say on these matters. In his *Anabasis* (3.2.18), he downplays the effectiveness of the harassing Persian cavalry by observing that it was difficult simply to remain mounted on a horse let alone attack firmly grounded infantrymen. Even as special pleading, there is a good deal of truth in his remarks. Anyone who has ridden bareback knows how immensely more difficult it is to stay mounted than with the aid of the simple device of stirrups. Galen, the Greek physician of the second century A.D., describes a series of physical problems resulting from hard horseback riding.[115] Theophrastos parodies the "late learner" who borrows a horse to ride in the countryside but falls off and cracks his head (*Characters* XXVII.10–11); and the orator Andokides suffers a fall from a colt while riding at the gymnasion at Kynosarges in 415 B.C.—he breaks his collarbone, fractures his skull, and has to be carried home on a litter (1.61). Xenophon (*Hipparch* 3.14) singles out the hard ground (*epikrotos*) around the Akademy as a place liable to cause horsemen to be thrown from their horses during cavalry displays.

Xenophon repeatedly cites the need to condition the hooves of the cavalry mounts of the Athenian cavalry. In the *Hipparchikos*, he suggests that stones be laid out for the horse to walk on continually (1.16; cf. *Peri Hippikes* 4.4), and emphasizes the benefits of constant practice on rough terrain.[116] In the purchase of a horse Xenophon advises the prospective buyer to examine the feet first—a warhorse is worthless if his hooves are not sound (*Peri Hippikes* 1.2–3; *Hipparch*. 1.4).

In sum, Attika was not a region that was well-suited to the pursuit of horse-breeding either for pleasure or for war. (Of course, for horse-racing the ground was deliberately leveled and prepared to decrease damage to the horses.) The number of Athenian families active as *hippotrophoi* must have been quite small—it is impossible to guess how many, considering our data base—but a cavalry force numbering ninety-six men (Pollux) does much, to my mind, to lend credence to this point. If this conclusion is correct, why then would Solon have created two census classes (out of four!) to recognize quite a small group of *hippotrophoi* whose military importance as cavalry at that date (and in that terrain) was relatively small? Its military value was

[115] Quoted by M. I. Finley and H. W. Pleket, *The Olympic Games; The First Thousand Years* (London, 1976), 31.

[116] 8.3; cf. *Oikonomikos* 11.14–18. The military manuals are filled with advice of this sort; Onasander 10.6 comments that the general should hold cavalry practice and skirmish in the plain, around the base of the hills and as far as is possible into the rough country.

probably limited to its role as a mobile force to patrol Attika. One even suspects that most of the *hippotrophoi* were technically not *hippeis*, but *pentakosiomedimnoi*.[117]

It should be recalled that the primary purpose (at least according to Aristotle) of the class division was to set forth minimal economic requirements for political participation (cf. *AP* 26.2). And it is generally agreed, at least for the Classical period, that thirty marked the minimum age requirement for high office.[118] Surely this was also the case for the Archaic period. Thirty seems to have marked a key turning point in the life of an ancient Greek male—it was the approximate age recommended for marriage,[119] producing children, and undertaking family and household responsibilities—in sum, maturity. J. K. Davies, in his *Athenian Propertied Families*, accepts the age of thirty as minimal for high political and military office, a veritable cornerstone for many of his prosopographical edifices.

Yet the prevailing image of the Athenian cavalry in the Classical period is that of youth, particularly, as I will elaborate upon more fully in the later chapters, youth in their twenties or perhaps early thirties. This is hardly surprising—who more appropriate than young men to have the skill, strength, and stamina to ride horses without stirrups in sometimes difficult cavalry maneuvers on uneven and hazardous ground?[120] The *hippeis* of Aristophanes' comedy *Knights* are identified as long-haired, aristocratic young men (580, 731, 1121), exemplified by Mantitheos, a litigant in the 390s.[121] And Xenophon tells his cavalry commander to recruit the young men (*neous*) not only by instilling in them an eagerness to join, but by persuading their "guardians" (*kyr-*

[117] For an explanation as to why only the *pentakosiomedimnoi* are specifically identified with agricultural production, see Rhodes, *CAP*, pp. 142–43, referring to an unpublished essay by G.E.M. de Ste. Croix.

[118] See Hignett, *HAC*, p. 224 and M. H. Hansen, "Seven Hundred Archai in Classical Athens," *GRBS* 21 (1980): 167–69. Recently, R. Develin, "Age Qualifications for Athenian Magistrates," *ZPE* 61 (1985): 149–59, has questioned this minimum age requirement, but his evidence and his arguments are not persuasive.

[119] Hesiod *Works and Days* 695–97; Ar. *Pol.* 1334b38–41 to 1335a1–4 recommends thirty-seven or a little younger; for discussion, see W. K. Lacey, *The Family in Ancient Greece* (Ithaca, N.Y., 1968) 129–30.

[120] In the recruiting of cavalrymen, physical endurance was as important as possession of wealth *(dynatôtatous kai chrêmasi kai sômasin*: Xen. *Hipparch.* 1.9; cf. *AP* 49.2). There is another physical phenomenon shared by ancient and modern man alike: men put on weight as they get older and settle into more sedentary lifestyles. The more weight, the greater the burden for the horse. The U.S. Cavalry accepted recruits only up to 165 lbs., the ideal range being 130 to 150 lbs.: W. H. Carter, *Horses, Saddles, and Bridles*, 3d ed. (Baltimore, 1906), 160.

[121] See Lys. *In Defense of Mantitheos.*

ious), (usually their fathers), that they will have to act as *hippotrophoi* for their sons, if not in that year, in some other (*Hipparch.* 1.11–12). And who indeed is gloriously represented in the cavalcade of the Panathenaic frieze on the Parthenon but young, vigorous Athenian horsemen (Figure 4)? Surely youth was similarly emphasized in the Archaic period.

Any cavalry that might have existed at the time of Solon would have been *numerically* dominated by young men who would not have been eligible for political office. I suggest that many of these young men would have still been subject to the authority (*kyria*) of their fathers, living in their parents' households (*oikoi*), and as yet unmarried. The number of young men who had already inherited their patrimony at the death or incapacity of their fathers or had already been allocated a large enough portion of the family estate by their fathers probably constituted a minority. And even here, their economic independence would not have ordinarily opened the door to political office—these wealthy young men would not have presumed to place themselves in competition for office until they had acquired more worldly experience—something that came with age.

An excellent exception that proves the rule comes from the Classical period. Alkibiades, barely thirty years old in 420 B.C. (*APF* 600, p. 18), is vying for political influence with Nikias and his faction. Thucydides (5.43.2) remarks that any other city would consider him still young (*eti tote ôn neos*), but because of his exceptional ancestry, the Athenians suffered him his boldness and listened to his suggestions on public policy. In fact, Thucydides pointedly comments that Alkibiades had taken offense that Nikias had dismissed him "because of his youth" (*kata te tên neotêta*).[122]

Would Solon have created such incredible ambiguity? I think not. The two systems, military and agricultural, need to be differentiated. There were those Athenians who served either as hoplites or cavalry or both—this military obligation and privilege was not, of course, unrelated to wealth. Hoplite armor and horses were the possessions of the well-off and the well-to-do. But if we can take a hint from the Classical period, there is no good evidence to indicate that the Solonian census classes provided the means of determining military responsibility for either hoplite or cavalry service. In the fourth century the state apparently took no interest in confirming what class one *claimed* to belong

[122] A similar case of political audacity has been made for Themistokles: D. H. Kelly, "The Athenian Archonship 508/7–487/6," *Antichthon* 12 (1978): 6.

to (though that may not have been true in the fifth century), yet, politically, the Solonian census classes continued to retain legal validity.[123]

I contend that the constitution of Solon was economic and political, not military. The lawgiver probably did choose the term *hippeis* for its symbolic, aristocratic value, but at the same time relegated it quite properly to secondary place in his system. In the end the information contained in *AP* 7.4 concerning the *hippeis* class cannot be used as proof, however attractive it might seem at first glance, that Solon was acknowledging the military importance of, or even the existence of, a cavalry and/or mounted hoplites in Archaic Athens.

I might also offer a suggestion to explain the confusion for *AP* 7.4. I have already stated my belief that the alternative explanation, i.e., *hippotrophia* for the second highest census class, did not belong to the original Solonian laws; I further suggest that it belongs to some writer, perhaps an Atthidographer, of the fifth or fourth century B.C., probably after the creation of a state cavalry in the mid-fifth century, who, misled by Solon's "hippic" term and the existence of a cavalry in his own day, assumed a similar military obligation, i.e., cavalry, extending back to Solon. This writer, impressed by the size of the 1,000-man Athenian cavalry, reasoned that the highest class, the *pentakosiomedimnoi*, could not alone supply enough young men to adequately sustain the cavalry quota. By necessity, the fifth-century cavalry represented a deeper penetration into the *hippeis* class,[124] a fact which perhaps generated the theory that the Solonian class had defined the lower limits of cavalry service. The top two classes, *loosely speaking*, did become the natural pool from which to recruit cavalry personnel, but the ambiguous economic meaning of those census classes in the Classical period[125] eliminated them as a basis of determining military obligations. That was left to the discretion of the recruiting officers, general perceptions of wealth, and specific traditions of military service within a given family.

Still other factors may also have contributed to Athens' failure to organize a respectable cavalry force in the Archaic period.

[123] See *AP* 7.4; 8.1; 26.2; 47.1; Isaios 7.39; *IG* I³ 46 (=ML 49); *IG* II² 30; Ps.-D. 43.54.

[124] As we might conclude from Xenophon's comments that a cavalry commander has to recruit personnel from not only the "most wealthy" (*tous dynatôtatous . . . chrêmasi*), but also the "less wealthy" as well (*Hipparch.* 1.9–10); see Anderson's comments on "rich" and "poor" troopers: *AGH*, pp. 134–39.

[125] At *AP* 47.1 Aristotle seems to be saying that though a man might be classed as a *pentakosiomedimnos*, he could in fact be "poor" (*penes*); see Rhodes, *CAP*, p. 551.

TYRANTS, HOPLITE WARFARE, AND ATHENIAN SOCIETY

I have already noted that the Peisistratidai maintained a close relationship with the Thessalians, those famed cavalrymen: one of Peisistratos's sons was named Thessalos and Hippias utilized the Thessalian cavalry against the Spartans in 510 B.C. During the tyrannies of Peisistratos and Hippias (546–510 B.C.), sons of the wealthy landowners of Attika, those natural *hippotrophoi*, probably did see cavalry service.[126] Early in his reign Peisistratos disarmed the populace (*AP* 15.3–4; Polyainos 1.21.2), presumably including both hoplites and horsemen, but he probably did not keep their weapons for long.[127] Both Aristotle (*AP* 16.8) and Thucydides (6.54.6) confirm the tradition that he preserved the existing laws and succeeded in gaining the support of the nobility (*tôn gnôrimôn*) as well as the majority of the people (*AP* 16.9). This accommodation with the nobility finds support in the inscription of the late fifth century B.C. found in the Athenian Agora which apparently records the Athenian eponymous archon list of the 520s.[128] The archons mentioned there include members of two of the most famous Athenian families, abundantly attested for their *hippotrophia*, the Philaidai (cf. also Hdt. 6.39.1) and the Alkmeonidai. We may assume that some members of these families, at least, were expected to perform cavalry service; the Thessalian cavalry could always be requested in the event of a serious military crisis—as in 510 B.C.

The assassination of Hipparchos, Hippias' brother, in 514 B.C. put an end to the cooperation between the tyrant and the aristocratic families.[129] The search for conspirators, the subsequent accusations against a number of "great families," and the resulting civil strife, evidenced by the skirmish (I hesitate to call it a battle) between Hippias and the Alkmeonidai at a place called Leipsydrion,[130] terminated the Athenian cavalry. By 510 B.C. Hippias had alienated, exiled, or executed too many Athenians to levy a citizen militia safely; only his mercenaries and Thessalian cavalry allies could be trusted against the Spartans. With the defeat and banishment of Hippias in 510 B.C. cavalry service was probably resumed.

We would dearly love to know more about the military campaigns that Athens waged against Chalkis and Boiotia in 506 B.C. (Hdt. 5.74,

[126] Anderson, *AGH*, p. 130: "the tyrants trusted Thessalians rather than the landowners of Attica, at least toward the end of their reign."

[127] A. J. Holladay, *Greece & Rome* 24 (1977): 52.

[128] ML 6, frag. c., pp. 11–12.

[129] Thuc. 6.56.2–3, 58.1–2.

[130] *AP* 18.4; 19.1; cf. Hdt. 5.62. See Rhodes, *CAP*, p. 234, for discussion of the controversy concerning the date and duration of their exile(s).

77), particularly since these two states had established cavalry traditions, e.g., the so-called Hippobotai of Chalkis, whose lands the victorious Athenians confiscated to settle 4,000 of their own citizens. The fact that Athens defeated both of them may have persuaded the Athenians that there was no compelling reason to reorganize and expand their modest cavalry force. And perhaps we should see in the glorious four-horse chariot monument,[131] paid for from the ransom of prisoners, a pointed comment on the ineffectiveness of cavalry against hoplites? But we ought not to press this point too far.

We naturally assume that cavalry service guaranteed great honor and glory, but modern perceptions have been influenced by the image of the Medieval knight. The truth of the matter is that in the period under consideration the infantryman faced the greater risk in battle—especially the one honored with the position in the front ranks of the phalanx—and consequently had the best opportunity to prove his excellence and to win honor. Mimnermos celebrates the exploits of a hero of his native Kolophon in Asia Minor who rushed forth with the front line to wreak havoc with his spear against the Lydian cavalry.[132] Tyrtaios, the martial poet of Sparta, lauds the young warrior who stands and dies "among the fore-fighters" (*eni promachois*).[133] And an inscription on the base of an Archaic *kouros* found at Anavyssos in Attika, dated to ca. 530 B.C.,[134] records the poignant epitaph of Kroisos, whom raging Ares, god of war, slew while he fought "among the fore-fighters" (*eni promachois*).

To be sure, the aristocrat who raised horses and raced them in the Panhellenic Games was envied and honored—the epinician (celebrating victory) poems of Bacchylides and Pindar bear this out—but this agonistic honor was in reality an appreciation and acknowledgement of his conspicuous wealth and social standing. The *hippotrophos* did not even do the actual racing, any more than the wealthy owners of thoroughbreds ride their own horses in the Kentucky Derby today. The hoplite warrior had to face the enemy at close range and stand his

[131] For this monument and accompanying epigram, see ML 15, pp. 28–29; Page, *Further Greek Epigrams*, pp. 191–93; and S. West, "Herodotus' Epigraphical Interests," *CQ* 35 (1985): 283–85. For translation, see Fornara, *Archaic Times*, no. 42.

[132] *Elegy and Iambus* 1, trans. J. M. Edmonds, frag. 14 (Loeb). It is possible that this man was mounted—Aristotle refers to Kolophon as one of those states in Asia Minor with early cavalry traditions and Polyainos 7.2.2 relates a strategem that assumes the reputation of the Kolophonian cavalry—yet there is nothing in the poem that says that he is a cavalryman; he could just as easily be a hoplite or even a mounted hoplite (on the latter possibility, see Greenhalgh, *EGW*, p. 93).

[133] *Elegy and Iambus*, no. 10, lines 1–2; see Pritchett, *GSW* 4, pp. 85–89.

[134] See G.M.A. Richter, *Kouroi: Archaic Greek Youths*, 3d ed. (London & New York, 1970), 115–16, 118–19, fig. 401 and *SEG* x.461; xxi.157; xxii.70; xxiv.70.

ground; he knew all too well that if his lines broke and he was forced to flee, his back became an easy target and his corpse an object of disgrace.[135] The horseman, on the other hand, unless he confronted an opposing cavalry force, could approach the enemy lines, hurl his javelin, turn and gallop to safety.

The celebrated victory at Marathon was but a link in the long chain of the heroic hoplite tradition which extended from Tyrtaios in the seventh century B.C. to the young Athenian aristocrat Mantitheos who petitioned to be relieved of cavalry duty to join the hoplite ranks in the expedition to the Haliartos River in Boiotia in 395 B.C., knowing that it was common opinion that the cavalry would be safe (*asphaleian*), while the hoplites would have to face danger (*kindynon*).[136]

Even if the Athenians had conceived of such a "democratic" practice, the Archaic Athenian state could not have financed the cost of an efficient, year-round, war-ready cavalry. Whenever an emergency arose, the state expected the wealthiest members of society—and that surely meant those who raised horses as a matter of aristocratic tradition—to furnish their own horses, their own equipment, and their own rations.[137] Besides these expenses, we can assume that the cavalryman, like his Classical counterpart,[138] took along a mounted attendant. Even if the attendant or groom only rode a pack animal, e.g., a mule as suggested by Roman practice,[139] the attendant and his mount still needed to be fed. In the sixth century, at any rate, those who can be identified as attendants on vase paintings are riding *horses*.

Since the Athenian aristocrats probably raised horses primarily for pleasure (e.g., racing and hunting) and not for war, their response to cavalry service, especially in view of our earlier comments on the topography of Attika, may have been less than enthusiastic. Not only did the state not provide military pay, but it simply did not have the financial means nor the incentive to reimburse the cavalryman for the death or crippling of his mount; the hapless cavalryman had to absorb the loss and subject yet another horse to the fortunes of war.

I contend that some aristocrats preferred to fulfill their military obligations as hoplites, perhaps riding to the battlefield and dismounting to fight, as some have argued, rather than as cavalrymen. This spared

[135] So forcefully described in the poems of Tyrtaios, e.g., no. 11, lines 19–20.

[136] Lys. 16.13.

[137] Pritchett, *GSW* I, p. 23, has shown that before the mid-fifth century B.C., a Greek soldier was expected to maintain himself on campaigns (usually of short duration).

[138] Attested during the Sicilian campaign in 413 B.C., e.g., Thuc. 7.75.5 (*akolouthôn*); in 404/3 B.C. in Attika (*hippokomoi*: Xen. *Hell.* 2.4.6); Xenophon recommends the *hippokomoi* to be intermixed with the cavalrymen in order to exaggerate the size of the force: *Hipparch.* 5.6.

[139] See Anderson, *AGH*, p. 137.

one of their most valuable possessions, their horses—an issue of particular importance for *hippotrophoi* of fine-bred, exceedingly expensive, racehorses. L. H. Jeffrey has aptly observed, "Horses, unsuited in any case to the rough terrain of much of Greece, were basically status symbols, too precious to be sacrificed through breaking legs in a massed frontal attack or sustained shoving."[140] There were legal means to force a man to serve in the Athenian cavalry in the fourth century B.C.,[141] but we cannot assume the same held true for the cavalry of the sixth century. Even in Xenophon's day, persuasion appears to have been the usual method of recruiting rather than threats of legal action. The cost of raising a horse or committing a horse to war so greatly exceeded the cost of hoplite armor that the state could not impose cavalry service on any except the very rich—and that fact severely limited the size of the cavalry force.

The naucraric system described by Pollux probably designated which two of the available *hippotrophoi*—or much more likely their young sons who had the leisure time to acquire equestrian skills through practice and training—were to serve as cavalrymen for the year in what was probably a border and coastal patrol. One can only guess how the system operated after the institution of the Kleisthenic reforms in 508/7 or the establishment of the office of the ten generals (*strategoi*) in 501/0.[142] The paucity of available—and perhaps willing—cavalrymen, combined with the relative ineffectiveness of such a force in hoplite warfare, explains in great part why Athens had not developed a respectable cavalry by the time of the Persian Wars. Why did the situation change thereafter? The "regular" Athenian cavalry was a creation of the Athenian empire—an empire whose resources could furnish sufficient tribute to allow Athens to increase the size of its cavalry to 1,000 men, to provide the means of keeping that cavalry war-ready throughout the year, and to reimburse a cavalryman for the loss of his horse in combat. Analogously, is it possible to explain the astonishing rise of Athens to naval preeminence in the late 480s without speaking of the state's decision to commit the newly discovered silver vein at Laureion to the building of a fleet?[143]

[140] Jeffrey, *Archaic Greece*, p. 67. For a similar suggestion by Greenhalgh, with regard to the banning of missile weapons in the Lelantine War in the late eighth century B.C., *EGW*, p. 92. Recently, E. Wheeler ("The Prohibition of Missiles: An Archaic Treaty?" a paper delivered at the American Philological Association annual meeting in San Antonio, Dec. 28, 1986) has questioned the historicity of this prohibition.

[141] Xen. *Hipparch.* 1.10.

[142] *AP* 21.5; 22.2; see Rhodes, *CAP*, pp. 264–266, on the *strategoi.*

[143] See Hdt. 7.144 and *AP* 22.7.

CHAPTER TWO

Cavalry of Empire

"Then for the first time we established 300 horsemen."
(And. 3.5)

SOMETIME between 479 and 431 B.C. the Athenians chose to create a cavalry that was radically different from that of the Archaic period. For one thing, the size of that cavalry is attested variously at 300,[1] 600,[2] and 1,000 men (with an additional 200 mounted bowmen).[3] One thing is clear: after 431 B.C., at the latest, the canonical size of the Athenian cavalry was 1,000 men for the rest of the Classical period. The most logical (and surely correct) explanation is that the Athenians increased the size of the cavalry in stages during the period commonly referred to as the *pentekontaetia*, "the fifty-year period" between the end of the Persian Wars (479) and the beginning of the Peloponnesian War (431).[4]

Our task in the first part of this chapter is to analyze the historicity of these increments, their probable dates, and the reasons for the apparent necessity to increase the cavalry at those times. The second part of this chapter will be devoted to a discussion of the organization and internal operation of this new-style cavalry. The third part addresses certain political considerations.

ESTABLISHMENT OF AN IMPERIAL CAVALRY

Only the two figures of 300 and 1,000 have strong literary support. The scholiast to Aristophanes' *Knights* 627 informs us that "at first the number of Athenian horsemen was 600. Afterwards, after the city had grown in size, the number was increased to 1200." Helbig suggested that perhaps the scholiast had read the next line in Andokides referring to 300 Skythian bowmen and had mistakenly assumed that they were mounted and simply added their number to the 300 Athenian horse-

[1] And. 3.5 (cf. Aischin. 2.173)

[2] Scholiast to Ar. *Knights* 627.

[3] Thuc. 2.13.8; Ar. *Knights* 225 (and scholion to 627); And. 3.7 (cf. Aischin. 2.174); *AP* 24.3 and Philochoros (*FGrH* 328 F 39).

[4] For an excellent account of this period, see R. Meiggs, *The Athenian Empire* (Oxford, 1972).

men.[5] Alföldi, for one, assumed the Skythians were mounted.[6] But Andokides says *toxotai* (bowmen) not *hippotoxotai* (mounted bowmen), and Plassart has convincingly argued that the 300 Skythians were foot soldiers, not mounted archers.[7] An intermediary stage of 600 men is not impossible (though that is not what the scholiast intends by his use of *prôton*, "at first"). A gloss of Hesychios s.v. *hippês* on Aristophanes' *Knights* 225 reports that the Atthidographer, Philochoros, recorded in his fourth book that time when the 1,000-man cavalry was established. Hesychios adds, "for the number of Athenian horsemen was different *from time to time*" (*kata chronon*), i.e., prior to the enrollment of the 1,000.[8] But that need not refer to a stage of 600; it could mean simply 300 and even 96 or some other number unattested in our sources. In the main, the case for 600 is not strong and I consider it only a scholiast's conjecture or simply a mistake.[9] It might be worth noting that the *Suda* gloss, s.v. *hippeis*, (Adler) repeating the scholiast to Aristophanes' *Knights* 627 (Koster) almost verbatim, conspicuously omits this stage of 600 cavalrymen (troubling, however, is the fact that the *Suda* also reports that "at first, the number was 1,000, then after the city had grown, 200 were added to the 1,000"—an incremental sequence otherwise unattested.) In any event, the 200 must certainly be the *hippotoxotai*, here to be included within the general category of cavalrymen.[10]

As for the other two figures, I find no compelling reason to reject the testimony of Andokides (despite the confused nature of his account) to the effect that Athens had at first a cavalry of 300, and later one of 1,000. What remains is to propose probable dates for the two stages and to suggest reasons for the need for a new cavalry, as well as for the increases in it. One fact stands out: military exigencies were primary.

Our sources seem to affirm that military activities during the first seventeen years of the Delian League focused mainly on the fleet and at times on the infantry, e.g., in the battle on the Eurymedon River (ca.

[5] Helbig, *Hippeis*, p. 11.

[6] "Reiterei," p. 30. See M. F. Vos, *Scythian Archers in Archaic Vase-Painting* (Groningen, 1963).

[7] "Les archers d'Athènes," *REG* 26 (1913): 204–5.

[8] So Helbig, *Hippeis*, p. 10. This has some support in the fact that the fourth book of Philochoros, cited specifically by Hesychios, appears to have covered at least the latter part of the *pentekontaetia*: *FGrH* 328 F 39.

[9] Keil, *AA*, p. 144, n. 1 rhetorically asks, "wenn jedes uniformierte Glied auf dem Parthenonfries . . . eine Schwadron vertritt, gab es da um 440 etwa 600 Cavalleristen? 300, 600, 1,000 wäre hier eine natürliche Progression." I respond only by saying that counting specially clothed figures in the Parthenon cavalcade is a risky business.

[10] As in Thuc. 2.13.8; And. 3.7; *AP* 24.3; see Appendix D.

467 B.C.) and in the Messenian Expedition to Mt. Ithome (ca. 463/2).[11] The Athenian cavalry is not recorded to have participated in any of these expeditions.[12] Perhaps for logistical reasons. Thucydides 2.66.2 can be interpreted to mean that the Athenians did not make use of horse-transports until the early years of the Peloponnesian War.[13] This is not to say that the Athenians might not have embarked horses on ships in the past, e.g., before Salamis in 480 B.C., but that due to the special problems involved with transporting horses by sea and accommodating them on ships designed for other uses, the numbers transported were probably quite small and relatively insignificant.

The change in the importance of the cavalry was a direct result of Athens' involvement in the affairs of mainland Greece, starting with Megara ca. 460/59.[14] It is possible that Athens organized a new cavalry sometime between 460 and 457, but the military exigencies probably did not demand it until the pivotal battle of Tanagra in 458/7. During the battle the Thessalians deserted to the Spartans (Thuc. 1.107.7). Diodoros 11.80.1–6 offers a much expanded account of this treachery. We should not dismiss the entire account simply because Diodoros, *contra* Thucydides 1.108.1, claims that the battle was a draw (11.80.6) and that the Athenians even laid claim to the victory. Thucydides furnishes the very briefest of accounts of the battle, and Diodoros' source—probably Ephoros—may have been correct in at least some of the details. But we do know that the Thessalians made a dedication to Apollo at Delphi from the spoils of Tanagra.[15] That should settle the issue of who won, at least.

According to Diodoros, the Thessalians deserted in the thick of the battle. After the battle was over, the Thessalians attacked a large supply train at night coming from Attika. The Athenian escort troops, unaware that the Thessalians were now enemies, naturally welcomed them. Confusion resulted when the Thessalians slew the unsuspecting guards. The main force of the Athenians, having already retired to their camp, received news of the supply train's desperate situation,

[11] Thuc. 1.100.1; cf. Diod. 11.61.4. On the Messenian expedition, see Thuc. 1.102.2.

[12] Demosthenes (13.23; 23.119) recounts that at the siege of Eion (ca. 476 B.C.) a Thessalian nobleman, Meno of Pharsalos, provided a private army of 200 or 300 cavalrymen levied from his own serfs. This does not mean that Athens contributed any cavalry.

[13] Cf. Pliny, *HN* 7.56.209.

[14] See Thuc. 1.105.3ff; Diod. 11.79.1–4. For recent discussion, see R. Legon, *Megara: The Political History of a Greek City-State to 336* B.C. (Ithaca, N.Y., 1981), 183–99.

[15] *SEG* XVII.243; XVIII.207. The Spartans made a thank-offering at Olympia for their part in the victory at Tanagra: Paus. 5.10.4; ML 36, pp. 78–79; for translation, see Fornara, *Archaic Times*, no. 80.

sped to the rescue, routed the Thessalians, and proceeded to slaughter the Thessalians (11.80.5–6). Thereupon the Lakedaimonians came to the rescue of their new allies. A pitched battle ensued, and many were slain on both sides. It is to this second confrontation that Diodoros connects the Athenian claim to victory or at least a draw. Since the matter was still in dispute as night fell, a truce was concluded for four months (11.80.6).

Thucydides speaks of only one battle at Tanagra. It seems probable that the day-long battle is the same battle which Thucydides relates, while the evening action with the supply train was more of a skirmish than Diodoros would like us to believe.[16] The skirmish was probably emphasized as a face-saving device for the Athenians, and it may mask Diodoros' silence about the outcome of the main day's fighting.[17]

The Thessalians' support of the Athenians had not been particularly steadfast in the past either. While an Athenian might not have recalled that the Thessalians had fled the plain of Phaleron in 510 B.C., he certainly would have remembered, prodded by this recent treachery, that the Thessalians had gone over to the Medes in the Persian Wars.[18] Can there be any doubt that the Athenians' subsequent attempt to restore Orestes, the exiled king of the Thessalians, a few years later was fueled in part by their desire for revenge?[19]

The ramifications of the battle of Tanagra are not limited to retribution. Athens now faced the very real possibility of a combined infantry and cavalry (both Boiotian and Thessalian) coalition attacking Attika. Later in the Peloponnesian War this situation was dramatized when the Athenian general Hippokrates exhorted his troops before the battle of Delion in Boiotia (424 B.C.), explaining to them that if they defeated the Boiotians, the Peloponnesians, deprived of cavalry support, would not risk an invasion of Attika again (Thuc. 4.95.2). And in the 360s Xenophon recalls the same danger of an enemy invading Attika supported by its own cavalry and that of allies; he also provides

[16] Paus. 1.29.9 speaks of two engagements on two days with the Thessalians deserting on the second day. Like Diodoros, he claims that Athens did well on the first day, then lost on the second because of the desertion of the Thessalian cavalry.

[17] Unfortunately, Diodoros does not inspire us with confidence. For example, Diod. 11.78.1–2 claims that the Athenians were victorious at Halieis in 459 B.C., yet Thuc. 1.105.1 reports that they were defeated. In addition, note the apparent doublet on the battle of Oinophyta in Diod. 11.81–82 and 11.83; for discussion, see Gomme, *HCT* 1, p. 317.

[18] Hdt. 7.132. The cavalry contribution by Meno of Pharsalos at Eion in 476 B.C. represents the generosity of an individual Thessalian and should not be generalized.

[19] For this expedition to Thessaly, see Thuc. 1.111.1; Diod. 11.83.3–4.

grudging respect for the Boiotian cavalry by declaring that the cavalry tradition of Athens was just as noteworthy.[20]

Unlike any other specifically mentioned battle during the *pentekontaetia*, Tanagra has a decent claim to have witnessed Athenian horsemen in action.[21] In his stroll through the Kerameikos, Pausanias saw a stele on which horsemen were depicted in battle (1.29.6). Pausanias reports that the names inscribed on the stele were those of Melanopos and Makartatos who died fighting the Lakedaimonians and Boiotians on the borders of Eleon and Tanagra.

However, the generally accepted view that Pausanias is referring to the battle of Tanagra in 458/7 has been challenged by B. D. Meritt on the basis of a fragmentary inscription recovered in 1934 from a Turkish wall in the Athenian Agora.[22] The name Melanopos is clearly visible; Meritt has restored the rest of the line to include the second name Makartatos. He argues that by reason of its Ionic letters and "its general appearance" a date near the end of the fifth century rather than the middle is to be preferred. He conjectures that this inscription commemorates some unrecorded battle fought after the Spartan occupation of Dekeleia in 414 B.C. (p. 148). Meritt believes that this epitaph agrees so well with what Pausanias says about the stele of Makartatos and Melanopos that this is in fact the stele that Pausanias describes in 1.29.6.[23] If Meritt is correct in this identification, then the two *hippeis* killed near Tanagra would have no significance for dating the reorganization or the existence of the Athenian cavalry in the mid-fifth century B.C.

However attractive Meritt's interpretation is, a few difficulties remain: (a) while the name Melanopos is plain to see on the stone, Makartatos is only a restoration; (b) Pausanias does seem to be associating the two men with the well-known action near Tanagra in Boiotia in 458/7, so Meritt must assume that Pausanias' guide misinformed him—which is of course not impossible; and (c) Melanopos may not be the same man in both places. In addition, though it is a small point, Pausanias records the two names in the order Melanopos and Makar-

[20] *Hipparch.* 7.2–4.

[21] See Gomme, *HCT* 1, p. 316, for comments.

[22] *Hesperia* 16 (1947): 147–48, no. 36, pl. 23 (=*SEG* x.426; xx.64).

[23] Hölscher, *Historienbilder*, pp. 110–11, accepts the date and identification proposed by Meritt. Davies, *APF* 2921, p. 85, hesitates to commit himself except to note the prosopographical ramifications for the stemma of Makartatos if Meritt is correct. L. H. Jeffrey, "Lykios, Son of Myron: The Epigraphic Evidence," in *Stêlê. Festschrift in Memory of Nikolaos Kontaleon* (Athens, 1980), 51–54, accepts Meritt's conclusions. For bibliography see Pritchett *GSW* 4, p. 168, n. 161.

tatos, yet the inscription has to be restored with the names in the reverse order. Would not Pausanias have been more likely to record the names as they appeared on the stele, especially since the name Melanopos, as Meritt observed (p. 148), (mnemonically and) metrically ends a hexameter verse?

If Pausanias 1.29.6 refers to Tanagra in 458/7, could these two horsemen have been members of the new corps of 300? Busolt and Swoboda[24] thought that the 300 horsemen were organized before the battle of Tanagra, but were careful to add "wohl" and "wahrscheinlich." I am inclined to believe that they belonged to the old cavalry order. The presence of the Thessalian cavalry, allied since 460 B.C. (Thuc. 1.102.4), would have made a radical reorganization of the Athenian cavalry militarily superfluous. After Tanagra, on the other hand, the need for a larger and more effective cavalry force would have been imperative in anticipation of the Boiotian-Thessalian-Spartan alliance.

Another possible proof of the presence of Athenian cavalry at Tanagra is contained in a very fragmentary inscription, *IG* I² 946. A. Wilhelm[25] discovered that the fragmentary lines belonged to an epigram in the corpus of Simonides (*Anthologia Palatina* 7.254). The inscription was then securely restored to read:

> Farewell brave peers of Battle, names renown'd
> Young cavaliers of Athens! for whose peace
> Your own sweet lives ye lost in battle-stormed
> Against almost the complete host of Greece.[26]

Two dates have been suggested for the action so memorialized. Wilhelm connected it with Tanagra, von Domaszewski[27] with the cavalry engagement in 431 B.C. between the Athenian cavalry and their restored Thessalian cavalry allies (Thuc. 2.22.2). Pausanias refers to this action: "And there is a tomb of Thessalian horsemen who had come because of an old friendship, when Archidamos and the Peloponne-

[24] *GS* 2, p. 978 and n. 3; p. 1128 and n. 5.

[25] *Österreichische Jahreshefte* 2 (1899): 221ff; for text and commentary, see D. L. Page, *Further Greek Epigrams* (Cambridge, 1981), pp. 274–76, who argues for its association with the battle of Tanagra; in agreement, see Ch. Clairmont, *Patrios Nomos: Public Burial in Athens during the Fifth and Fourth Centuries B.C.* (Oxford, 1983), nos. 21c, e, pp. 138–41 and Pritchett, *GSW* 4, pp. 180–81.

[26] Trans. by H. T. Wade-Gery, "Classical Epigrams and Epitaphs," *JHS* 53 (1933) 79.

[27] *Sitz. Berichte* (Heidelberg, 1917), Abhand. 7, and *Der Staatsfriedhof*, p. 18, as cited by Wade-Gery, "Classical Epigrams," p. 79.

sians first invaded Attika with their army" and several lines later to the corresponding monument dedicated to "those horsemen who died when they shared the danger with the Thessalians."[28]

Those who favor the later date point out that the presence of the key letter form, the four-bar sigma, at least places the dedication some years after Tanagra, i.e., after 446 B.C. Wade-Gery, however, suggests that the temporal *pote* in line 3 indicates that although the inscription was cut some years after the event, it still refers to Tanagra. He also points out that a four-bar sigma appears in a public document as early as 452/1.[29]

Those who favor the earlier date suppose that the last words "against almost the complete host of Greece" are more appropriate to Tanagra than to the brief cavalry engagement (*hippomachia bracheia*) in 431 B.C.[30] Wade-Gery is reluctant to place much weight on this particular argument as it might have seemed to an Athenian cavalryman that he had encountered "most of the Greeks" when he faced the Boiotian cavalry and the enemy hoplites coming to its relief.

All things considered, the case for Tanagra appears feasible, but not conclusive. In any case, there is nothing in the poem that sheds light on the nature of the cavalry force so praised. Worth noting, however, is the clear reference to the fact that these brave horsemen were youths (*koroi*).

We also possess an important epigraphical document specifically citing the Athenian cavalry from this period. *IG* I² 400 is a dedication by the Athenian *hippeis* which was inscribed on the base of a bronze monument of a young man with a horse erected at the entrance to the Acropolis. In the Augustan period the base was flipped over and reused, and the same inscription was copied on it in archaizing letter forms. Presumably at this time another monument was set up on the opposite end of the entrance, whose base also bore the inscription.[31]

[28] Paus. 1.29.6. This date is followed by A. E. Raubitschek, "The First Athenian Casualty List of the Peloponnesian War," *Hesperia* 12 (1943): 25–26, no. 5 connecting it with *IG* I² 944 and by D. W. Bradeen and D. M. Lewis, "Notes on Athenian Casualty Lists," *ZPE* 34 (1979): 244 (= *SEG* XXIX.60).

[29] Wade-Gery, "Classical Epigrams," on *IG* I² 335, line 14. For evidence of the four-bar sigma as early as the 450s, see M. Walbank, "Criteria for the Dating of Fifth-Century Attic Inscriptions," in *Phoros. Tribute to B. D. Meritt*, ed. by D. W. Bradeen and M. F. McGregor (Locust Valley, N.Y., 1974), 161–69.

[30] Page, *Further Greek Epigrams*, pp. 274–75, n. 1.

[31] For bibliography and discussion of these three inscriptions, see Raubitschek, *DAA*, nos. 135, 135a, b, pp. 146–152 and Anderson, *AGH*, p. 131. See now *SEG* XXX.24 (cf.

Pausanias saw these two monuments in the second century A.D., and conjectured from the name Xenophon preserved in the inscription that the young horsemen might be the sons of the famous Xenophon.[32]

The original inscription has been variously dated: before 445 B.C.,[33] mid-fifth century B.C.,[34] and shortly after Oinophyta in 457 or shortly after the suppression of the Euboian revolt in 446 (Thuc. 1.114.3).[35] The text reads:

> The horsemen [dedicated this monument] from
> the spoils of war; the hipparchs were
> Lakedaimonios, Xenophon, and Pronapes.
> Lykios, son of Myron, from Eleutherai made it.

We are fairly well informed about these hipparchs: Lakedaimonios is none other than the son of our much discussed Kimon. Xenophon, not related to the historian, served as *strategos* during the Samian War (441–438) and again at Poteidaia in 430/29 and at Spartolos where he lost his life in 429. His daughter married the father of the speaker of Lysias' *On the Property of Aristophanes*, who is described as rich, a horseman, and winner of equestrian events at the Isthmian and Nemean Games (14). Xenophon is probably the paternal grandfather of a certain Xenophon attested as a cavalryman shortly after the mid-fourth century B.C. Finally, Pronapes dedicated a monument on the Acropolis to commemorate his four-horse chariot victories at Nemea, Isthmia, and the Panathenaia in the period 450–430 B.C.[36] Clearly, these men belong to families with strong equestrian traditions. We should expect no less from commanders of the cavalry.

Although no Athenian cavalry is attested as participating in the battle of Oinophyta or on Euboia, Raubitschek has commented that "it is conceivable that the Athenians employed a strong detachment of knights in either of the two campaigns" (p. 151). G. F. Hill agrees to the effect that the monument with its inscription "possibly commemorates Oenophyta."[37] If Athens had some sort of cavalry at Tanagra

---

26) and Jeffrey, *Stêlê*, pp. 51–54. For translation and brief notes, see Fornara, *Archaic Times*, no. 83.

[32] 1.22.4; so Anderson, *AGH*, p. 131.

[33] G. F. Hill, *Sources for Greek History*, B.C. 478–431, 2d ed. by R. Meiggs and A. Andrewes (Oxford, 1951), 288, no. 20 and the editor of *IG* I² 400.

[34] Dittenberger, *SIG*³ 51 (following A. Wilhelm).

[35] See Raubitschek, *DAA*, p. 151. Jeffrey, *Stêlê*, pp. 52–53, argues for the latter date and on the basis of letter forms dates it ca. 450–430 B.C.

[36] For the primary evidence and prosopography, see *APF* 8429, XIII (Lakedaimonios); *APF* 5951 and Kroll, "Archive," no. 24 (Xenophon); and *APF* 12250 (Pronapes).

[37] Hill, *Sources*², p. 288; for sources on the battle of Oinophyta, see, p. 342, no. 8.

two months earlier, then it is inconceivable that cavalry would not also have been present at Oinophyta.

Though the exact date of the cavalry reorganization still eludes us, I favor the relatively brief period between Tanagra (458/57) and Oinophyta (457). The victory at Oinophyta, by which the Athenians asserted their authority over Boiotia, would well justify the claim "from the spoils of war" (*apo tôn polemiôn*). Plutarch may confirm the military urgency at hand: he reports that after the Athenians were defeated at Tanagra, they fully expected the Peloponnesians to invade Attika in the coming spring (*Kim.* 17.6; *Per.* 10.2). What better occasion for a reform of the cavalry?

The information provided by Plutarch is linked chronologically, however, with the return from exile of Kimon, a vexed historical issue. Andokides (3.4–5) leaves the impression that the creation of the 300-man cavalry took place in the period of the Five-Year Truce between Athens and Sparta (Thuc. 1.112.1), commonly dated to ca. 451–446 B.C. However, because the text of Andokides goes on to say that the peace lasted thirteen years, modern editors, puzzled by the logical inconsistency and following the lead of Aischines (2.172) who supplies the figure "fifty" instead of "five," have corrected the text of Andokides accordingly. But another ancient tradition maintains that Kimon, ostracized from Athens in 462/1, was recalled in the fifth year of his exile to negotiate a peace treaty with Sparta.[38] W. E. Thompson has ingeniously argued that Andokides meant exactly what the best manuscripts preserve, a five-year peace lasting beyond the stated terms of the treaty, i.e., to thirteen years.[39] According to this interpretation and following the variant tradition, Kimon was recalled from exile to negotiate a peace in 458/7; war between Athens and Sparta did not break out again until 446/5. Taken inclusively, the dates result in a thirteen-year span of peace. Thus, on this line of argument, the creation of the 300-man cavalry could be placed at least in the early 450s. If Kimon was recalled after Tanagra as a peacemaker, it would be tempting to see in the election to the hipparchy of his son Lakedaimonios a vote of approval for the measure. But chronological worries persist.

In any event, the participation of the cavalry at Oinophyta may not

[38] Theopompos (*FGrH* 115 F 88); Nepos *Cimon* 3.3, and discussion, A. E. Raubitschek, "Kimons Zurückberufung," *Historia* 3 (1954–55): 379–80; but see criticism by W. R. Connor, *Theopompus and Fifth-Century Athens* (Washington, D.C., 1968), 24–30, n. 17, pp. 149–151; D. Kagan, *The Outbreak of the Peloponnesian War* (Ithaca, N.Y., 1969), 91–94; and H. Bengtson, *Griechische Staatsmänner des 5. und 4. Jahrhunderts v. Chr.* (Munich, 1983), 103–4.

[39] "Andocides and the Peace of Cimon," *Phoenix* 38 (1984): 216–20.

have directly affected the outcome of the battle, and therefore the silence of the literary tradition about their presence is not conclusive.[40] One should also not forget who commissioned the monument and dedication—the cavalrymen themselves. Their estimation of their importance during the battle might not have been entirely realistic. In addition, though I believe that he neglects the importance of *apo tôn polemiôn*, Anderson may be near the mark when he suggests "perhaps we may suppose that [the monument] celebrated the foundation of the new corps as much as its first success."[41]

It is of course quite possible that the monument celebrates the success of Euboia. After Perikles had led an army across to Euboia to quell a revolt, he received word that Megara had in turn revolted, had destroyed the Athenian garrisons stationed there, and had called on Korinth, Sikyon, and Epidauros for aid. To meet this emergency, Perikles hastily returned to the mainland with his army, only to find that the Spartan king, Pleistoanax, had already invaded Attika with this combined force of Peloponnesians and had laid waste the Eleusinian and Thriasian plains. Pleistoanax withdrew his army (under suspicious circumstances), allowing Perikles to return to Euboia and subdue the island (Thuc. 1.114.1–2). Plutarch (*Per.* 23.2) reports that Perikles banished the famous *Hippobotai* from Chalkis (apparently they had regained power sometime after 506 B.C., probably during the Persian Wars; see Hdt. 6.100.) If the *Hippobotai* actually fought as cavalry, then the Acropolis monument may dramatize the victory of the 300-man cavalry over the famous "horse-breeders" of Chalkis.[42] But this scenario still presupposes that the new-style cavalry with its three hipparchs had existed prior to the revolt of Euboia. The events move too quickly for a reorganization and deployment during the hectic campaigns. But how far back? In my view, the military demands brought on by the Athenian hegemonial designs in mainland Greece and the

[40] However, Polyainos 1.35.2 reports an incident, presumably on the eve of the battle of Oinophyta, where Myronides orders his troops to plant their weapons (*ta hopla*) and look about them. Stressing the wide extent of the Theban plain, he tells them that if they flee, the pursuing Theban cavalry will cut them down, but if they stand fast, there is hope for victory. This Athenian army appears to be exclusively infantry (*ta hopla*), but his observation may be simply the acknowledgement of the numerical superiority of the Theban cavalry who are fighting on their own ground. In any case, Myronides does not say that the Athenians have no cavalry of their own.

[41] *AGH*, p. 131.

[42] Analogous, perhaps, to the restoration of the quadriga dedicated from the spoils of the Athenian victory over Boiotia and Chalkis, the date of that restoration being associated with Oinophyta or the Euboian revolt in 446 B.C.; see ML 15 for a brief discussion and earlier literature.

hard lesson learned from Tanagra should move the date of the establishment of the 300 back into the 450s. In any case, the fact that the Spartan king was able to ravage the plains of Attika in 446 may have convinced the Athenians that 300 horsemen were not sufficient to help defend Attika and may have prompted their decision to increase their numbers to 1,000, as was certainly the case by 431 B.C. at the latest. The renewed threat of an independent and hostile Boiotian cavalry after the signing of the Thirty-Year Peace in 445 B.C. would have made this decision even more imperative. Certainly the events in the Archidamian phase of the Peloponnesian War (as we shall see in the next chapter) bear out this concern.

There is one last piece of evidence that may influence one's decision whether to associate *IG* I² 400 with Oinophyta or with Euboia. We have mentioned that Xenophon, one of the three hipparchs, died while serving as *strategos* in 429 B.C. If we agree that the highest military offices in Athens ordinarily required a candidate to be at least thirty years old, then, if associated with Oinophyta, Xenophon would have been at least in his late fifties at the time of his death. This is approaching the upper limits of normal military service, though it is certainly not unheard of for commanding officers, e.g., Phokion in the fourth century, elected general forty-five times (Plut. *Phok.* 8.2). On the other hand, if he were hipparch ca. 446 B.C., the issue of age would bear little concern. The same consideration applies to Kimon's son, Lakedaimonios: he served as *strategos* to Kerkyra in 433/2.[43]

One of the most interesting aspects of *IG* I² 400 is the number of hipparchs mentioned. It can be accepted that these three hipparchs held office simultaneously, since their names are joined in a dating formula.[44] One is led to conclude, therefore, that ca. 457–445 (the approximate date of the inscription) the commanders of the Athenian cavalry numbered three each year. As a result, scholars have cautiously approved the neat arithmetic equation of three hipparchs commanding three hundred horsemen, adding support to the testimony of Andokides.[45] Martin, incorrectly I believe, rejected the Andokidean account and consequently the existence of an Athenian cavalry numbering 300 in the mid-fifth century B.C.[46]

[43] ML 61, line 8; Thuc. 1.45.2; Plut. *Per.* 29.1; see *APF* 8429, pp. 305–6.

[44] Raubitschek, *DAA*, p. 151.

[45] Keil, *AA*, pp. 144–45; Busolt-Swoboda, *GS* 2, p. 1128 and n. 5; Gomme, *HCT* 1, p. 328 and n. 1: "probably about this time [post 454 B.C. following Diodoros for the date of the Peace of Kimon], the first regular cavalry corps of three hundred was established," and Anderson, *AGH*, pp. 130–31.

[46] *Cavaliers*, p. 367.

A. Wilhelm[47] ingeniously suggested from information in the *Athenaion Politeia* that the three hipparchs represented the two hipparchs elected from all the citizens (61.4) plus the hipparch sent to the island of Lemnos (61.6). B. Keil considered Wilhelm's suggestion "unprobabel."[48] Keil is surely correct. There is no evidence earlier than the fourth century B.C. that Athens even dispatched a hipparch to Lemnos.[49] Besides, the two hipparchs who commanded the later 1,000-man cavalry are always distinguished from the hipparch to Lemnos.[50] This applies to the Hellenistic period as well, when the Athenian cavalry had been reduced to 200–300 men.[51]

All things considered, there is no good reason to reject the testimony of Andokides concerning the 300 cavalrymen, and, at the same time, the case for the three hipparchs having served in the same year. It makes good sense to associate these two pieces of information with one another.

Although the battle of Oinophyta is the best claimant for this cavalry monument, which is so impressive both in itself and because of its placement at the entrance to the Acropolis, still, the words *apo tôn polemiôn* might also refer to other known military actions, e.g., to Myronides' subjugation of the Opuntian Lokrians and the Phokians in 457 B.C. (Thuc. 1.108.3) or even to Athens' punitive expedition against Thessaly ca. 456/5 (Thuc. 1.111.1; Diod. 11.83.3–4).

The Thessalian expedition fails to attract our attention because, quite frankly, it is described as an Athenian military failure. The Athenians besieged Pharsalos but could not take the city. Consequently, they could not restore Orestes to the throne. Thucydides explains the reason for their failure:

> The Athenians were masters of the land as long as they did not venture much beyond their camp (for the Thessalian cavalry prohibited their movement), but they did not take the city nor achieve any other

[47] *Göttingische Gelehrte Anzeigen* 165 (1903): 773.

[48] *AA*, p. 144, n. 1.

[49] See Appendix B.

[50] For the Classical period, see *AP* 61.4, 6 (over and against Demosthenes' ambiguous implication that the hipparch to Lemnos was one of the two regularly elected hipparchs: 4.26–27); *IG* II² 102 (= Tod, *GHI* 2, no. 129, 375–373 B.C.); *IG* II² 104 (= Tod, *GHI* 2, no. 134, 368 B.C.); E. Vanderpool, "Victories in the Anthippasia," *Hesperia* 43 (1974): 311–13 (two inscriptions, late fourth–early third century B.C.).

[51] For the Hellenistic period: Threpsiades-Vanderpool, p. 104, lines 7–10 (cavalry increased from 200 to 300 men), lines 25–27; p. 110, lines 12–13 (two inscriptions, early third century B.C.); *IG* II² 957.II.39–41, 158–157/6 B.C.

goal for which they had undertaken the expedition. In the end they returned home with Orestes, having failed in their efforts. (1.111.1)

We should not be put off by Athens' apparent failure to achieve its goals in Thessaly, prominent among which was revenge for Tanagra because of the Thessalian treachery (Diod. 11.83.3).[52] The Athenians did make themselves master of the land—if ever so briefly and of modest compass—and Diodoros 11.83.4 includes this expedition among Myronides' rapid successes.[53] As masters of some part of the land, the Athenians must have acquired some "spoils of war." Analogously, although Agesilaos in his campaign against the Akarnanians in 389 B.C. was accused of having accomplished nothing because he had failed to capture any city (Xen. *Hell.* 4.6.13), we are told earlier in the narrative that he ravaged the land, captured cattle, horses, stock, and great numbers of slaves, made a public sale of the booty (4.6.6) and even set up a trophy after an engagement with the Akarnanians (4.6.12). In other words, victory and defeat can be relative.

Although there is no evidence to confirm it, the Athenian cavalry may have proved its bravery during the withdrawal from Thessalian territory. Just as the Thessalian cavalry caused considerable consternation to Agesilaos' rear guard on his return to mainland Greece in 395/4,[54] it no doubt harassed the Athenian forces all the way to Thermopylai. Since we hear of no extensive Athenian losses—though there must have been some—during the expedition or on its return,[55] can we not suppose that a combined Athenian-Boiotian-Phokian cavalry,[56]

[52] See A. E. Raubitschek, "Menon, Son of Menekleides," *Hesperia* 24 (1955): 286–89: Menon of Thessaly was probably ostracized immediately following Tanagra.

[53] Diodoros conflates all the successes from Oinophyta to Thessaly into one continuous campaign, though he does separate the battle of Oinophyta from the other events with the temporal expression, *meta tauta* (11.83.2). Thucydides split his material between 1.108 and 1.111, interposing the final phases of the Egyptian campaign (454 B.C.). The Thessalian expedition is portrayed as a separate campaign, perhaps one or two years after Oinophyta.

[54] Xen. *Hell.* 4.3–7, *Ages.* 2.2–5; Plut. *Ages.* 16.5; and Paus. 3.9.12–13. Similarly, who can forget what the Syracusan cavalry did to the Athenian army on its retreat inland in 413 B.C. (Thuc. 7.78.3, 6–7)?

[55] Paus. 10.15.2 offers a teasing piece of information: to Apollo at Delphi, the Thessalians of Pherai dedicated statues of certain mounted cavalry commanders to commemorate their victory over the *Athenian cavalry*. Is it too bold to connect this victory with the Athenian invasion of Thessaly and thus to confirm the presence of Athenian cavalry? Or can it somehow be associated with the dedication also offered to Apollo at Delphi by the Thessalian cavalry from the spoils of Tanagra (see above)?

[56] Since the Boiotians and Phokians accompanied the Athenians as allies (Thuc. 1.111.1), it is possible that they contributed horsemen as they were to do for the Lakedaimonians in the Peloponnesian War (Thuc. 2.9.4).

though not numerous enough to match the awesome Thessalian cavalry in a major clash outside the walls of Pharsalos, might have prevented heavy Athenian losses during the withdrawal by delaying tactics in the rear? We should note in this respect that it was the cavalry of Agesilaos that prevented the Thessalian cavalry's approach to his main army.[57]

It seems unlikely, then, that after the treachery at Tanagra the Athenians would have failed to include some cavalry in their expedition against the foremost cavalry power in Greece, particularly if, as I have argued, they now possessed a cavalry 300 strong. And would the Boiotians and Phokians have been any more trustworthy than the Thessalians had been at Tanagra?

To sum up this section, immediately after the Persian Wars, despite the telling effect of the Persian cavalry during the Plataian campaign, Athens entered a period in which the navy lorded it over the other military arms. A change in the old cavalry system was simply not justified by military demands. When the Athenians began to intervene in mainland affairs in the 450s, they did not hesitate to enlist the Thessalian cavalry, as the Peisistratidai had done in the past. But after Tanagra, they must have recognized the need for an effective cavalry corps, one composed of their own citizens.[58] This is the time most appropriate for the new organization of the 300 cavalrymen.

Organization of an Imperial Cavalry

Up to this point, we have been referring to a new-style cavalry of 300 men without providing much by way of detail. What was different about its organization? Who were to be its recruits?

The wealthiest Athenians—the ones most likely to raise horses as a matter of interest and tradition—were expected not only to participate in the political life of the city but to perform various liturgical duties, most noteworthy of which was the trierarchy, which exempted them at least for the year from the new, continuous cavalry service. Also, some wealthy families, particularly those whose wealth lay not in land, probably did not engage in *hippotrophia*, e.g., Demosthenes the orator (Dem. 18.320; see *APF* 3597, pp. 123–38). We cannot simply assume

[57] Xen. *Ages.* 2.5; Plut. *Ages.* 16.5 reports how proud Agesilaos was of his cavalry (which he had personally organized in Anatolia) that it had defeated the Pharsalians, "so reputed for horsemanship."

[58] See T. Dohrn, "Phidias, Perikles, und Athen," in *Symbola Coloniensia* (Cologne, 1949), 71–99 and repeated by J. A. Davison, "The Date of the Prometheia," *TAPA* 80 (1949): 93.

they did because they belonged to the liturgical class. Their sons, however, in their twenties and early thirties, many without an independent means of income, like Pheidippides in Aristophanes' *Clouds*, became prime candidates for the new corps—as in fact they had been in the old system. The difference was that the young men were now expected to serve throughout the year, not just during an emergency. This is exactly the reason Xenophon cites to justify the state spending forty talents a year on its cavalry in the fourth century (*Hipparch.* 1.19). The additional burden placed upon their fathers, especially during years of liturgical obligations, cannot have been welcomed by some. Consequently, to defray the cost imposed upon the cavalryman expected to provide year-round service and to compensate for the risks associated with the demanding Attic terrain, the Athenians introduced a new loan called *katastasis*. And consistent with the recently introduced policy of pay for hoplite service, the state was to furnish a food allowance, called the *sitos*, to cover the daily sustenance expenses for man and horse.

What follows is a reconstruction of the operation of recruitment, the grant of state loans and subsidies, the scrutinies of the cavalrymen and their mounts, and the probable length of cavalry service. It draws upon evidence from both the Classical and Hellenistic periods, but care is taken not to assume that every aspect of the process originated in the mid-fifth century B.C. In this regard I mention pointedly the information contained in *AP* 49.1–2.

In the fourth century B.C., a special board called the *katalogeis* had the responsibility of enrolling new recruits for cavalry service (*AP* 49.2). I have argued elsewhere that this was a recent innovation of the 340s or 330s and that before this period the commanding officers performed this function.[59] During the period of the 300-man cavalry, this meant the three hipparchs; when the cavalry was increased to 1,200 (1,000 *hippeis* and 200 *hippotoxotai*) sometime between 445 and 431 B.C., this responsibility was assumed jointly by the newly created board of two hipparchs and their ten tribal commanders, the phylarchs.[60] The cavalry commanders, like the *strategoi*, were elected by

[59] Glenn R. Bugh, "Introduction of the *katalogeis* of the Athenian Cavalry," *TAPA* 112 (1982): 23–32.

[60] Xen. *Hipparch.* 1.9–12 seems to place most of the recruiting duties personally on the hipparch, but his earlier remarks (1.8) that the phylarchs are elected to relieve the burdens of office from the hipparchs make it clear that all the officers participated in the process, for example, see Pamphilos, hipparch during the Korinthian War, (Lys.) 15.5 (*APF* 9667, p. 365); Orthoboulos, hipparch or phylarch also during the Korinthian War, Lys. 16.13 (*APF* 9667, pp. 364–65; though listed by Davies in his *Wealth and the Power of Wealth in Classical Athens* [New York, 1981], 153 as phylarch); and even at a time after the introduction of the *katalogeis* (attested only once after *AP* 49.2: *SEG* XXI.435,

popular vote from the wealthiest and most powerful Athenian families, and not surprisingly, received no pay. The two hipparchs were elected from the whole citizen body; the phylarchs one from each tribe.[61]

The new recruits assembled along with the existing cavalrymen before the Council of Five Hundred, presumably at one predetermined time in the year, probably shortly after the new hipparchs and phylarchs had taken office (i.e., after 1 Hekatombaion, around July 1). According to *AP* 49.2, the hipparchs and phylarchs opened the tablet (*pinax*) on which were inscribed the names of existing cavalrymen. The names of those who requested release from service on the basis of bodily incapacity and were willing to swear an oath to that effect, were deleted from the cavalry register. The new hipparchs and phylarchs, in consultation with their retiring counterparts, surely must have had prior knowledge of approximately how many *hippeis* had died in the course of the year and also of those who hoped to secure release from service in order for the necessary number of recruits to have been found prior to the scrutiny. Despite this, the two sides of the equation probably rarely matched up perfectly. In the period of the *katalogeis*, these men may have routinely enrolled a greater number of potential recruits to cover any discrepancies in the tally. Perhaps some young men in each tribe escaped service because they were not needed in a given year. There was a general expectation, however, that the system would catch up with them eventually (Xen. *Hipparch.* 1.11). The newly enrolled were summoned before the council. Those who swore that they were incapable of serving in the cavalry did so at that time on two grounds: bodily incapacity or lack of sufficient wealth.[62] These

line 27, 187/6 B.C.), see eleven phylarchs (twelve tribes in 282/1 B.C., but one name, belonging to the Aigeis tribe, is missing from the stone) and the two hipparchs praised for "establishing" 100 additional cavalrymen to join the existing 200-man force in 282/1 B.C.: Threpsiades-Vanderpool, p. 104, lines 6–11. A cooperative effort is only logical; after all, the phylarchs knew their tribesmen better than the hipparchs—except for hipparchs' own tribesmen, of course. It is also possible that since all the military officers were elected after the sixth prytany of the Athenian year (*AP* 44.4), i.e., around January–February of the Julian calendar, yet presumably did not take office until 1 Hekatombaion (around July 1), the hipparchs- and phylarchs-elect participated in the process of recruiting men who would in fact serve under them, not under the current officers.

[61] See [Xen.] *Athenaion Politeia* 1.3 and *AP* 61.4–5; cf. 4.2 ("Drakontian constitution").

[62] Xen. *Hipparch.* 9.5 implies that (a) rich men who were physically unfit, (b) orphans holding large estates, and (c) resident aliens (cf. *Poroi* 2.5) were exempt from cavalry service. But this information is not altogether helpful—unfit men, rich or otherwise, were naturally exempt; orphans because they had not yet attained their majority, i.e., eighteen years old (thus no military responsibilities) and had not yet come into full possession of their patrimony (thus no civic responsibilities); and metics who were in fact barred from

were released. The council then judged the qualifications of the remaining recruits; those whom they approved were duly inscribed on the *pinax*.

In addition to the *pinax*, called a *sanidion* in 403 B.C. (Lys. 16.6), we know that the *hippeis* were identified as such on the deme registers, *lexiarchika grammateia*, administered by the demarchs. *IG* I³ 138 (= *IG* I² 79), dated to the 440s, records that the cavalrymen and hoplites listed on the deme registers are to pay a tax to Apollo Lykeios of two drachmas and one drachma, respectively.[63] These registers may have served as a working tool for recruitment purposes, establishing precedents and patterns of cavalry versus hoplite service.[64]

---

the cavalry, though expected to serve in the infantry. This last exemption must have something to do with aristocratic elitism associated with the cavalry, but legally to do with the inability of metics (resident aliens) to own land in Attika (unless granted *enktesis*), therefore voiding the traditional link between landed estates and *hippotrophia*. For brief remarks, see D. Whitehead, *Ideology of the Athenian Metic* (Cambridge Philological Society, 1977), 82, 127–28.

[63] *SEG* XXX.5; for text and discussion, see M. Jameson, "Apollo Lykeios in Athens," *Archaiognosia* 1 (1980): 213–23 and D. Whitehead, *The Demes of Attica 508/7–ca. 250 B.C.* (Princeton, 1986), 35, n. 130; 135-36. On *lexiarchika grammateia* for military purposes, compare the famous "Themistokles Decree" of 480 B.C.: ML 23, lines 29–30 and M. Jameson, "The Provisions for Mobilization in the Decree of Themistokles," *Historia* 12 (1963): 385–404. On the question of a central hoplite register in Athens, see M. H. Hansen, "The Number of Athenian Hoplites in 431 B.C.," *SO* 56 (1981): 19–32. For its connection with ephebes, see Lyk. *Against Leokrates* 76; Tod, *GHI* 2, no. 204; and P. Siewert, *Der Eid von Plataiai* (Munich, 1972). For translation, see Fornara, *Archaic Times*, no. 57.

[64] More on the *lexiarchikon grammateion*. The information contained in Pollux 8.104 s.v. *lexiarchoi* to the effect that "one who has been inscribed on the *lexiarchikon grammateion* has already inherited his patrimony (*ta patrôia*); for family property is called a *lêxis*," cannot really mean what it implies (although from a reading of Is.7.27 one can see how it might be so construed)—many of the young *hippeis* (and hoplites for that matter) listed on this register (*IG* I³ 138) cannot have yet inherited their father's estates. The gloss in the *Suda* s.v. *lexiarchikon grammateion* (Adler, 462) has yet other problems: "the names of those Athenians who have reached an age to hold office (*archein*) were inscribed on this register, with their demes being added. And from these registers, they were allotted the offices." But this would imply that they were not registered until they were thirty, the minimum age for most *archai*. This would make more sense with respect to a proper age for inheriting property, but it no doubt has to do with the alternate use of *lêxis* in the sense of election to office by lot. But a second entry on *lexiarchikon grammateion* in the *Suda* (463) may provide the solution: "on this register were inscribed the names of those who had reached maturity, for whom it was possible to assume charge of their patrimony." This surely means eighteen years old, the age at which an Athenian *was allowed* to inherit and assume formal control over his patrimony—particularly important for orphans, e.g., Demosthenes. For others, it was the official register guaranteeing right of inheritance upon the death or incapacitation of their fathers, whenever that happened to occur.

The next step in the process must have been the granting of the *katastasis* to the new recruits.[65] Within the last decade there has been a revolution in our understanding of this unique state loan to Athenian cavalrymen. Prior to the discoveries in the Kerameikos and Agora in Athens of hundreds of lead tablets belonging to the cavalry archives of the fourth and third centuries B.C.,[66] the exact nature of *katastasis* could only be conjectured. The first reference appears in two lines from a comic play by Eupolis entitled *Philoi* ("Friends")[67] whose production is dated between 429 and 425 B.C.[68] The lines translate: "You have not displayed good sense, old man, by rashly accepting the *katastasis* before even learning horsemanship."

The most informative evidence for *katastasis* comes in the Lysianic speech (16), *In Defense of Mantitheos* 6–7 on which the lexicographer Harpokration again provides comment. In this speech, dated between 393 and 389 B.C., we learn that at his examination for admission to the Council of Five Hundred, Mantitheos was charged with having served as a cavalryman under the Thirty Tyrants (404/3). In the course of his defense, Mantitheos mentions the fact that after the restoration of the democracy in 403, the phylarchs were ordered to determine who had actually served in the cavalry that year in order to force them to return their *katastaseis* (16.6). Harpokration explains *katastasis*: "it seems to be money (*argyrion*) which the 'established' (*katastathentes*) *hippeis* received from the public moneys for the *katastasis*." He goes on to quote a lost work by the comic poet Plato entitled *Syrphax* ("Litter") which supplies the information that the *argyrion* was given back when the cavalrymen left service and were replaced by other men (who would in turn receive the *argyrion* for *katastasis*).

These are the earliest references to the noun form, *katastasis*, but I have argued elsewhere that this grant is also implied in a number of cavalry-related testimonia in which the verb form, *kathistêmi*, appears, most notably Andokides when speaking of the "establishing" of 300 horsemen and the later 1,200 *hippeis*.[69] In fact, I am convinced that

[65] An ancient grammarian mistakenly called the *dokimasia* of the *hippeis* by the council the *katastasis*: I. Bekker, ed., *Anecdota Graeca*, 3 vols. (1814–21) 1:270; they are indeed related, but not technically the same.

[66] See Braun, "Funde," pp. 197–269 and Kroll, "Archive," pp. 83–140.

[67] Contained in a gloss by Harpokration to Lysias 16.6–7 s.v. *katastasis*.

[68] For date, see J. M. Edmonds, *The Fragments of Attic Comedy*, vol. 1 (Leiden, 1957), frag. 268, p. 407, note "e"; (a new edition [Kassel] of Eupolis is in progress) Ch. Fornara, "Kleon's Attack Against the Cavalry," *CQ* 23 (1973): 24, has detected a possible reference to *katastasis* in a fragment of Philochoros (*FGrH* F 93) concerning Kleon in the 420s.

[69] 3.5, 7; for other possible references, see G. R. Bugh, "The Three Hundred Athenian

the creation of the state grant called *katastasis* led to the technical adoption of *kathistêmi* to mean "enroll" or "enlist" when specifically applied to the Athenian cavalry.[70]

But what exactly did it mean to be granted "establishment money"? Kroll has cogently argued that it was "a loan made by the state to each recruit when he was formally enrolled ("established") in the cavalry primarily to assist him in the purchase of his mount."[71] By an exhaustive study of the lead tablets from both the Agora and the Kerameikos, Kroll suggests that the state grant probably had a maximum limit of 1,200 drachmas.[72] This amount causes some concern (see later, Chapter Five): it matches the price of a very fine horse, of race quality, not a hardy war-horse, in the Classical period.[73] A very respectable sum, indeed, approximating four years' wages for a skilled workman and further emphasizing the fact that cavalry service was not for poor men, nor even the middle classes. And one point cannot be forgotten—*katastasis* was a loan, not a gift; it had to be repaid upon retirement from service.

Interestingly enough, the lead tablets of the Athenian cavalry archives do not record, technically speaking, the *katastasis*, but rather another closely related procedure known as *timêsis*, "evaluation," also perhaps attested in two third-century and one second-century B.C. inscriptions.[74] The lead tablets themselves (Figure 5), approximately 100 from the Agora dating to the fourth and third centuries and approxi-

---

Cavalrymen," *Phoenix* 36 (1982): 309–11. Kroll, ("Archive," p. 99) on the other hand, connects the first use of *katastasis* with the creation of the 1,000-man Athenian cavalry.

[70] To mean here, the full execution of the process: initial enrollment, scrutiny before the Council of Five Hundred, and the grant of *katastasis*, not simply "enroll" in expectation of service, as those called the *katalogeis* (="enrollers") would perform later (*AP* 49.2); see Bugh, "Athenian Cavalrymen," p. 311.

[71] "Archive," pp. 97–98.

[72] "Archive," p. 99.

[73] Attested by Ar. *Clouds* 21–23; Lys. 7.10; and Xen. *Anab.* 7.8.6 (here Xenophon owns a horse worth 50 darics, approximately 1,200 drachmas, but he apparently brought several horses on this Persian campaign [3.3.19], so perhaps this one was intended more for show/riding than for actual combat. It is clear that he dearly loved this horse: having been forced to sell it for the money, his friends bought it back and gave it to Xenophon as a gift [7.8.6]). For discussion of horse prices, see Anderson, *AGH*, p. 136 and Kroll, "Archive," p. 89. For a serviceable mount, a hack, 300 drachmas appears to have been sufficient: Is. 5.43.

[74] Threpsiades-Vanderpool, p. 104, lines 14–15, 282/81 B.C. (=*SEG* XXI.525; cf. L. Moretti, *Iscrizioni storiche ellenistiche*, vol. 1 [Florence, 1967], no. 16); *IG* II² 672, lines 22–23 (280/79 B.C.), as restored by Ch. Habicht, "Neue Inschriften aus dem Kerameikos," *AM* 76 (1961): 135 (for comments, see J. and L. Robert, "Bulletin épigraphique," *REG* 77 [1964]: no. 132); and Habicht, "Neue Inschriften," p. 129, lines 27–28, 188/7 B.C. (=*SEG* XXI.435).

mately 570 from the Kerameikos dating to the third century, provide valuable detail on the administration of the cavalry: the owner's name, the color of his horse, the brand marked on his horse, and the monetary value in drachmas of his horse, sometimes introduced by the term *timêma*, "value," or some abbreviation of the word, e.g., *tim* or *ti*. It is clear that the tablets are related to the *timêseis* recorded in the inscriptions cited above.

Kroll's analysis of the tablets has shown that they were a record keeping device to render account of the depreciating value of the cavalry mounts over the years of service. An exact and up-to-date record of a horse's worth allowed the state to reimburse the cavalryman the actual price of his horse if it were incapacitated or killed while in service. This guaranteed that the cavalryman did not claim the loss of a horse worth more than was justified; the state in turn demonstrated its fairness in light of the expected hazards of war and full-time service. Long ago, Martin had suggested this possibility on the analogy that the rich citizens responsible for equipping warships (*triremes*) of the Athenian fleet were not held financially culpable for their ships should their charges be lost in bad weather or war.[75] The fact that the Council of Five Hundred had close supervision over both the cavalry and trierarchies only strengthens this comparison.[76] Again, when the trooper retired from service, he was still obligated to repay the *katastasis* in full in order to finance his replacement. Therefore, the *timêsis* procedure guaranteed fair reimbursement for loss to a cavalryman while engaged in cavalry service; it was related to *katastasis* only in its initial and maximum amount.

In addition to the ongoing *timêseis*, no doubt carried out by the cavalry commanders and their secretaries (*grammateis*)[77] on their own authority—as befitting the day-to-day bureaucratic demands—the cavalrymen and their mounts underwent several examinations, called *dokimasiai*, overseen by the Council of Five Hundred.[78] The two op-

[75] *Cavaliers*, p. 345.

[76] See Kroll, "Archive," p. 98, n. 39 and P. J. Rhodes, *The Athenian Boule* (Oxford, 1972), 175.

[77] Attested in three inscriptions: (a) in the cavalry inscription of 282/1 B.C., they are referred to in the plural *grammateis tôn hipparchôn*: Threpsiades-Vanderpool, p. 105, lines 41–42; (b) only one *grammateus* accompanied the cavalry escort for the Pythaid procession to Delphi in 106/5 B.C.: *FD* III.2, no. 28.III.38–40; and (c) in a second-century B.C. inscription only one *grammateus* is assigned to the hipparchs (*[gramm]ateus hipparchois* phy[etôn?]: M. Crosby, "Greek Inscriptions," *Hesperia* 6 (1937): 460–61, no. 8.)

[78] For brief discussion, see Kroll, "Archive," pp. 85–86. For discussion of the process of *dokimasia*, see F. Borowski, "Dokimasia: A Study in Athenian Constitutional Law"

erations, *timêsis* and *dokimasia*, are clearly differentiated in the cavalry inscription of 282/1 B.C. (lines 14–15). In the fourth century, the council examined the horses to see if they had been cared for properly. If they had not, the guilty owners were fined their grain allowances (*sitoi*). The council also observed the horses in action to see if they could keep pace with their fellows or if they refused to submit to formation drills. Those horses which exhibited these traits failed the *dokimasia* and were barred from future cavalry service by being branded with the sign of a wheel—the mark of rejection (*AP* 49.1). It was in the cavalryman's own interests to feed and train his mount properly, though that must not always have been the case.[79] If his horse were rejected, he would have had no choice but to sell it, certainly at a much reduced price, and purchase another more expensive one which could pass the next inspection. The *timêma* reimbursement would not have been applicable in such a case of apparent malfeasance.

We know that the cavalry displays (*epideixeis*) that formed part of the *dokimasiai* took place at various locations, e.g., in the Akademy, in the Lykeion,[80] at Phaleron, and in the hippodrome (*Hipparch.* 3.1). One particularly famous cavalry display was the *anthippasia*, which Xenophon places in the hippodrome.[81] It was a mock cavalry battle; the cavalry split into two squadrons of five tribes commanded each by a hipparch. The two sides charged one another and rode through each other's ranks (*Hipparch.* 3.10–13). A fair bit of epigraphical evidence from the fourth and third centuries attests to the popularity of this display and confirms its place at the festivals of the Greater Panathenaia and Olympieia.[82] There is no reason to doubt that this was true for the fifth century as well.

(Unpubl. diss., University of Cincinnati, 1975), 27–49, though certain sections are now out-of-date due to Kroll's publication of the lead tablets.

[79] See Anderson, *AGH*, p. 139. Cf. Plato *Hipparchos* 226 A: a horseman who feeds his horse poor grain will ruin his horse.

[80] On the cavalry parades at the Lykeion, see R. E. Wycherley, "Xenophon *Hipparchicus* 3.6–7: Cavalry at the Lyceum," *CR* 13 (1963): 14–15 and Kyle, *AAA*, p. 80.

[81] See the lexicographical literature: Hesychios; the *Suda*; and I. Bekker, ed., *Anecdota Graeca* 1:404. For general discussion, see Martin, *Cavaliers*, pp. 261–63; Reisch, "anthippasia," *RE* I.2 (1894) 2378ff; and Kyle, *AAA*, pp. 189–90 (*anthippasia*) and 95–97 (hippodrome and its suggested location).

[82] See *IG* II² 379 (321/20 or 318/7 B.C.); *IG* II² 3079 (ca. 280/79); *IG* II² 3130, mid-fourth century B.C. (cf. L. Moretti, *Iscrizioni agonistiche greche* (Rome, 1953), 66–68, no. 28); B. D. Meritt, "Greek Inscriptions," *Hesperia* 15 (1946): 176–77, no. 24; W. K. Pritchett, "Tribal Decree for Anthippasia Victor," *Hesperia* 9 (1940): 111–12, no. 21 (although apparently associating it with the Olympic Games; it properly refers to the Olympieia at Athens: H. W. Parke, *Festivals of the Athenians* [Ithaca, N.Y., 1977], pp. 143–44, 198–99 with n. 186; Kyle, *AAA*, p. 46; and M. B. Walbank, "Regulations for

We also possess two fine artistic representations from the fourth century which celebrate victories in the *anthippasia*: the first is the now well-known Bryaxis base whose sculptured reliefs depict three horsemen and their prize tripods (Figure 6).[83] The accompanying inscription (*IG* II² 3130) identifies the celebrants as Demainetos, son of Demeas, of the deme Paiania, and his two sons, Demeas and Demosthenes, all of whom had been phylarchs at one time or another.[84] This monument is exhibited in the National Archaeological Museum of Athens. The second, a sculptured relief, recently discovered in the Agora, shows a defile of galloping young horsemen led by a bearded horseman, probably the phylarch (Figure 7).[85] On the back of the relief are inscribed the words: "The tribe of Leontis was the winner." The *anthippasia* is the most likely choice for the equestrian event depicted here, but we cannot be absolutely certain. The relief is on display in the portico of the Stoa of Attalos in the Agora.

To return to the subject of state support, Xenophon informs us that in the fourth century the Athenians spent forty talents a year on the cavalry (*Hipparch.* 1.19). This expenditure must represent the grain allowance, the *sitos*. We have already mentioned that it was withdrawn from a horseman if he had not properly cared for his horse. The *sitos* for the cavalry is first attested in 410/9 in irregular allotments from the Treasurers of Athena Polias and Athena Nike to the Hellenotamiai,[86] but it was probably first introduced in conjunction with the

---

an Athenian Festival," *Studies in Attic Epigraphy, History, and Topography, Hesperia Supplement* XIX, presented to E. Vanderpool [Princeton, 1982], 179, n. 28; for the festival in the 330s, see *IG* II² 1496, lines 88 and 113); T. Leslie Shear, Jr., "The Athenian Agora: Excavations of 1970," *Hesperia* 40 (1971): 271–72, pl. 57, c; and E. Vanderpool, "Victories in the Anthippasia," *Hesperia* 43 (1974): 311–13.

[83] For a suggested reconstruction of the trophy above the base, see N. Pharaklas, "Ti istato epi tês Baseôs tou Bryaxidos?" *AD* 24 (1969): 59–65.

[84] That is, not in the same year from their tribe. I interpret the clause in the cavalry inscription of 282/1 (Threpsiades-Vanderpool, p. 104, lines 11–14) which gives evidence of an apparent scarcity of phylarchs prior to this legislation (so Threpsiades-Vanderpool, p. 106) and which abolishes the law that prevented phylarchs from being elected from "all the horsemen" to mean that before that date only one phylarch could come from a given tribe (his own), but that now more than one phylarch could come from a tribe in a given year (see Chapter Six for further discussion). This may parallel the change in the election of the generals (*AP* 61.1), i.e., from "all the Athenians."

[85] *Hesperia* 40 (1971): 271–72, pl. 57, c.

[86] *IG* I³ 375 (= I² 304A), lines 4, 8, 9, 11–12, 24 (see ML 84). Here the *sitos* allotments appear in prytanies 1, 3, 4, and 7, totaling over sixteen talents (Anderson, *AGH*, p. 138; the twenty-two talents mentioned in ML 84, p. 259 cannot be correct); for brief discussion see W. K. Pritchett, "Loans of Athena in 407 B.C.," *AncSoc* 8 (1977): 37. Recently, I. G. Spence, "Athenian Cavalry Numbers in the Peloponnesian War: *IG* I³ 375 Revisited," *ZPE* 67 (1987): 167–75, has argued that (a) the *sitos* payments for prytanies III

grant of *katastasis* in the mid-fifth century. In 300/299 the *hippeis* honored the Treasurers of Athena for having assisted the hipparchs in securing the requisite grain allowance from the state, along with other unnamed good services.[87] But we should not assume that in either case this involvement was customary.[88] In the detailed cavalry inscription of 282/1 the hipparchs and phylarchs are praised for having seen to the distribution (*dosis*) of the *sitos* "in each prytany" of their year of service.[89] In 351 B.C. Demosthenes in his *Fourth Philippic* (28) recom-

and VI (restored) have been doubled to compensate for the lack of payments in prytanies II and V respectively; (b) *sitos* payments can be recovered in their approximate amounts from the payments for unspecified purposes listed in prytanies VIII–X; (c) consequently, the prytany by prytany *sitos* allotments for the cavalry in 410/9 are now sufficiently complete to suggest that at a rate of one drachma *per diem* per horseman "cavalry numbers fell below the 1,000 strong establishment figure and that by 410/09 B.C. at least they had dropped to slightly over half their total in 431 B.C." (p. 175). Spence's argument is not persuasive. I do not doubt that the number of Athenian horsemen serving in the years after the Sicilian Expedition had fallen from the canonical 1,000, but there is no evidence that it was as low as ca. 500–600 men. In fact Diod. 13.72.7, recording a battle in Attika in 407 B.C., would lead us to believe that the Athenian cavalry had scarcely diminished in size (see further remarks, Chapter Three, footnote 17). This evidence, albeit inexact, cannot be dismissed. And *IG* I³ 375 cannot be used to substantiate Spence's claim. First of all, it still remains that the inscription records *hippois sitos* in only four of the ten prytanies. But more importantly, these *irregular* payments should be interpreted as loans from Athena to the state at a time of fluctuating fiscal distress (see footnote 88 below)—as Pritchett ("Loans of Athena," p. 37) has observed, this document does not detail the "normal expenditures for maintenance of the military establishment." Therefore, the *sitos* allotments recorded (or unrecorded) in *IG* I³ 375 were never intended to be a complete tally and do not accurately reflect the size of the Athenian cavalry in that year.

87 *IG* II² 1264, lines 2–11.

88 The year 410/9 was the first year of the restored democracy after the Revolution of 411 B.C., therefore probably reflecting troubled financial times (ML 84, p. 258). The year 300/299 appears to be a time of peace, but 301 B.C. had marked the beginning of significant political and financial changes in Athens, one of which apparently resulted in the abolition of the office of the Treasurers of Athena: W. S. Ferguson, *HA*, pp. 129–30; Ferguson, *The Treasurers of Athena* (Cambridge, Mass., 1932), 126–27 (*IG* II² 1264 marks the last appearance of the Treasurers of Athena in our records: p. 126). Perhaps of relevance, in an inscription, *IG* II² 1272, dated 283/2, we find reference to the secretary of the "Treasurers of the Sitonikon" being honored by the soldiers serving in the frontier garrisons of Attika for doing a fine job with "the distribution of the grain" (*peri tên tôn sitôn dosin*; cf. wording in Threpsiades-Vanderpool inscription, p. 104, lines 19–20: *tês tôn sitôn doseôs*.) And a newly discovered and unpublished inscription from the 1982 excavations from the northwest corner of the Athenian Agora can be plausibly restored to have similar wording: I 7551, line 11. I wish to thank John Camp, Assistant Director of the Agora Excavations, for directing me to it and T. L. Shear, Jr., Director, for permission to examine it in Athens in the summer of 1985.

89 Threpsiades-Vanderpool, p. 104, lines 18–21. Also, Ch. Habicht, "Neue Inschriften," p. 129, lines 30–31, restores the word *sitos* in the inscription honoring a hipparch in 188/7.

mends that Athens send an expeditionary force to oppose Philip of Macedon; that force was to include 200 cavalrymen, each of whose grain allowance (*sitêresion*)[90] would amount to thirty drachmas a month, i.e., one drachma *per diem*. Some scholars have accepted this figure as standard for the Classical period,[91] but the issue is complicated by a reference to a four-obol rate in a fragment of a Lysianic speech dated to 403/2.[92] I will have occasion to deal with this problem in more detail in Chapters Four and Five.[93]

We cannot leave our discussion of the administration of the Athenian cavalry without addressing the question of the length of cavalry service. There is no doubt that Athens recruited its cavalrymen from well-to-do young men (Xen. *Hipparch.* 1.9–12); the *hippeus* Dexileos, who died fighting at Korinth in 394 B.C. at the age of twenty, is a good example.[94] More problematic, however, is the normal length of cavalry service. There is a body of evidence from the Classical and Hellenistic periods which seems to indicate that at least some of the cavalrymen continued to serve into their middle age despite the rigors of this branch of the military. On the basis of the prosopography of the third-century cavalry documents of the tribe Erechtheis, Kroll observes that "most of the cavalrymen . . . were still active more than a decade afterwards, and that a cavalry career of 20–45 was probably the norm." Kroll goes on to say that "the names of as many as five Athenian *hippeis* who participated in the Pythais at Delphi in 128/7 appear 22 years later in the list of cavalrymen who attended the Pythais of 106/5; although some of the repeated names may belong to sons of the 128/7 knights, one or more may equally be of cavalrymen who participated in both festivals, being at least in their 40's at the time of the second

[90] For a thorough treatment of "provisioning" for ancient Greek armies, see Pritchett, *GSW* 1, pp. 30–52.

[91] For references, see Kroll, "Archive," p. 97, n. 36; cf. *IG* I³ 83 (= I² 86; Tod, *GHI* 1, no. 72; Thuc. 5.47.6) where the *sitos* payment for the cavalrymen of the quadruple alliance of 420 B.C. is listed at one Aiginetan drachma (worth slightly more than an Athenian drachma): see Pritchett, *GSW* 1, pp. 16–17.

[92] Lysias, frag. 6, lines 72–81 (B. P. Grenfell and A. S. Hunt, *The Hibeh Papyri* 1 [London, 1906], 49–55); see L. Gernet, *Lys.* frag. 6, pp. 257–59 (Budé); Pritchett, *GSW* 1, pp. 21–23. For a related inscription with discussion and date, see R. S. Stroud, "Greek Inscriptions, Theozotides, and the Athenian Orphans," *Hesperia* 40 (1971): 280–301 (= *SEG* XXVIII.46) and M. H. Hansen, "*Nomos* and *Psephisma* in Fourth-Century Athens," *GRBS* 19 (1978) 320.

[93] For brief remarks, see my "Introduction of the *katalogeis*," pp. 26–27.

[94] See *IG* II² 6217; Davies fails to include him in his *Athenian Propertied Families*, presumably because his family cannot be proved to have belonged to the liturgical class (worth 3–4 talents).

festival."[95] However, inasmuch as 62 *hippeis* participated in the Pythais of 128/7 and 92 *hippeis* in 106/5, out of a possible 200 to 300 *hippeis* (see Chapter Six) on each occasion, the repetition of at most five names argues for the exception rather than the rule. In fact, we know nothing of the mechanism of selection: did they volunteer or were they chosen? The crossover evidence between 128/7 and 106/5, therefore, is clearly not reliable. On the other hand, on the basis of a prosopographical study of the three Pythaides of 128/7, 106/5, and 98/7, I have determined that out of a total of 158–168 *hippeis* attested, it is possible to calculate the ages of forty-five (thus approximately 30%), though only twenty-nine of these forty-five can be securely dated. In my study I have assumed that the men married between twenty-five and thirty-five. Table 1 presents the results (for a list of the specific horsemen examined, see corresponding Appendix A at back). The figures are a little surprising inasmuch as we should expect more *hippeis*

TABLE 1. Age Distribution of the Horsemen of the Pythais

Age	20s	>20s	30s	>30s	40s	>40s	50s
Hippeis	8	2	11	10	5	8	1

in their twenties than appear here. But from these data, the prime age for cavalry service appears to be the thirties, thus lending support to Kroll's conclusions. What I could not determine exactly was when these *hippeis* entered service, but presumably the emphasis on youth still prevailed; why should we assume otherwise?

My primary objection, however, rests with applying the paradigm of cavalry service from the Hellenistic period—when the cavalry numbered at most 300—to the Classical period when it numbered 1,000. The thrust of my argument is that 1,000 cavalrymen represent an artificially high number of men performing cavalry service; that the 200 or 300 attested in the third and second centuries B.C. reflect more closely the number of Athenians who were at once wealthy and also traditionally engaged in horsemanship and cavalry service. Therefore, these Hellenistic horsemen would by reasons of wealth, tradition, and interest be more likely to serve longer than many of their Classical counterparts (see also Chapter Six).

No one can deny that there were older Athenian cavalrymen in the Classical period. Xenophon warns his cavalry commander that he must be alert to recruiting needs, as "it is inevitable that some men will leave service due to old age (*gêra*) and others for other reasons" (*Hip-*

[95] Kroll, "Archive," p. 103.

parch. 1.2). And the hipparch should station the "oldest and most sensible" (*presbytatôn kai phronimôtatôn*) in the rear ranks (*Hipparch.* 2.3). He also observes that some of the "older" *hippeis* (*presbyteroi*) will require their grooms to give them a leg up on their horses, a technique called the "Persian method" (*Hipparch.* 1.17; *Peri Hippikes* 6.12). This would have been acceptable for practice sessions, but what about a *presbyteros* who fell off his horse in the heat of battle? And Meidias, whom Demosthenes so bitterly attacked, served as a hipparch (less likely a simple *hippeus*) when he was over fifty years old.[96] But this very fact may have contributed to Meidias' reluctance to perform his military duties with proper enthusiasm. More on this later. Finally, *AP* 49.2 reports that a horseman could retire from the cavalry only if he swore that he was no longer physically able to serve—the implication being that he was either physically disabled or simply too old.

The comic fragment of Eupolis about the old man requesting the *katastasis* before becoming familiar with horsemanship is problematic; we cannot be sure of the exact intent of the barb—his age, or his inexperience, or both. Without the context it is difficult to say. Certainly, Xenophon assumes in his equestrian treatises that new recruits are not yet skilled horsemen and therefore need to undergo training (e.g., *Hipparch.* 1.12). In the fourth century the phylarch Pheidon (Figure 8) could be seen training his young cavalrymen to mount and dismount near the Herms in the northwest corner of the Athenian Agora,[97] a location frequented by the cavalry officers and no doubt near their headquarters, the *Hipparcheion*.[98]

Despite the fact that the Athenian cavalry was composed of both younger and "older" *hippeis*, the emphasis in Aristophanes' *Knights* is on the youthfulness of the horsemen. Demos asks the Paphlagonian slave (a.k.a. Kleon the demagogue) who it is that does him injustice. He replies: the Sausage-seller and the young horsemen (*hoi neaniskoi*) (731). And the artistic representations of the Athenian *hippeis* consistently depict them beardless, a sign of youth.

All of this leads me to my main point. In the Classical period, at the time when the cavalry numbered 1,000, most of the cavalrymen were

[96] Dem. 21.154, 162.; see *APF* 9726, pp. 386–87.

[97] In the Middle Comedy play, *The Horse-Breeder* (*Hippotrophos*) by Mnesimachos (frag. 4) as reported by Athenaios 9.402. A number of clay sealings inscribed with the name Pheidon as "Hipparch to Lemnos" were uncovered with the lead tablets of the cavalry archive; for analysis, see J. H. Kroll and F. W. Mitchell, "Clay Tokens Stamped with the Names of Athenian Military Commanders," *Hesperia* 49 (1980): 86(89)–96 (and *SEG* XXX.14).

[98] See Appendix C.

in their twenties and early thirties. In addition, I suggest that it was common for these *hippeis* to retire from service after ten to fifteen years. The reasons vary. First of all, the physical demands of cavalry service began to take their toll by the thirties, as with today's professional athletes. Both Xenophon (*Hipparch.* 1.9–10) and Aristotle (*AP* 49.1–2) stress the importance of physical fitness for cavalry service. Lykophron tells his audience that in his eager, civic-minded involvement with *hippotrophia* (here meaning cavalry service), he has continuously taxed his strength (*dynamis*) and his estate (*ousia*; Hypereides 1.16). Secondly, until their early thirties, many of the young horsemen were relatively free from the encumbrances of family, estate, liturgical responsibilities, and political life, for example, like the young men enamored of expensive racehorses (*Hipparch.* 1.11–12) as parodied in the character of Pheidippides in Aristophanes' *Clouds*,[99] and thus available for active duty throughout the year. It is interesting that in the Roman system, a cavalryman was required to complete ten years of service, the minimum military requirement for eligibility for political office—in other words, he could not stand for political office until his twenty-seventh year (Polybios 6.19.2–5).

To return to the point, Xenophon informs us that the state expected a war-ready cavalry force for its money, not one just for seasonal campaigns; to this end he recommends training, individual practice (cf. Ischomachos, Xen. *Oik.* 11.17; *Peri Hippikes*, *passim*), maneuvers in the field, and preparation for the requisite cavalry displays (*dokimasiai*) throughout the year (cf. Xen. *Mem.* 3.3.1–7). What a marvel to see "cavalry advancing by squadrons (*kata taxeis*) in orderly ranks" (Xen. *Oik.* 8.6; *Mem.* 3.3.4)! Practice was essential. It is not surprising that Xenophon felt obliged to counter the charge that all this equestrian practice would be burdensome (*polla echein an pragmata: Hipparch.* 8.5–7). But nowhere does he say that it will not be time-consuming, however pleasant or glorious cavalry service may be envisioned. It might be appropriate to mention here that, unlike the operating norm for the U.S. Cavalry, the Athenian cavalrymen were ordinarily not quartered together in Attika, but lived at home, e.g., Ischomachos. Perhaps only on garrison duty on the Attic frontiers and abroad, e.g., Lemnos, Salamis (see Appendix B), were groups of them actually "barracked" together.

Xenophon may provide some support for the differentiation be-

[99] For an extensive survey of the evidence for the pursuits of the youth of Athens, see A. A. Bryant, "Boyhood and Youth in the Days of Aristophanes," *HSCP* 18 (1907): 73–122.

tween the pursuits of young men and those appropriate to the mature man. In his *Peri Hippikes* (2.1), when discussing whether it is best to break a colt oneself or have it done by a professional horse breaker, Xenophon comments:

> For states lay the duty of cavalry service upon those who are best provided with money (*hoi tois chrêmasi te hikanotatoi*) and who play the major part in politics, and rather than be a horsebreaker, a young man (*neos*) should see to his good health and horsemanship, or if he is already a good rider, should practice riding. And a grown man (*presbyteros*), rather than breaking in horses had far better pass his time in looking after his estate and friends and political and military affairs.[100]

This passage confirms that older, more mature men did participate in the cavalry—this fact is not disputed—but juxtaposed with the typical young recruit still under the authority of his father or guardian (*kyrios*: *Hipparch.* 1.11–12), without independent "money" nor yet playing a "major part in politics"; this passage simply observes (as with *Hipparch.* 1.9–10) that Athenian cavalrymen come from the politically powerful upper-class families. Moreover, the contrast between the young *hippeis* and their guardians as *hippotrophoi* (*Hipparch.* 1.11–12), offers a realistic dichotomy, i.e., *neos* = *hippeus*, *presbyteros* = *hippotrophos.* Of course, we should not press this equation too far.

The nature and purpose of the *katastasis* grant may also have affected the normal length of cavalry service. The 1,000-man cavalry was a composite of very wealthy and less wealthy members (*Hipparch.* 1.9–10), certainly including the liturgical class whose wealth ranged from three to four talents and up.[101] If the cavalry had been maintained at 300 men, the truly wealthy families might have been sufficient to sustain the quota each year,[102] but even here the state recognized the need of providing a loan for the purchase of a sturdy and reliable cavalry mount (as opposed to showy race horses), a practice, significantly, that the Athenians continued into the Hellenistic period. But with the increase to 1,000 men, the state was compelled to recruit from the less

[100] Trans. by Anderson, *AGH*, p. 159; cf. his *Oik.* 3.9–10 for the same advice on horse-breaking.

[101] *APF*, p. xxiv; for a handy categorization (perhaps overly schematic) of Athenian classes by wealth, see L. Casson, "Athenian Upper Class and New Comedy," *TAPA* 106 (1976): 32–33.

[102] Casson, p. 59 comments that "only the very rich, the members of the Three Hundred [elite of Athens' liturgical class] had no financial worries: they could afford to go in for horses and other forms of conspicuous consumption. . . ."

wealthy, from families which did not raise horses normally and which had no genuine equestrian traditions. No matter what the level of your wealth, you still had to repay the *katastasis* to the state.

Kroll has observed that in the third century tablets there appears to be an average depreciation in the value of the cavalry mounts of 100 drachmas a year.[103] If a new recruit spent his entire 1,200-drachma loan on a horse, and it depreciated on the average of 100 drachmas a year, then within twelve years or so his horse would have depreciated in value to a point where it could no longer pass the *dokimasia* of the council and would consequently have been branded and rejected. This is an important point. It would have been impossible to fob off a branded horse on an unsuspecting buyer. The *hippeus* who continued to serve in the cavalry will have had to purchase another mount at his own expense (there is no evidence that the state granted a second *katastasis*, which in any case would also have had to be repaid upon retirement) and yet still have to repay the original 1,200-drachma loan upon eventual retirement.

How many of the less wealthy families could afford or would want to undertake this double cost? Admittedly, there are a lot of variables in the figures: Kroll comments that some horses "evaluated at the maximum 1,200 drachmas retained the same evaluation for as much as three years in succession, doubtless, as we have seen, because most of them were actually worth more than this sum" (p. 94). But this would not have been the case for the less wealthy who would have had no reason to spend more than their state allotment for a horse. Even with astute and active horse-trading or selling, as can be interpreted from the tablets (p. 94), inevitably the amount of the original loan would be exhausted and the *hippeus* would find himself the less-than-proud owner of a horse which could not pass the scrutiny. Only the input of additional private capital for the purchase of a more expensive horse would prevent this diminution, a case perhaps best represented by Arkesas of the Erechtheis tribe who replaced his chestnut valued at 600 drachmas with a black valued at the maximum allowable 1,200 drachmas, and, as Kroll has suggested, probably worth much more.[104] But is Arkesas representative of cavalry personnel? In the third century perhaps: he was the son of Nikogenes, son of Arkesas, of the deme Euonymon who is attested as one of the hipparchs of 282/1 B.C.[105]—in other words, a man with certain wealth and established cavalry traditions. But even Kroll has conceded:

[103] "Archive," pp. 93–94. [104] "Archive," p. 90.
[105] Threpsiades-Vanderpool, p. 104, lines 25–26.

> Though small, the 4th century sampling implies that during the century that separates the earlier and later groups of tablets the over-all values of the horses used by the Athenian *hippeis* had increased. This could be evidence that the cost of war horses had risen in the early Hellenistic period. But the more probable explanation may be simply that a higher standard was maintained for the horses of the 3rd century cavalry, which was a much smaller, more exclusive, and presumably *per capita* wealthier body than the cavalry of the preceding century. (p. 89)

What of the active service careers of the horses themselves? Xenophon comments at the end of his treatise *Peri Hippikes* that "if a man buys his horses well, trains them so that they can stand work, and uses them properly in the training for war, in the exhibition rides (*pros epideixin hippasiais*) and on the battle-fields, what is there then to hinder him from making horses more valuable than they are when he takes them over?"[106] Aristotle informs us that a mare reaches her full length and height in five years, a stallion in six; and that in the following six years they reach full body bulk (*to plêthos*) and continue to improve until their twentieth year (*HA* 576b4–8). And modern writers have observed that "horses ten years of age and older seem better able to withstand hardships than younger horses."[107] But I am inclined to

[106] *Peri Hippikes* 11.13, Loeb trans. (passage curiously omitted by Anderson in his translation). In a conversation between the wealthy Kritoboulos and Socrates in his *Oikonomikos* (dated 390 to 371 B.C.: P. Chantraine, *Xenophon Economique* (Paris, 1949), 5–6, Budé), Xenophon has Socrates point out to Kritoboulos that some fall into poverty by keeping horses, while others profit from it and that as long as the state "has compelled him (i.e., Kritoboulos) to maintain horses" (*hippotrophia*, referring back to *Oik.* 2.6; cf. Hyp. 1.16), the more knowledgeable he is about them, the more likely his horses will be useful and will bring a profit when he sells them (3.9). Socrates goes on to say that horses and human beings share something in common—upon reaching a certain age, they become useful and continue to improve. It is interesting to note that Kritoboulos, like his famous father and friend of Socrates, Crito, seems to have married quite young, before twenty-five years old (*APF* 8823, pp. 336–37), and at the time of the setting for this dialogue (420s), he will still have been in his twenties. The many civic liturgies, e.g., choregia, gymnasiarchy, prostateia "to which he was or might be subjected" (Davies, *APF* 8823, p. 337), listed at *Oik.* 2.5–6, cannot have been required of him in one year and must be a conflation of his liturgical career over decades; thus we cannot say how many of his liturgies were synchronous with his cavalry service. Moreover, when Xenophon says that Kritoboulos will be required to be a trierarch and also to pay war-taxes (*eisphora*), "if war breaks out," he cannot be speaking about a time in the Peloponnesian War (431–404) unless we assume it occurs during the short-lived and ill-observed Peace of Nikias, 421–415 B.C. So problems remain.

[107] F. L. Devereux, Jr., *The Cavalry Manual of Horse Management* (New York, 1979), 49.

think that Aristotle is exaggerating on the upper range (and he is not talking about war-horses) and that Xenophon's optimistic remarks apply to the exceptional horse rather than to the typical one. Obviously, much depends on the quality of care—and that on the interest, time, and money committed to the task. Not every *hippeus* had a keen eye for horse flesh; after all, Xenophon wrote *Peri Hippikes* not for the expert, but for the novice.

Both ancient and modern authorities on horses concur that the average duration of a horse was twenty years.[108] Since horses do not fully mature until their fifth to sixth year,[109] the U.S. Cavalry, logically, applied that age as the minimum for the purchase of cavalry mounts; it was considered unsound and risky practice to fully train cavalry horses much younger.[110] That being the case, "the period of usefulness will average about ten to eleven years.[111] It is possible that ancient practice permitted training at a younger age (as Xenophon seems to suggest: *Peri Hippikes* 1.2–2.5, 3.2, 9.1), but surely not too much younger; otherwise, the serviceable lifetime of the horse would have been correspondingly reduced. Another modern expert has estimated that the average serviceable career of a U.S. Cavalry mount was thirteen to fifteen years, but that during the Late Indian Wars, it dropped to ten years because of "long marches, often poor feed, shortage of grain, and abuse."[112] And we should not forget that these modern cavalry horses wore horseshoes, which cut down on the damage to their hooves on

[108] Arist. *Historia Animalium* 576a27–28; Devereux, *Cavalry Manual*, p. 49; General W. H. Carter, *Horses, Saddles, and Bridles* (Baltimore, Md., 1906), 223–24.

[109] At a time when they shed their milk teeth: Arist. *Historia Animalium* 576b12–13. Up to this point in their development one can accurately judge the age of a horse on the basis of his teeth. Thus Xenophon's remarks at *Peri Hippikes* 3.1 concerning the risk of purchasing a cavalry mount beyond this age—you may get a good buy, but it becomes extremely difficult to determine the age of a horse and thus the length of serviceability; this makes the prospects of reselling a gamble.

[110] Of an age "not less than five nor more than nine," according to the Quartermaster General for the U.S. Cavalry in 1876–77 as quoted by R. Steffen, *The Horse Soldier 1776–1943*, vol. 2 (Norman, Okla., 1978), 204–5; cf. J. Boniface, *The Cavalry Horse and his Pack* (Kansas City, Kans., 1903), 53–54 who states that the average age of European and American horses at purchase was (3)4 to 8, but he goes on to say "while it is good to buy the new cavalry horse between three and a half or four years of age to perhaps six or seven, in order that while still young he may be taught the cavalry work he must perform, yet horses under five are rarely sufficiently developed to stand what field service demands of him" (p. 119).

[111] Devereux, *Cavalry Manual*, p. 49.

[112] These figures generously supplied *per litteras*, July 24, 1984, by Henry B. Davis, Jr., former member of the U.S. Cavalry before World War II, former director of the Cavalry Museum at Ft. Riley, Kansas, and sometime consultant on cavalry matters for the Center of Military History.

rough terrain. Can we assume that the mounts of the Athenian cavalry fared better than this on comparably rocky and uneven terrain? It is very unlikely.

Drawing upon ancient and modern practice, therefore, I suggest that an Athenian *hippeus* could not have reasonably expected his mount to remain serviceable beyond, say, fifteen years, and probably some years less than that. Of course, for a splendidly wealthy *hippeus*, like Ischomachos,[113] the question of the serviceable longevity of his mount(s) was trivial—he could continue to perform his civic and military duties beyond the "lifetime" of his *katastasis* loan. But surely the same cannot be claimed for the majority of *hippeis* whose family wealth was not unlimited, and who at best could afford one horse.

As a result of these considerations, I am tempted to suggest that organizers of the new cavalry system in the mid-fifth century B.C. established the maximum allowable loan of 1,200 drachmas on the basis of the known average serviceable lifetime of a cavalry mount, and that they could not have been unaware of the social dynamics of cavalry service. And we should not be surprised to discover, considering the artificiality of the Athenian cavalry numbering 1,000, most of whose members, I contend, came not from the very wealthy classes and did not have equestrian traditions, that there would be a natural propensity to seek retirement at an age (early to mid-thirties?) when they became distracted by other commitments, e.g., marriage, family, estate, etc.; at a time when they would need to purchase a new mount at their own expense, if they continued in cavalry service; at a time when full-time military service lost some of its glamorous and youthful appeal, when they themselves could no longer identify so closely with the carefree, perhaps frivolous interests of the young recruits. The image of the Athenian cavalry as youthful is not the device of artists. Most of the Athenian *hippeis* in the Classical period, I suggest, were exactly that.

A question will naturally arise. Was it even possible for a *hippeus* in his early thirties legally to retire? If one takes the testimony of *AP* 49.2 at face value, a cavalryman could only retire from service if he claimed physical disability. This contrasts with the information provided in the later passage concerning eligibility—to wit, that a potential recruit could gain exemption if he swore he had not the financial means or physical health to serve. Aristotle seems to imply that protestations by current *hippeis* to the effect that they had fallen on hard times while serving were not taken into account. This implication gains some support in the earlier recommendations by Xenophon (*Hipparch.* 9.5) to

[113] See *APF* 7826, XIII–XIV, pp. 265–68.

allow those who object to serving in the cavalry to buy their way out, as if they could not retire otherwise. But Xenophon does not supply any information as to the age of those who might want to retire—his concern is to rid the cavalry of disinterested, and therefore useless, personnel.

In any case, Aristotle cannot be giving us all the facts—even if his information about restrictions on retirement is applicable not only to his own time, but to the fifth century as well.[114] Certainly a cavalryman who had lost his fortune or estate could not have been expected to continue serving; in this respect, analogies with liturgical responsibilities are apt. If an Athenian felt that he had been unfairly burdened with a liturgy when another possessing greater wealth had escaped, he could challenge the decision by offering to exchange property, a process called *antidosis*.

It seems clear to me that Aristotle has given us only an abbreviated account; he assumes most of the existing cavalrymen will not have suffered a significant financial setback during the tenure of their cavalry careers. Moreover, a *hippeus* who had been appointed to the trierarchy,[115] or elected to one of the archonships, the Council of Five Hundred, or the courts and perhaps many other offices was not obligated to serve in the cavalry during his year in office, thereby gaining at least a year's exemption.

Furthermore, we do not know what constituted a legitimate case of physical disability nor how often such a claim would have been contested or tested. There is at least no denying the fact that cavalry service was typically more arduous than hoplite service, and so the range of acceptable excuses was probably broader. Finally, we do know that some *hippeis* retired from cavalry service on grounds other than old age—we have already mentioned that Xenophon in his *Hipparchikos* (1.2) notes that some men retire from service because of old age *or for other reasons*, unspecified. I suppose that the last could simply mean, for reasons of physical disability, i.e., a specific medical problem (or even death), as opposed to a general deterioration of physical health concomitant with old age. But this seems too narrow.

[114] See my "Introduction of the *katalogeis* of the Athenian Cavalry," pp. 23–32 for discussion of elements in Aristotle's account that appear to be recent innovations.

[115] Although P. J. Rhodes, "Problems in Athenian *eisphora* and Liturgies," *AJAH* 7 (1982): 1–19, esp. 4–5, believes that Athenian cavalrymen "ought to have been exempt from the trierarchy," whereas "there is no reason why they should not have performed festival liturgies." On the last I would not disagree, but there is no reason why a rich Athenian could not have been temporarily withdrawn from cavalry service to perform the trierarchy. How often this happened we cannot say.

In fact, we know of two relatively young men, Mantitheos and Phainippos, who succeeded in leaving cavalry service with no hint that they had claimed exemption on the basis of bodily incapacity. Moreover, these were wealthy young men, which only goes to confirm that not all members of the wealthiest classes of Athens saw cavalry service as a long-term career option.

Mantitheos, we recall, petitioned to be removed from the cavalry in order to serve as a hoplite in the campaign at the Haliartos River in Boiotia in 395 B.C. (Lys. 16.13), and was still serving as a hoplite in the following year in campaigns around Korinth (16.15–17). And on the occasion of this Lysianic speech involving his scrutiny for admission to the Council of Five Hundred sometime between 392 and 389 B.C., Mantitheos gives no indication that he had rejoined the cavalry in the interim. Unfortunately, there is no evidence one way or the other that he did after this speech. It is worth noting that he was "certainly little if at all over 30 years of age at the time (392–390)."[116]

The case is admittedly somewhat unusual, but the probability remains that the young aristocrat Mantitheos, having performed cavalry duty in his twenties, resigned from that service, not on the basis of physical infirmity or even loss of financial means, but from commitments of a different and varied sort. Why should he have been the target of a recruiting officer again after a lapse of as many as five years and after entrance into the civic responsibilities presumed of a mature man of thirty? Was there not an ever-expanding pool of available young men each year, just as was the case with the modern draft in the United States in the early 1970s? The older a man gets, the less likely he is to be drafted; and in the case of Mantitheos, he had already seen some cavalry service.

The second case involves Phainippos, whom Pseudo-Demosthenes in the 320s describes as young (certainly in his twenties),[117] an avid horse breeder (*agathos hippotrophos*),[118] rich (*plousios*), and physically fit (*ischyros*; Ps.-Dem. 42.21).[119] Yet, Phainippos sold his warhorse and secured his retirement from cavalry service after only a few

[116] *APF* 9667, p. 365.

[117] Probably born ca. 355–350 and "nearer 20 than 30" at the time of this legal dispute: *APF* 14734, p. 553.

[118] Cf. Dem. 18.320 where he sarcastically describes Aischines and his pro-Macedonian faction each as "a great and splendid horse-breeder" (*megas kai lampros hippotrophos*). For same tone, cf. Lyk. *Against Leokrates* 139.

[119] See also G.E.M. de Ste. Croix, "The estate of Phaenippus (Ps-Dem. XLII)," in *Ancient Society and Institutions. Studies presented to Victor Ehrenberg on his 75th Birthday*, ed. by E. Badian (Oxford, 1966), 109–14.

years (42.24). How he accomplished this is not explained. Wealth can buy many things. But the fact remains that he apparently accomplished something which *AP* 49.2 states that one was not supposed to be able to do. Did he swear falsely before the council on this point?

Though the relative chronology is uncertain, perhaps the brief cavalry career of Phainippos mentioned in Pseudo-Demosthenes 42 fell before the formal regulation governing resignations from cavalry service in *AP* 49.2; could this case have prompted its introduction or formalization? But, as we have noted earlier, Xenophon, writing in the 360s, also seems to say that a cavalryman could not leave service simply because he desired to do so; otherwise, the author would not suggest that disinclined cavalrymen might want to pay a monetary sum (*argyrion*) in order to buy their way out of service (*Hipparch.* 9.5).[120] But again, at what age? Xenophon may be thinking primarily of young men in their twenties who had no legitimate excuse not to be serving in the cavalry. Dikaiogenes III, the subject of Isaios' lawsuit *On the Estate of Dikaiogenes*, dated to 389 B.C., if he had been given the chance would have jumped at Xenophon's suggestion. Despite the fact that his grandfather (by adoption), Menexenos I, had died while serving as phylarch at Spartolos in the Chalkidike in 429 (Thuc. 2.79), Dikaiogenes III, around thirty at the time of the speech, concerned himself so little with cavalry service while a *hippeus* that he had purchased a war-horse worth no more than 300 drachmas—a mere hack;[121] nor had he deplenished his resources in *hippotrophia*, in this case, chariot-racing.[122]

But inasmuch as Phainippos was in possession of both his father's and grandfather's properties—though he is accused of having performed no liturgies nor paid the war-tax (*eisphora*) expected of the rich—it may be, as Davies suggests, that we should assume that "Phainippos had not long come of age in the early 320s and knew that

[120] In Demosthenes' speech *Against Meidias*, we learn that the hipparch Meidias repeatedly tried to avoid cavalry service when fighting broke out on Euboia in 349/8 and to that end he introduced something called the "two percent cavalry tax" (*hippikê pentekostê*, 21.166). But there are many uncertainties as to the exact meaning of this innovation; see Chapter Five.

[121] On the cavalry tablets, "the lowest normal evaluation is 300 drachmas": Kroll, "Archive," p. 89.

[122] Is. 5.43. The charge leveled at Dikaiogenes III in section 46 that he had not served in the military at all, if true, surely refers to campaigns (*strateiai*), probably in the Korinthian War as the context implies, not service *per se*; for discussion and prosopography, see *APF* 3773, pp. 145–46; W. Wyse, *The Speeches of Isaeus* (Cambridge, 1904), 470–81; and R. F. Wevers, *Isaeus, Chronology, Prosopography, and Social History* (The Hague, 1969), 42.

the tax-structure would catch up with him sooner rather than later, but in the meanwhile was doing nothing about it" (p. 553). Phainippos may have appealed to his cavalry officers or to the council arguing that the onerous obligations of assuming control of property holdings of his patrimony precluded his active involvement, at least temporarily, and that in any case he would be serving the public sector in the near future (whether true or not) in a liturgical capacity because of his wealth. If this is true, it would suggest that the length of cavalry service was dependent for some *hippeis* on the increasing complexity and extent of private and public commitments. The difference is that Phainippos assumed those personal commitments at an unusually early age and thereby shortened his cavalry tenure.

To summarize, I believe that it was legally possible for a *hippeus* to retire from cavalry service on grounds other than physical incapacity. First of all, the prosopographical evidence, albeit sparse, supports such a conclusion. Secondly, the information in Xenophon's *Hipparchikos*, particularly at 9.5, is open to interpretation and too ambiguous to be decisive. Thirdly, *AP* 49.2 either misleads by its brevity or reflects a period proximate to the composition of the *Athenaion Politeia* (320s) when the Athenians may have imposed new restrictions governing exemptions from cavalry service, perhaps in conjunction with the introduction of the board of *katalogeis* (see Chapter Five). In my view, for most of the Classical period at least, Athenian horsemen could gain release from the cavalry after a reasonable term of service, say ten–fifteen years, and many of them did so as they approached their mid-thirties. A variety of socially comprehensible reasons contributed to this phenomenon, e.g., lack of wealth, lack of genuine equestrian traditions, and increasing civic and personal commitments.

Athenian Politics and the Imperial Cavalry

Having directed our attention to military considerations that prompted the creation of a new-style cavalry and to the organization of that cavalry, it is relevant to note certain political factors which might have played a supporting role in the decision, that is to say, the establishment of the cavalry corps of 300 horsemen may have been an attempt to reconcile the aristocracy to the new democratic political order after 461 B.C.

From Thucydides 1.107.4,6,[123] we learn of an oligarchic plot at the time of the battle of Tanagra. The creation of a state-supported cavalry

[123] Cf. Plut. *Kim.* 17.4–7, *Per.* 10.1–3.

may have been intended as such to win support or at least to split the ranks of the oligarchic opposition. What the aristocrats were losing in the political arena might be compensated for by military and monetary concessions. Along these lines, Gomme has stated:

> This was by no means a democratic measure. . . . Just as, in spite of Salamis and the radical democracy, the hoplites preserved their social as well as their military distinction, so did the nobility survive, and its status, or at least its wealth, was recognized by the institution of cavalry.[124]

B. Keil, on the other hand, interprets the creation of this cavalry as a democratic measure. He argues in the following manner:[125] Because of the land warfare after 461 B.C., the state could no longer depend on the half-voluntary service of a class with possible sympathies for Sparta. Thus it became necessary to form a cavalry which was unconditionally available to the state and possibly free from political tendencies (p. 141). By supplying the grant of *katastasis*[126] and *sitos*, the state induced obedience and gained complete control over the cavalry (p. 142). These grants opened the door to a certain number of individuals of the cavalry census for whom it was too costly to maintain a cavalry mount.[127] Thus a democratically disposed cavalry became possible. However, with the increase in the size of the cavalry from 300 to 1,000 sometime between 445 and 431, more of the old cavalry class became interested and joined the corps. This led to a reassertion of the old political *esprit de corps* of the cavalry class, best portrayed in Aristophanes' *Knights* and the die-hard loyalty to the Thirty Tyrants to be evidenced in 404/3. Thus, ironically, out of a cavalry corps developed a cavalry class.

This thesis has merit, but it cannot be accepted as it stands. For instance, there is no evidence to support the claim that the political temperament of the 300 horsemen initially recruited was significantly different from that of the 1,000 horsemen a few years later. Aristophanes' comedy and the actions of the cavalry in 404/3 support no recent convergence of disparate political attitudes. Of course, the addition of 700 horsemen further widened the field of eligibility. More moderately

[124] *HCT* 1, p. 328.

[125] *AA*, pp. 141–44.

[126] For Keil, as for his contemporaries, *katastasis* meant some sort of state loan for equipment, not the substantial loan for the purchase of a war-horse, as we now know.

[127] Cf. A. Andreades, *A History of Greek Public Finance* (Cambridge, Mass., 1933), 218.

well-to-do, as opposed to wealthy, Athenian families were expected to purchase and maintain a war-horse year-round, as I have argued.

But Keil may be correct inasmuch as this new cavalry system resulted in an increased number of "democratically disposed" personnel in the cavalry. It may have been conceived as a measure of compensation and reconciliation aimed at the conservative upper-class elements of Athenian society. Hippias seems to have attempted a similar policy when he allowed opposing aristocratic families to occupy the archonship in the 520s.

The Athenian cavalry was increased to 1,200, which included 1,000 cavalrymen and 200 mounted archers, sometime between 445 and 431 B.C. As I noted briefly earlier, this significant increase in strength may have been prompted by military exigencies such as Pleistoanax' easy invasion of Attika in 446 B.C., by anticipation of invasions of the same type if war broke out again,[128] and by the hostility of neighboring Boiotia, independent since 446 B.C. The Athenians may have been forced to acknowledge that the 300 Athenian cavalrymen could not possibly stop a serious Spartan invasion, particularly if supported by Boiotian cavalry.

Perikles himself may have anticipated the political benefits from the expansion of the 300-man cavalry. He and his *hetairoi*, who incidentally were accused by comic critics of being the new Peisistratidai,[129] sought to gain aristocratic support, or perhaps neutrality, by inviting aristocratic participation in the dividends of empire. This concession is consistent with Perikles' own aristocratic heritage. We recall that Perikles had his two sons Paralos and Xanthippos taught to be the foremost *hippeis* in Athens.

In particular, during the struggle for supremacy between Perikles and Thoukydides son of Melesias, Perikles may have personally argued for the needed increase in the size of the cavalry, not unaware that it might garner for himself additional upper-class support—not only by the increase in the corps, but by the requisite multiplication of cavalry command positions. In essence, twelve officer positions were created out of three, that is to say, two hipparchs and ten phylarchs out of three hipparchs and possibly a hipparch to Lemnos.[130] Clearly, a sig-

[128] Plut. *Per.* 22.2–3 expressly states that Perikles was fearful of battle with the Spartans in 446 B.C., so he had to bribe the young Spartan king. Pleistoanax had ravaged as far as Eleusis and Thria (Thuc. 1.114.2; 2.21.1, Plut. *Per.* 23.1).

[129] Plut. *Per.* 16.1. Plutarch, however, dates the changes after the ostracism of Thoukydides, son of Melesias, in 444/3 at a time when Perikles no longer had effective opposition.

[130] See Appendix B.

nificantly larger number of popular positions of power, i.e., positions filled by direct election (*cheirotonia*; *AP* 61.4–6) now became available to the upper classes. The Athenian *demos* may have gained access to almost the whole political and judicial system but it continued to elect members of the upper classes to the important military offices.[131]

This new cavalry system, financed by the democracy and the empire, was certain to win some aristocratic support and possibly divide Thoukydides' partisans. Still, we cannot ignore the possibility that it represented an attempt at reconciliation *after* the ostracism of his opponent. In *Perikles* 14.2 Plutarch claims that as a result of the ostracism Perikles effected the dissolution of the opposing faction (*hetaireia*); political differences were removed and the city became united under Perikles' leadership—his controlling hand touched upon all important issues, e.g., imperial tribute, the army, the navy, etc. (*Per.* 15.1). The picture is no doubt oversimplified, but the subsequent fourteen years of apparent acquiescence to Perikles' authority confirms his overall success.

The magnificent Panathenaic cavalcade on the outer wall of the cella of the Parthenon may glorify the new 1,000-man cavalry (Figures 3, 4, 9). It was completed around 440 B.C.,[132] permitting the chronological possibility, though tight, that Perikles suggested the theme as a visual celebration of this expanded force.[133] Still, the increase in the size of the cavalry may not have occurred until the 430s, a measure anticipating Perikles' defensive strategy in the war to come. On balance, however, it is probably safer to assume that the frieze glorified the recently established corps of 300 cavalrymen, along with its military success (*IG* I^2 400). The theme of the frieze was probably sketched out before the revolts in Boiotia, Euboia, and Megara in 447/6, thus before the earliest date suggested for the creation of the 1,000-man cavalry. In any event, the grandeur and aristocratic bearing of the frieze must surely have been related to the new cavalry,[134] whatever its size may have been at the time.[135]

[131] Davies, *Wealth*, pp. 122–31.

[132] Robertson & Frantz, *Parthenon Frieze*, p. 9.

[133] See Dohrn, "Phidias, Perikles, und Athen," pp. 71–99, esp. 75–79.

[134] O. Reinmuth, *The Ephebic Inscriptions of the Fourth Century* B.C. (Leiden, 1971; Mnenosyme Supplement 14) 137, calls these young riders "representatives of the early ephebeia in the Panathenaic procession." J. Boardman, "The Parthenon Frieze—Another View," in U. von Hoeckmann and A. Krug, eds., *Festschrift für F. Brommer* (Mainz, 1977), 39–49, has imaginatively suggested that the horsemen depicted on the frieze represent, to the exact number, the hoplite heroes who died on the field of Marathon, shown mounted to indicate heroic stature. This theory has little to recommend it; for critical treatment, see E. Simon, *Festivals of Attica: An Archaeological Commentary* (Madison,

Finally, that the impetus for the theme of the frieze may have come from considerations of internal politics is only a suggestion. The beauty of youths and prancing horses is sufficient justification for such an honored place on the Parthenon. Xenophon may be thinking of this very frieze when he praises the spectacle of prancing horses:

> This is the attitude in which artists represent the horses on which gods and heroes ride, and men who manage such horses gracefully have a magnificent appearance. Indeed, a prancing horse is a thing so graceful, terrible, and astonishing that it rivets the gaze of all beholders, young and old alike.[136]

Wis., 1983), 58–60. An excellent *comparandum* was carved only two decades earlier: E. Harrison, "The South Frieze of the Nike Temple and the Marathon Painting in the Painted Stoa," *AJA* 76 (1972): 353–78 (esp. 373–74), has suggested that the south frieze of the Nike Temple depicted the battle of Marathon with the Athenian hoplites engaging Persian cavalry. Neither the Athenian sculptor nor painter found a problem in depicting the *Marathonomachoi* in their historical role as hoplites.

[135] Robertson & Frantz, *Parthenon Frieze*, commentary to West Slab VIII.15, suggests that since there exist only two cavalrymen on the frieze with beards, West Slab IV.8 and West Slab VIII.15, these may be the two hipparchs (Figure 9). If this is correct, then it would seem to confirm that the canonical size of 1,000 Athenian horsemen was achieved in the 440s as it is to be associated with the reduction of the three hipparchs to two. However, since approximately fifty feet of the frieze are now lost and fifty more are preserved only in drawings, who can confidently claim that another bearded horseman did not exist on one of the missing slabs? Furthermore, although it would necessitate an even larger number of bearded horsemen, it is possible that one or both of the bearded horsemen and even the bearded marshallers are phylarchs. The victory monument commemorating the *anthippasia* of the Leontis tribe in the early fourth century B.C. depicts a bearded horseman heading a seried line of beardless youths on horses: *Hesperia* 40 (1971): 271–72, pl. 57 c (my Figure 7). This bearded horseman should be the commander of the tribe's cavalry unit, its phylarch. Also, note the famous Bryaxis base showing the three bearded phylarchs from the same family (*IG* II² 3130; see Figure 6). I repeat these examples only to caution that the bearded horsemen on the Parthenon frieze might not necessarily be the hipparchs. If they are phylarchs, then this too would confirm the existence of the 1,000-man cavalry by the late 440s, because the ten phylarchs belong to that system.

[136] *Peri Hippikes* 11.8 (Loeb trans.). Cf. Ar. *Frogs* 653: "I see the horsemen," surely reflecting the excitement among the spectators at the Panathenaic or other cavalry processions.

CHAPTER THREE

The Peloponnesian War

"One thousand good horsemen"
(Ar. *Knights* 225)

IN HIS Funeral Oration Perikles extolled the efficacy of the Athenian military system. According to Thucydides (2.39.1–4) he claimed that Sparta and her allies had yet to encounter the unified land forces of Athens because her troops were always scattered on many expeditions. This explanation was intended primarily for domestic consumption. Perikles had a healthy respect for the Spartan hoplite and was very reluctant to risk the outcome of the war on a single pitched battle. The Spartans would be allowed access to Attic territory, while the Athenian fleet would carry on offensive campaigns against the shores of Lakonia itself.

Perikles, however, was not content to permit the Spartan forces uncontested mobility in Attika. His main defensive weapon was the cavalry. He and his successors regularly dispatched the Athenian cavalry to harass the Spartans and prevent them from approaching the walls of Athens. Since Athens alone was to furnish a cavalry (except perhaps for Thessaly in the early years, Thuc. 2.22.2–3), her allies contributing ships, infantry, or money (2.9.6), the 1,000 Athenian horsemen and the 200 mounted archers (2.13.8, *AP* 1.24.3) held the key to Perikles' defensive strategy. A recent theory[1] argues that Perikles' war policy of withdrawing the rural Athenians behind the city walls depended upon his ability to convince the Athenians that the cavalry and the garrison outposts of Attika would "limit the damage done by the Spartans to rural property." Although I think that Ober has correctly emphasized the important function of containment expected of the Athenian cavalry in the war, it is unlikely that Perikles believed the cavalry (albeit bolstered by the Thessalians) and the garrisons could really protect the rural farms.[2] If he did not, we must accuse him of risking political su-

[1] J. Ober, "Thucydides, Pericles, and the Strategy of Defense," in *The Craft of the Ancient Historian: Essays in Honor of Chester G. Starr*, ed. by J. W. Eadie and J. Ober (Lanham, Md., 1985), 171–88, esp. 181. Cf. P. A. Brunt, "Spartan Policy and Strategy in the Archidamian War," *Phoenix* 19 (1965): 255–80 and T. Kelly, "Thucydides and Spartan Strategy in the Archidamian War," *AHR* 87 (1982): 25–54.

[2] Xen. *Hipparch.* 7.4 does seem to imply that protection of the rural farms of Attika was possible and refers specifically to the strategy of the Peloponnesian War. But Xeno-

icide with a promise which he knew would be revealed as misleading soon enough (as it turned out, by the second year). And it is difficult to believe that any Athenian would have accepted such an argument in light of the dominance of hoplite warfare over cavalry in that age.[3] Moreover, the Spartans had cavalry allies, most notably the redoubtable Thebans, whose numbers, skill, and tradition more than matched those of the Athenians. Would not most of the experienced Athenian soldier-farmers have known that the Peloponnesian army could operate more or less at will in Attika unless confronted by a combined force of hoplites and cavalry? On the other hand, if Perikles was the architect of the 1,000-man cavalry, as I believe, it is not unreasonable to suggest that Perikles stood before the Athenian assembly and predicted that the war would soon prove how wise had been the decision to establish the system of state support, i.e., *katastasis* and *sitos*, and thereby create an effective cavalry force with which to make life very uncomfortable for the Spartans and encourage them to withdraw. To what extent Perikles might have promised that the cavalry could save some or most of their farms from destruction is another matter.[4]

The activities of the cavalry were not restricted to the defense of Attika. The cavalrymen were to accompany offensive expeditions against Athens' hostile neighbors, Megara and Boiotia, to the Peloponnesos, the northern Aegean, Sicily, and Anatolia.

The Peloponnesian War also witnessed a heightened political unity among the cavalrymen. Fast upon the reorganization of the Athenian cavalry in the 450s, the cavalrymen were already acting as a single body. We recall the equestrian dedication on the Acropolis discussed in the previous chapter. And with the grant of *katastasis*, the *hippeis*, unlike the hoplites, stood in a unique contractual relationship with the

phon speaks of it as if it is a near impossible achievement requiring "aid by the gods and a brilliant hipparch." Furthermore, Ober may be reading too much into the key phrase, *ta te ektos tou teichous diasôzein*, "to preserve all that lies outside the walls" (Loeb). Delebecque in the Budé edition translates *ta* with *ce qui* "what," leaving it indefinite. Also, Thuc. 2.22.2 says that Perikles sent out horsemen to stop outrunners (*prodromoi*) of the enemy army from falling upon and doing harm to "the fields near the city" (*es tous agrous tous engus tês poleôs*). Is this not what Xenophon is more likely referring to, not all of Attika?

[3] Only when a hoplite phalanx was drawn out of its close serried formation could cavalry inflict real damage; cf. Polyainos 2.5.2; 7.14.3; and Asklepiodotos 7.1.

[4] Recently, V. D. Hanson, *Warfare and Agriculture in Classical Greece* (Pisa, 1983) 105–6, 134, has also remarked on the importance of the Athenian cavalry and garrisons in the war, and, from the agricultural context, has argued that real damage to the farms has been exaggerated (pp. 109–43); see review article by J. Ober in *Helios* 12 (1985): 91–101.

state. The constant contact during the requisite cavalry exercises, the numerous festivals, processions, sacrifices, and equestrian displays[5]—reenforced by the common social and economic backgrounds of certainly the majority of the cavalry personnel—encouraged a unanimity of political outlook. But it was during the Peloponnesian War that the Athenian cavalry developed a true *esprit de corps*, the result in part of their newly acquired military importance and in part of a bitter and personal conflict with the perceived champion of the *demos*, Kleon.

Military Activities[6]

Attika

Archidamos and the Lakedaimonian army invaded Attika for the first time in 431 B.C. The Athenian cavalry met them in the Thriasian plain and suffered their first recorded defeat (Thuc. 2.19.2).

The Spartan army proceeded to the deme of Acharnai and began ravaging the countryside. Despite the terrible sight of Athenian land and crops being destroyed, Perikles refused to commit the infantry. Instead he continually (*aei*) dispatched the cavalry to attack raiding parties from the main army who were ravaging the fields near the city. It was during this operation that a cavalry battle (*hippomachia*) at a place called Phrygia took place between a unit (*telos*)[7] of Athenian

[5] We know of certain sacrifices associated with the cavalry processions, called *hippades*: Sch. Ar. *Knights* 627 (Koster); the *Suda* s.v. *hippeis* (Adler); Harpokration s.v. *hippas*. On a gate of the city of Athens called *Hippades Pylai*, see S. G. Miller, *Hesperia* 41 (1972): 72, lines 141–42, with discussion, p. 85. Xen. *Hipparch*. 3.1 sketches the hipparchs' duties in sacrifices and processions: "first he must sacrifice to propitiate the gods on behalf of the cavalry; secondly, he must make the processions during the festivals worth seeing (*axiotheatous*)" (Loeb). In honor of Athena, there was the famous Panathenaic festival (cf. *IG* II² 2316–2317, second century B.C.; see also, Ar. *Knights* 581–94; Plut. *Kim*. 5; Paus. 1.30.4 (altar to Athena Hippias, cf. Paus. 1.31.1). In honor of Zeus, see Plut. *Phok*. 37.1, and for participation in the Olympieia, see Chapter Two; and in close association with their patron god Poseidon Hippios (see Ar. *Knights* 551–60; cf. *IG* I² 310, lines 60–61; II² 4691; and Chapter One); and an involvement with the cult of Apollo, at least financially: *IG* I³ 138 (I² 79; *SEG* XXX.5; see Chapter Two). The cavalry also participated in the festival of Theseus, though their presence is not attested before the second century B.C.: *IG* II² 956–61. For recent discussion of these festivals, see Kyle, *AAA*, pp. 32–55, *passim*.

[6] For a recent discussion of the Peloponnesian War in a straightforward chronological survey, see the books by D. Kagan, *The Archidamian War* (Ithaca, N.Y., 1974); *The Peace of Nicias and the Sicilian Expedition* (Ithaca, N.Y., 1981); and *The Fall of the Athenian Empire* (Ithaca, N.Y., 1987).

[7] It is not known what exactly is meant by this term. The Athenian cavalry was divided into tribes, *phylai*. Gomme, *HCT* 2, p. 77 suggests a *telos* may be a tribal squadron.

horsemen, aided by some Thessalian allies, against the Boiotian cavalry (Thuc. 2.22.2–3).[8] According to Thucydides the Athenian and Thessalian horsemen held their own until hoplite reinforcements arrived to rally Boiotians to victory. A few of the Thessalians and Athenians were killed. Pausanias records having seen the monument dedicated to these fallen Thessalians and Athenians on his guided tour through the Kerameikos.[9]

During the course of the Archidamian War, the Spartans invaded Attika again in 430, 428, 427, and 425. Cavalry action is recorded only once during these years, in 428, but it is clear from Thucydides' use of "as they were accustomed to" that the Athenian cavalry routinely harassed the Lakedaimonians during every invasion whenever the opportunity presented itself (3.1.2).[10]

War returned to Attic soil in the spring of 413 when the Lakedaimonians established a permanent garrison at Dekeleia. As with the yearly invasions of the 420s, the Spartans sought to dominate and devastate the Athenian plain and the other fertile parts of Attika. This time, however, they could operate on a continuous basis (Thuc. 7.19.1–2).[11] At this point the Athenian cavalry was put to the supreme test. Not only did they have to cope with invading armies, but now had to combat a permanent garrison which sent forth raiding parties and provided encouragement and refuge to deserting slaves (7.27.4–5). The cavalry sallied forth every day against Dekeleia and patrolled Attika. The hazardous and unrelenting duty on the rocky Attic soil

[8] The Spartans also received cavalry contributions from Phokis and Lokris (Thuc. 2.9.3).

[9] Paus. 1.29.6. Gomme, *HCT* 2, p. 77 expresses skepticism that the monument recorded by Pausanias is connected with those who fell at Phrygia in 431. He prefers to associate it with the Thessalian allies of Hippias, aided by Athenian nobles (perhaps on horseback or as mounted hoplites) who fought against Kleomenes in 510 B.C. His skepticism is unwarranted and his argument that "we have no evidence that the cavalry of Athens had separate monuments in the Kerameikos—we should expect that their dead would be named with the others on a single monument for all the casualties of the year—"is weakened by the fact that the cavalrymen who died at Koroneia and Korinth in 394, even though perhaps only of one tribe (see my Chapter Four), received a war memorial in the Kerameikos: see *IG* II^2 5222 (=Tod, *GHI* 2, no. 104).

[10] See Hanson, *Warfare and Agriculture*, p. 106, n. 6.

[11] For the year 430, Thuc. 2.65.2 records that the "great men" (*dynatoi*) of Athens were distressed at Perikles' strategic policy because it had cost them their fine estates (*kala ktêmata*) in the country, with buildings and costly furniture. The historian of the *Hellenica Oxyrhynchia* (12.5), however, claims that Attika had not really suffered until the Dekeleian occupation. For comment, see I.A.F. Bruce, *An Historical Commentary on the "Hellenica Oxyrhynchia"* (Cambridge, 1967), 115–16 and Hanson, *Warfare and Agriculture*, pp. 111–43 (and footnote 4 above).

took its toll on the horses of the cavalrymen (7.27.5). It is not surprising that the cavalry was not asked to perform guard duty on the city walls at night (7.28.2); their daylight duty was sufficient.

In 411, despite the overtures of peace from the short-lived Athenian oligarchy of the Four Hundred to the Spartan king at Dekeleia, Agis gathered reinforcements from the Peloponnesos and together with his garrison marched on Athens, hoping to catch them in a state of military unpreparedness because of the internal political crisis. He was deceived in his hopes. The Athenians sent out the cavalry, a portion of the hoplites, light-armed troops, and bowmen. Some of Agis' men were killed when they approached the walls. Agis was forced to withdraw (Thuc. 8.71.2–3).

In 410, Agis again marched on Athens. Thrasyllos led forth "the Athenians and all others in the city" and stationed them near the Lykeion. Upon seeing the size of the force, Agis beat a hasty retreat, losing some of his rear guard to Athenian light-armed troops (Xen. *Hell.* 1.1.33). In spite of Xenophon's silence, it is quite likely that the Athenian cavalry, as in 411 B.C., participated in this show of arms. What follows may confirm this.

It is possible that among the cavalry involved was the youngest son of Polystratos.[12] In the twentieth speech of the Lysianic corpus, *For Polystratos*, the speaker, the second son of Polystratos, relates that when the exiles—who had not only brought ill in Athens, but also were attacking the Athenians from the wall (= "walled place": *apo tou teichous*)—made a descent against the city, his younger brother advanced alone beyond the cavalry lines and killed one of them (28). The date of the speech is 410/9. The exiles referred to must have been the deposed supporters of the oligarchy of the Four Hundred who had fled to Dekeleia in 411 B.C. (Thuc. 8.98.1). Consequently, we should understand the "walled place" as referring to Dekeleia.[13] Furthermore, it is quite likely that this "walled place" is not explicitly identified because no Athenian would have thought it to be any place other than Dekeleia. Since we know that the attack by Agis from Dekeleia in 411 took place while the Four Hundred were still in power (Thuc. 8.71.2), the information supplied by Pseudo-Lysias cannot apply to this invasion. It is natural, then, to connect the bravado of Polystratos' youngest son with

[12] Polystratos was commissioned to draw up the list of 5,000 citizens in 411, but enrolled 9,000 instead ([Lys.] 20.13). On involvement with the Four Hundred in 411, see A. Lintott, *Violence, Civil Strife, and Revolution in the Classical City 750–338* B.C. (London, 1982), 149–50 and Andrewes and Dover, *HCT* 5, pp. 201–6, whose authors doubt the reliability of the figure 9,000.

[13] So the editor of the Loeb edition of Lysias, p. 469.

the Spartan show of force recorded by Xenophon for 410 B.C. It is less likely that the information in *For Polystratos* relates to some otherwise unrecorded cavalry skirmish in 410 B.C.

The speaker of this oration states further that while he was in Sicily serving as a cavalryman (24), his two brothers were in Boiotia (4). It is natural to assume that the speaker is trying to highlight the patriotism of himself and his brothers. One might speculate that the youngest son (and possibly the eldest) was stationed as a cavalryman at Oropos (on the Boiotian border) when the city was betrayed to the Boiotians in 412/11.[14] He may have been assigned to the garrison at Oropos in deference to his father who held office there at the time (see 6). The fact that the youngest son was a cavalryman in 410/9 does not, however, automatically mean he was one two years earlier, any more than Alkibiades was a *hippeus* at Poteidaia (431) simply because he was at Delion (424). The brother who is speaking participated in the Sicilian Expedition as a horseman, but shortly after the disaster he was both a cavalryman and a hoplite in the service of Katana, a city in Sicily (25). The cavalry experience of the sons of Polystratos did not end with the Peloponnesian War. A certain Lykios, son of Polystratos, who commanded the fifty horsemen organized by Xenophon for the march of the Ten Thousand back from Cunaxa (401 B.C.), has been identified as one of these sons.[15]

Diodoros 13.72.3–73.2 records a third military action between Agis and the Athenians in the archonship of Euktemon, 408/7. However, his own narrative relates in chronological order Alkibiades' triumphant return to Athens, his assumption of the office of *strategos* (Xen. *Hell.* 1.4.10,20), and his subsequent departure on an expedition to Andros prior to the attack by Agis. Thus the latter engagement should fall sometime in autumn 407 and should be dated to the archonship of Antigenes, 407/6, not that of Euktemon.[16] On this occasion Agis gath-

[14] Thuc. 8.60.1. Cf. Lys. 20.6.

[15] See Xen. *Anab.* 3.3.20; 4.3.22, 25; 4.7.24. For discussion of the family stemma, see *APF* 12076, pp. 467–68; the three sons, though which is uncertain, have been identified as Philopolis, Polystratos II, and Lykios.

[16] The scholiast to Ar. *Frogs* 1422 places Alkibiades in Athens in the archonship of Antigenes; his return to Athens is dated to the spring of 407 by N. Robertson, "The Sequence of Events in the Aegean in 408 and 407 B.C.," *Historia* 29 (1980): 285. Since Alkibiades participated in the procession of Eleusis (September) of that year (Xen. *Hell.* 1.4.20; Plut. *Alk.* 34), his departure should be dated some time in the fall of 407/6; Xen. *Hell.* 1.4.21 indicates that Alkibiades set sail for Andros in the fourth month after his arrival in Athens. If this is true, his return has to be placed in very late spring 407. W. J. McCoy, "Theramenes, Thrasybulus, and the Athenian Moderates" (Unpubl. diss., Yale University, 1970), 139, argues for summer 407.

ered a large force, including 1,200 horsemen, the Boiotians contributing 900 of that number. The Athenian generals sent out the cavalry, approximately equal to that of the enemy (13.72.7), and a fierce battle ensued for some time.[17] Since we know that Alkibiades took along 150 Athenian horsemen when he set sail for Andros,[18] there must have remained in Attika 850 horsemen (and some *hippotoxotai*, perhaps), if we assume that the Athenian cavalry was at full strength. The difference in the numbers of the opposing cavalry forces could not have been less than 350 men. Nevertheless, the Athenian cavalry was victorious and pursued the Boiotian horsemen as far as the Spartan lines. Upon encountering the infantry, they withdrew to the city. The Athenians even erected a trophy. Agis failed to entice the Athenians to come out and fight on the next day and finally withdrew, content merely to ravage the land.

Megara

As I have said earlier, the role of the Athenian cavalry was not limited to home defense. It was actively enlisted to carry the war to the enemy. Just as the neutralization of Megara had played an important part in Athens' military strategy in the 450s and 440s, so now Athens sought to occupy Megara and block Sparta's approach to Attika. In 431 B.C., as soon as Archidamos had withdrawn from Attika, the Athenians levied their "whole military force" (*pandêmei*), the largest army that the Athenians had assembled in one body (*stratopedon megiston*; Thuc. 2.31.2), and invaded the Megarid. Thucydides reports that there were 10,000 hoplites, 3,000 heavy-armed metics, and a considerable number of light-armed troops. They ravaged the Megarid and then retired.

Although Thucydides does not mention the Athenian horsemen in

[17] Ch. Clairmont, *Patrios Nomos: Public Burial in Athens during the Fifth and Fourth Centuries* B.C. (Oxford, 1983), 317, n. 16, refers to a lecture by D. L. Page on Sophokles *Oidipous at Kolonos* 1044–1095, suggesting that the battle scene envisioned by the chorus between Theseus and Kreon's men trying to steal away Antigone and Ismene was inspired by the present battle. Diodoros' information that the Athenian cavalry numbered nearly 1,200 in this battle has been unjustifiably dismissed by I. G. Spence, "Athenian Cavalry Numbers in the Peloponnesian War: *IG* I³ 375 Revisited," *ZPE* 67 (1987): 168, n. 10, who argues in his article that in 410/9 the Athenian cavalry numbered approximately 500–600 men. If his analysis were correct (I do not believe it is, see Chapter Two, footnote 86), it would signal a marvelous feat of arms for the Athenian cavalry to have bested the redoubtable Boiotian horsemen, particularly in light of the fact that Alkibiades was campaigning elsewhere with 150 Athenian horsemen at the time. One must conclude that either the numbers of Athenian horsemen had been dramatically increased immediately after 409 B.C. or the Athenian cavalry was victorious with only approximately 350 to 450 men. The former is unsubstantiated, the latter unbelievable.

[18] Xen. *Hell.* 1.4.21; Diod. 13.69.4; Plut. *Alk.* 35.1.

connection with this invasion, their presence should be accepted. First of all, their absence would have been conspicuous in an expedition of this size and inexplicable against such a close neighbor whose terrain was suitable for cavalry maneuvers. Secondly, we know that the Athenian cavalry had suffered two defeats that same year (see above) and it would have been natural for them to have wanted to recover their honor at the expense of Sparta's ally Megara.

Thirdly, the *hippeis*, paid for year-round service and potentially always on call, can occupy an ambiguous position with respect to the irregular and seasonal levy of infantry. For example, Thucydides 2.31.3 records that later in the war (*hysteron en tôi polemôi*) Athens invaded the Megarid every year, both with the cavalry and the full military levy (*esbolai . . . kai hippeôn kai panstratiai*) until Nisaia was captured in 424 B.C.[19] This should mean that other incursions into the Megarid took place between 430 and 424,[20] even though Thucydides does not mention them in the course of his annual narrative. He was no simple chronicler of events; here he strove to avoid superfluous repetition.

Describing the events of 424, Thucydides supplies us with the specific information that two invasions into the Megarid occurred each year "with the full military levy" (*panstratiai*, 4.66.1).[21] Taken in conjunction with the statement in 2.1.3, the language seems to imply that the cavalry is to be considered distinct from the forces included in *panstratiai* since we learn elsewhere that there were in fact 600 Athenian horsemen at Megara in 424 (4.68.5). I therefore suggest that Thucydides intended the *pandêmei* and *stratopedon* in 2.31.2 and the *panstratai* in 2.31.1,2 and 4.66.1[22] to represent the seasonal and temporary levies of infantry forces. This would explain why he listed only the infantry present in 431 B.C. Thus Thucydides' statement in 2.31.3 that later invasions included both cavalrymen and the whole army was not intended to stand in contrast to the invasion of 431—Thucydides 2.31.3 merely filled out the complete and normal military complement.

After the Athenians gained possession of Nisaia, the port of Megara,

[19] See Thuc. 4.66.69.

[20] It is not clear how long a span of time is meant by "later in the war" (Thuc. 2.31.3), so it cannot be stated with absolute assurance that the Megarid invasions took place in unbroken succession from 431 B.C. on. For discussion, see Gomme, *HCT* 2, p. 93 and R. Legon, *Megara: The Political History of a Greek City-State to 336 B.C.* (Ithaca, N.Y., 1981), 229, who appears to accept the idea of repeated attacks during the years 431–424.

[21] See also Plut. *Per.* 30.3 and Paus. 1.40.4.

[22] Cf. Thuc. 4.90.1; 4.94.1 for similar military terminology for the expeditionary force to Delion in 424 B.C.; for discussion, see A.H.M. Jones, *Athenian Democracy* (London, 1969), 178.

in 424, Brasidas arrived with troops from several Peloponnesian states. The Boiotians, already prepared to march to Megara's defense, hastened to join him with hoplites and 600 cavalry. The Boiotian cavalry initiated the battle with a devastating attack on Athens' light-armed troops. The Athenian cavalry promptly engaged the Boiotian horse and a lengthy cavalry battle ensued (Thuc. 4.72.3). In almost a repeat performance of the Boiotian-Athenian cavalry battle in 431, neither side was willing to admit defeat. The Athenians killed the Boiotian hipparch and a few others. Notwithstanding the trophy which the Athenians set up, Thucydides indicates that neither side gained a sure victory (4.72.4). Nevertheless, the battles of 431 and 424 demonstrate that the Athenian cavalry could effectively neutralize Sparta's principal cavalry arm, Boiotia. But when a pitched battle was imminent with Brasidas and his infantry (clearly a larger force), the Athenian generals, like Perikles before them, refused to risk destroying the "best of their hoplite force" (4.78.4), and withdrew without engaging.

Diodoros 13.65.1–2 records another cavalry engagement on Megarian soil during the Peloponnesian War. He dates it to the archonship of Diokles, 409/8. Apparently the Megarians had retaken Nisaia, and the Athenians responded with a force of 1,000 hoplites and 400 horsemen. The Megarians met them *en masse*, bolstered by some troops from Sicily, near the hills aptly called The Horns. According to Diodoros, the Athenians, although outnumbered, fought brilliantly, routed the enemy, and slew many Megarians.

Boiotia

The one major battle during the war on Boiotian soil resulted in disastrous defeat for the Athenians. The battle of Delion took place in winter, 424 (Thuc. 4.76–77, 89–101.1–2), shortly after the events in Megara. Hippokrates, the Athenian general, levied a mixed force of citizens, metics, and aliens. The cavalry probably numbered at least as many as had participated in the Megarian campaign. Three hundred cavalrymen were left to guard the sanctuary of Apollo at Delion (4.93.2), on the northern coast of Boiotia; the battle itself took place some distance inland.[23]

In the battle an unspecified number of cavalrymen were stationed on either wing of the Athenian army (4.94.1). It is hazardous to speculate on their numbers, but in light of the fact that Hippokrates felt confident enough to leave 300 horsemen at the sanctuary surely knowing

[23] For a detailed examination of the topography and its relation to the battle, consult Pritchett, *Studies* 2, pp. 24–36.

that he would be confronting Boiotian horse (1,000 strong) and Lokrian, the Athenian cavalry force present in the battle must have been considerable.

Hippokrates' exhortation to his men before the battle is a telling statement on the importance of the Boiotian cavalry to the Spartan cause. He declared that if the Athenians won, the Peloponnesians would be deprived of the Boiotian cavalry and consequently would be unable to invade Attika again.[24] As it turned out, ironically, the battle was decided by the sudden appearance of two squadrons (*telê*) of Boiotian cavalry (4.96.5).

The actual part played by the Athenian cavalry in the battle is disputed. Thucydides' account appears to eliminate them from an active role with the observation that the extremities of both forces, where the horsemen were stationed, did not, or rather could not, engage because of "rushing streams" (*rhyakes*). The exact meaning of *rhyakes* is a matter of debate, but it appears certain that some topographical obstacle prevented the outer flanks from engaging. Diodoros 12.69–70, on the other hand, recounts a cavalry engagement in which the Athenian cavalry gave a good account of itself before the actual hoplite lines met. It would be interesting to know where he obtained this information. Both Gomme and Pritchett simply reject the testimony of Diodoros on this initial cavalry skirmish.[25] It is not impossible that a cavalry battle occurred before the main lines met, as at Megara in 424, but Thucydides' understanding of the battle and the topography argue against it.

There is, however, a personal touch connected with Athenian defeat at Delion. Plato relates in his *Symposium*[26] that in the general retreat from the field of battle, Alkibiades, participating as a cavalryman, protected his mentor Socrates, who was serving as a hoplite.

Peloponnesos

As early as 430, Athens was able to extend the range and usefulness of its cavalry by means of horse-transports. While the Peloponnesians were ravaging Attika as far as Laurion (and presumably being harassed

[24] Thuc. 4.95.2; cf. Xen. *Hipparch.* 7.3–4; *Mem.* 3.5.4: Socrates is speaking with the son of Perikles in 411 B.C. about the consequences of the defeat at Delion, but the information probably reflects the military environment following the rise of Theban power in the 370s and 360s—the young Perikles claims that the Boiotians, who would not dare to invade Attika without aid from Sparta and the Peloponnesians before the battle of Delion, now threaten to invade and plunder Attika by themselves.

[25] *HCT* 3, p. 568 and *Studies* 2, p. 34, n. 38, respectively.

[26] 221 A–C (repeated in Plut. *Alk.* 7.4; cf. Pl. *Laches* 181B and Athenaios 5.215 C–216 C; the account in Diog. Laert. 2.22–23 about Xenophon the historian falling from his horse in this battle is chronologically impossible).

by flying columns of Athenian cavalry) and the plague was sweeping through Athens—to the loss, ultimately, of 300 *hippeis* (Thuc. 3.87.3), Perikles organized an expeditionary force against Epidauros consisting of 100 ships, 4,000 Athenian hoplites, and 300 horsemen transported "in horse-transports used for the first time from old ships" (*en nausin hippagôgois prôton tote ek tôn palaiôn neôn poiêtheisais*, Thuc. 2.56.2). While it is certain that the Persians possessed horse-transports in the Persian Wars (Hdt. 6.48.2; 7.97), there is no evidence that the Athenians did before 430.

Gomme suggests three possible meanings for *prôton tote*[27] used here: that horse-transports were used (a) for the first time in this war, (b) for the first time in Athens (or Greece), and (c) for the first time made of reconstructed triremes.[28] He rejects (a), considers (b) less probable, and prefers (c). I suggest that (b) and (c) may not be mutually exclusive, that the reference to old ships stands as an explanatory footnote to a military innovation.[29] The Athenians must have had some means of transporting horses before 430,[30] but not, I believe, in a state-organized capacity. The Athenians had no need for horse-transports until the real organization of the Athenian cavalry in the 450s and had no need of annually maintained horse-transports until the war with Sparta forced them to bypass their hostile neighbors, Megara, Boiotia, and Korinth. Perikles, the architect of Athens' war strategy, appreciated the advantage of mobility in the raids on the coast of the Peloponnesos and probably suggested their introduction. In fact it was the first time they were employed in the war, so even (a) is technically correct.[31]

[27] On uses of *prôton tote* in Thucydides, see Gomme, *HCT* 2, p. 163 and J. G. Griffith, "A Note on the First Eisphora at Athens," *AJAH* 2 (1977): 3–7. Nothing exemplifies better the difficulty of this phrase than Thuc. 8.97.1 which reports that an assembly met on the Pnyx *tote prôton* in 411 to remove the government of the Four Hundred. But Thucydides explicitly tells us in the next line that the Pnyx was where they had been accustomed to meet at other times (*allote eiôthesan*), thus apparently meaning here "for the first time since the revolution of 411"; see Andrewes and Dover, *HCT* 5, p. 323.

[28] *HCT* 2, p. 163.

[29] As the Loeb edition seems to translate: "four (read 'three') hundred cavalry in horse-transports, then employed for the first time, which had been made out of the old galleys."

[30] For example, perhaps horses were transported, along with family members and possessions, to Salamis, Troizen, and Aigina in 480 B.C. when Xerxes invaded Attika (Hdt. 8.41). It is also quite possible that the horses were abandoned as spoils of war for the Persians. And horses would have simply been ferried across to Euboia for Perikles' invasion of the island in 446.

[31] And so Mme. de Romilly in the Budé edition explains her translation of the passage "sur des transports de cavalerie faits pour la première fois avec les anciens navires de guerre" as meaning "pour la première fois à Athènes et au cours de cette guerre" (p. 99).

If one horse-transport ordinarily carried thirty horses and men (borrowing from Thuc. 6.43), then it would appear that Athens probably did not maintain many more than ten horse-transports each year. There was no seagoing expedition during the Peloponnesian War which included more than 300 Athenian cavalrymen. The number was probably limited for strategic reasons, e.g. Athens' understandable hesitancy to commit more than one-third of its cavalry force to distant foreign enterprises in a given year. The primary role of the Athenian cavalry throughout the long war was to defend Attika.

The Athenian cavalry received praiseworthy mention by Thucydides twice during the Peloponnesian War in engagements on Peloponnesian soil, first at Solygeia, near Korinth, in 425, and later at Mantineia, in 418.

In the summer of 425 the Athenians invaded the Korinthiad with 80 ships, 2,000 Athenian hoplites, and 200 cavalrymen transported in horse-transports (4.42.1). The Athenian force landed near the village of Solygeia, and the Korinthians hurriedly dispatched soldiers to protect the village. A fiercely contested battle ensued with neither side able to gain the advantage. Finally the Korinthians yielded. Thucydides states that the presence of the Athenian cavalry and the fact that the Korinthians had none (*ouk echontôn hippous*) proved to be the decisive factor in the Athenian victory (4.44.1).[32] When it was discovered that Korinthian reinforcements were on their way, the Athenians wisely departed, content with having won the right to set up a trophy.

In *Knights*, produced in 424 B.C. at the Lenaia, Aristophanes recalls the decisive role of the Athenian cavalry in this battle. In the play the Paphlagonian slave (who is really Kleon, the people's politician) cunningly tries to enlist the support of the chorus of cavalrymen by proposing to have a memorial erected on account of their bravery (266–68). In lines 565–80 the chorus praises its noble ancestors for their unfailing devotion to the native land and pledges similar devotion and bravery of themselves.

In lines 595–610, in more specific but thoroughly comic fashion,

[32] Note Thuc. 5.59.3 where, as late as 418, the Argives apparently possess no cavalry force either, *hippoi de autois ou parêsan*. See Gomme, Andrewes, and Dover, *HCT* 4, p. 83, for the evidence that Korinth and Argos did not have cavalries in the fifth century; yet they had had a tradition of horse-breeding in the Archaic period, for example, the famous Korinthian racehorse called "Breeze," owned by Pheidolas, who won his race riderless in 508 B.C. (Paus. 6.13.9) and Pindar who refers to Argos as *hippion*, "horse city" (*Isth.* 7.17). We do know that by the fourth century both states had acquired cavalries—Korinth: Xen. *Hell.* 6.5.51–52 (370 B.C.), 7.1.20 (369 B.C.) and Argos: Xen. *Hell.* 7.2.4 (368 B.C.).

Aristophanes has his chorus of cavalrymen praise their horses for their enthusiasm. In the Korinthian campaign their mounts leapt bravely aboard the horse-transports (599), and pulled the oars like humans.[33]

Two members of the chorus of cavalrymen are explicitly named, Simon and Panaitios (242–43). The scholiast plausibly identifies these men as the two hipparchs. If it is assumed that they held office during the civil year July 425–June 424 B.C., then one (probably not both in view of the relatively small cavalry contingent at Solygeia) may have participated in the battle which should have taken place sometime after mid-August 425.[34] It is generally agreed that this Simon is the same authority on horsemanship to whom Xenophon refers in his treatise *Peri Hippikes*.[35]

H. Fuhrmann has proposed that a dedicatory relief of the late fifth century B.C. found at Eleusis depicting a cavalry-infantry skirmish commemorates the battle at Solygeia (Figure 10).[36] The stele has two sculptured panels, vertically juxtaposed, depicting horsemen charging hoplites, some of whom defend themselves, while others flee. The terrain represented is uneven and hilly. As Hölscher points out,[37] the difficulty with Fuhrmann's suggestion that the relief recalls Solygeia is that Thucydides reports that the Korinthians had no cavalry (4.44.1),

[33] Note the double wordplay on *andrikôs*, "bravely" and "like a man." See Ar. *Knights* 604 and Sch. Ar. *Knights* 609 which explicitly connects this comic passage with the Korinthian expedition under the command of Nikias.

[34] Gomme, *HCT* 3, p. 478, calculates that the campaign at Pylos should have been over at the lastest by the second week in August (cf. esp. Thuc. 4.39.1). Since the Korinthian campaign did not commence until the Pylos affair had ended (4.42.1) and it is generally agreed that the military officers were inaugurated on Hekatombaion 1, the first day of the archon year, then it follows that the new hipparchs for 425/4 were already in office. On the date of entrance to office on Hekatombaion 1, see U. Kahrstedt, *Untersuchungen zur Magistratur in Athen*, pt. 2 (Stuttgart, 1936), p. 68; W. K. Pritchett, "The Term of Office of Attic Strategoi," *AJP* 61 (1940): 469–74; Hignett, *HAC*, pp. 347–48; and Ch. Fornara, *The Athenian Board of Generals from 501 to 404* B.C. (Wiesbaden, 1971), 40. See now also Rhodes, *CAP*, pp. 406, 537, 676. See footnote 55 below on the possibility that before ca. 407 the military officers entered office with the Council of Five Hundred, not with the archons. For the case under examination it makes no difference.

[35] 1.1 and 11.6. Xenophon reports that Simon dedicated the bronze horse in the Eleusinion in Athens and had his achievements sculptured in relief on the pedestal (1.1; see Pliny *Natural History* 34.19.76). A fragment of Simon's treatise on horsemanship is preserved in a manuscript at Cambridge: see *Hipp. Cant.* XCIII.1–11, pp. 228–31 in vol. 2 of E. Oder and Ch. Hoppe, *Corpus Hippiatricorum Graecorum*, 2 vols. (Teubner, 1924). On Simon, see also Pollux *Onom.* 1.190, 198. Pliny *Natural History* 34.19.76 claims that Simon was the first to write about horsemanship.

[36] *Philoxenos von Eretria* (Göttingen, 1931), 325–26. For discussion and bibliography, see Hölscher, *Historienbilder*, pp. 99–101.

[37] *Historienbilder*, p. 261, n. 516.

while there is clear evidence of a horse's leg in the left-hand corner of the relief. That horse must belong to the army of the defeated. It can be countered, however, that Greek artists are sometimes less concerned with accuracy than with heroic or pleasing anatomical representations.[38]

Another argument against the identification of the Eleusis relief with the battle of Solygeia rests on prosopographical grounds. A fragmentary inscription at the bottom of the relief (*IG* I² 816) names the dedicant (though only the final "s" of the *nomen* is preserved, the restoration is quite probable) and his rank:

> [Pythodoro]s, son of Epizelos, having served as hipparch
> [dedicated this to the gods].

Although we have only one other indisputable reference to this Pythodoros,[39] he has been plausibly identified with a member of the Peace Commission in 421 (Thuc. 5.19.2, 24.1) and one of the three generals on the Argolid campaign in 414.[40] The inscription proves that Pythodoros was one of the two hipparchs at the time of the battle depicted. He is thought to be the cavalryman located in the center of the upper panel of the relief.[41] The result is that we are encumbered with three

[38] Cf. e.g., the Dexileos grave relief and the Albani relief (Holscher, *Historienbilder*, pp. 102–4, 109, pl. 8.1) which depict the horseman nude, a practice that certainly was not customary for cavalrymen. On arms and armor for cavalrymen, see Xen. *Hipparch.* 1.22–23; *Peri Hippikes* 12.1–3. In the second century B.C. festival of the Theseia, one competitive event was judged on the basis of which Athenian tribe had the "most dashing equipment" (*tôn hippeôn euhopliai*, *IG* II² 956.1.58–60; 957.1.36–41; 958.1.56–59; 960, lines 22–24 [restored]; and 961, lines 22–24). In the Eleusis relief, the fully preserved cavalryman wears a corselet and *petasos* (cf. Figure 4). The hoplite in the lower panel was probably carved nude (except for helmet, shield, and spear) to demonstrate the artist's anatomical skills; the other hoplites are clothed. Nude hoplites similarly armed, some with flowing *chlamydes*, also appear on the south and west friezes of the Nike Temple on the Acropolis, see E. Harrison, "South Frieze," *AJA* 76 (1972): pls. 75–76 and E. G. Pemberton, "East and West Friezes," *AJA* 76 (1972): pl. 61.

[39] *IG* I² 770a celebrating his victory as *choregos* in the Dionysia of 414, belonging to the tribe Aigeis.

[40] Thuc. 6.105.2. See *APF* 12402, p. 481, for other testimonia. Hölscher, however, is hesitant to admit any secure identification with any other man than the *choregos* (p. 261, n. 507). To be sure, without a patronymic and demotic absolute certainty is impossible. The problem is compounded by the fact that Pythodoros is a common name. Most men, fortuitously, are eliminated by tribal affiliation. But if A. Andrewes and D. M. Lewis ("Note on the Peace of Nikias," *JHS* 77 [1957]: 178) are correct in their theory that Thuc. 5.19.2 and 5.24.1 recorded the proper tribal sequence, then the Peace Commissioner of 421, who belonged to the tribe Aigeis, is very likely to be our Pythodoros.

[41] This identification is based not only on his prominent position on the relief as foremost of the attacking cavalry, but by his beard, a good indicator of age and maturity. In

hipparchs for the year 425/24. In the period of the 1,000-man cavalry, of course, all three cannot have been serving in the same year. The most reasonable solution is to disassociate the relief from the historical event at Solygeia and to connect it with one of the Spartan invasions of Attika in the 420s[42] or possibly with the Megarian campaign in late summer 424.[43] In any event, if Simon and Panaitios were the hipparchs for 425/4, the hipparchy of Pythodoros must be dated to another year.

---

addition, Hölscher (p. 101) cites the authority of J. K. Anderson, *Military Theory and Practice in the Age of Xenophon* (Berkeley, 1970), 27, that the heavy armor of the Eleusis horseman may reflect his high office. But since the *torsi* of the other two cavalrymen are not preserved, we cannot be absolutely sure. Besides, a beardless horseman on the Parthenon frieze (Figure 4) wears a corselet and plumed helmet; should we identify him as one of the hipparchs, or perhaps phylarchs? The degree to which a cavalryman was armored, I believe, depended more on individual wealth and pride than on his status as an officer *per se*. To be sure, we should expect hipparchs and phylarchs to appear in full dress regalia (cf. Xen. *Hipparch.* 1.22)—they had been popularly elected by the people and that fact presupposed maturity and self-sufficience. But the conclusion that a wealthy young horseman could not and did not dress just as fully and splendidly is unfounded.

[42] So Hölscher, *Historienbilder*, p. 100. The possibility that the Eleusis relief is connected with some action during the Argolid campaign of 414 should be eliminated: A. Brückner, "Ein Reiterdenkmal aus dem Peloponnesischen Kriege," *AM* 14 (1889): 400–401 and Hölscher, *Historienbilder*, p. 261, n. 516. Thucydides reports that Pythodoros, Laispodias, and Demaratos were in command of this expedition (6.105.1). Despite the fact that he employs the general term *archontes* to describe their military status, *strategoi* are surely to be understood; see Fornara, *Strategia*, p. 65.

[43] Specifically, the cavalry action described in Thuc. 4.72.3–4. G. Busolt, *Griechische Geschichte*, 3 vols. (Gotha, 1893–1904; repr. Hildesheim, 1967), vol. 3, pt. 2, p. 1137 (followed by Fornara, *Board of Generals*, p. 59), states that the Megarian campaign took place "Anfang August." It cannot have occurred much later than this since we know that Brasidas, who participated in the action at Megara, was in the Chalkidike a little before "the vintage season" (Thuc. 4.84.1). Gomme, *HCT* 3, p. 551, indicates that September marks the "vintage season" in the north, so Brasidas is there "about the middle to the end of August." The Megarian campaign must have ended some time prior to that in order to accommodate the time required for Brasidas' activities recounted in Thuc. 4.74.1 (preparations in Korinth for the Thracian expedition), 4.78 (his crossing of Thessaly), and 4.83 (his attack with Perdikkas on the latter's neighbor, the king of the Lynkestrian Macedonians). Nevertheless, we need not suppose a date earlier than late July for the beginning of the Megarian campaign. In any event, Hippokrates (unlike Demosthenes) is not attested as *strategos* for 425/4 (Fornara, *Board of Generals*, p. 59) and the close association of these two generals in both the Megarid (4.66.3) and the Boiotian campaigns (4.76–77, 89.1) of summer and summer/winter 424, respectively, makes it almost certain that they were both elected *strategoi* for 424/3 (so Fornara, *Board of Generals*, p. 59). Since the new hipparchs presumably took office at the same time as the generals, i.e., Hekatombaion 1, it is possible for Pythodoros to have been one of the hipparchs for 424/3 and therefore to have participated in the Megarian campaign. Such a large cavalry force, 600, must have included at least one hipparch. All of this is speculation, to be sure, but at least the date of the Megarian campaign cannot be used to

During the year 425/4 the Athenian cavalry also saw action in Kythera and Lakonia. A small detachment of cavalry accompanied an Athenian force consisting of 60 ships, 2,000 hoplites, and an unspecified number of allies.[44] During the expedition the Athenians occupied Kythera, prompting the Spartans to make hurried preparations to defend Lakonia itself against the expected attack. The Spartans sent garrisons of hoplites throughout the country, fearing that their constitution might be subverted. With the recent disaster at Sphakteria (the unprecedented surrender of 292 of its warriors in 425 B.C.), and with both Pylos and Kythera now lost, they felt themselves encompassed by a swift and unforeseen war (Thuc. 4.55.1). As a result, the Spartans organized a cavalry, "contrary to their custom," numbering four hundred and a force of bowmen (4.55.2), no doubt partly to provide a mobile force between the garrisons and partly to counterbalance the cavalry advantage of the Athenians. Sparta simply did not have the time to wait for the arrival of her allies, especially her cavalry allies, Boiotia, Lokris, and Phokis, when faced with the hit-and-run attacks along her southern coast (4.54.4).

In 418 Argos found itself facing the army of Agis and the Peloponnesian allies. In accordance with the terms of the alliance of 420/19,[45] Athenian help was summoned. The Argives attached great weight to the Athenian cavalry since "they themselves had none," *hippoi de autois ou parêsan* (Thuc. 5.59.3). An Athenian force of 1,000 hoplites and 300 horsemen[46] under the command of Laches and Nikostratos arrived (5.61.1), only to discover that in the interim Argos had concluded a four-month truce with Agis, thereby removing the immediate threat. But in subsequent developments Argos was compelled to take the field at Mantineia alongside its allies, the Athenians and Mantineians. In the course of the battle the Athenians found themselves on the verge of being surrounded. Part of their force had already been routed (5.72.4) and now the rest were being outflanked by the right wing of

---

shift the hipparchy of Pythodoros back into 425/4, thereby positing the same surplus of hipparchs for that year as encountered in the identification of the Eleusis relief with the battle of Solygeia.

[44] Thuc. 4.53.1. This campaign is securely dated to May 424, see Gomme, *HCT* 3, p. 507.

[45] See Thuc. 5.47 and *IG* I³ 83 (=I² 86; Tod, *GHI* 2, no. 72). If the campaign had lasted over a month, Argos would have had to pay each Athenian cavalryman one Aiginetan drachma per day: Thuc. 5.47.6; *IG* I³ 83, lines 22–24.

[46] Diod. 12.79.1 records only 200 horsemen. Gomme, Andrewes, and Dover, *HCT* 4, p. 86, comment that the difference between the two figures represents "the type of variant common in Diodorus, for which we need not postulate a non-Thucydidean source." For an insightful analysis of this battle, see Pritchett, *Studies* 2, pp. 37–72.

the Lakedaimonians and Tegeates. At this critical moment Thucydides reports that "had not the Athenian cavalry been there and proved useful" *(ei mê hoi hippês parontes autois ôphelimoi êsan)*, the Athenians would have suffered more than the other allies.[47] Interestingly, the word *ôphelimoi*, "useful," repeats exactly Thucydides' description of the contribution of the cavalry at Solygeia (4.44.1). Fortuitously, Agis diverted his forces to aid his troubled left flank and did not engage the Athenians. The Athenians were thus allowed to escape, losing 200 men and both generals (5.74.2).

An unspecified number of Athenian horsemen was also at hand at Olympia during the Olympic Games of 420. The Eleians had denied the Lakedaimonians access to the sanctuary and, anticipating Spartan military intervention, had stationed some armed young men there. In addition, the Mantineians and Argives each sent 1,000 men to Elis in support. As for the Athenians, while it is natural to assume that their presence at Olympia fulfilled the terms of the alliance of 420, the relevant passage is not without ambiguity: "the Argives and Mantineians came to their aid with 1,000 men each, and some Athenian horsemen, who were at Harpine awaiting the festival" (Thuc. 5.50.3).[48] But does the relative clause refer to the Athenians alone or to all three allies? The authors of *HCT* 4, p. 66, tacitly approve the latter interpretation, if the textual emendation is correct. Thus, all three allies had been previously summoned, had arrived, and were waiting at Harpine when they received word to come to the sanctuary.

Let us consider the alternative, that the relative clause concerns only the Athenian horsemen. It is possible that the Athenians could have been waiting at Harpine when the Argives and Mantineians arrived because the time required for the cavalry force of the Athenians to cover the greater distance from Athens might have been less than the time required for the Argives and Mantineians to levy a hoplite force. But is it not also possible that Thucydides intended the relative clause to indicate that certain Athenian horsemen, perhaps a fairly small contingent, *happened to be domiciled* at Harpine waiting for the Games and answered the summons, in contrast to the Argives and Mantineians who came expressly to Olympia in considerable numbers to aid the Eleians? Perhaps a certain number of Athenian horsemen had been

[47] 5.73.1. It is interesting that whereas Diodoros describes a brilliant, but probably fictitious, cavalry action by the Athenians at the battle at Delion (12.70.2), he fails to note the role of the Athenian cavalry here, where Thucydides plainly comments on their importance.

[48] See Gomme, Andrewes, and Dover, *HCT* 4, p. 66 for arguments supporting the retention of the manuscript reading of *Argei*.

dispatched as an escort to the Athenian participants in the Games, or perhaps had intended to participate themselves in equestrian events, analogous to the role of the Athenian cavalry in the Pythaids to Delphi in the second century B.C.[49] Finally, why would the Argives and Mantineians, serving clearly as an emergency military force, be "waiting for the festival"? If Thucydides meant "waiting for the expected trouble at the festival," or "for the end of the festival" then his brevity has served to disguise his meaning.

I conclude this survey of recorded cavalry operations in the Peloponnesos with the Athenian expedition against Melos in 416 (Thuc. 5.84.1–2). It is here that we have one of the few references to Athenian mounted bowmen during the war, although Thucydides tells us that Athens maintained a force of 200.[50] In this expedition Athens transported twenty *hippotoxotai.* It is puzzling that mounted bowmen were included and not cavalrymen because the terrain posed the same problems for both. It has been suggested that this small force might have been designed to "raid isolated farms and hamlets."[51]

Northern Aegean

The northern Aegean also proved to be a frequented arena of war for the Athenian cavalry. In the summer of 430 Hagnon and Kleopompos assumed command of the force which Perikles had recently directed around the Peloponnesos, a force including 300 horsemen (2.56.2), and straightaway (*euthus*) set sail for Thrace, against the Chalkidians and Poteidaia, under siege since 432 (2.58.1). After unsuccessful operations around Poteidaia and oppressed by the plague, Hagnon assembled the fleet and returned to Athens, apparently without relieving the troops already stationed there (2.58.3).

Did the 300 Athenian cavalrymen return to Athens with Hagnon or did all or some of them remain at Poteidaia? There is no report of any force sent from Athens in relief having arrived in that area when the Athenians with 2,000 hoplites and 200 horsemen marched against the

[49] See *FD* III.2, pp. 29–46, esp. nos. 27 and 28, pp. 34–37. There are some difficulties with this analogy. The games held at Delphi at the time of the Pythais may have been limited to Athenians (p. 46), therefore constituting a special *national* event. Thus, it cannot be said that Athenian horsemen—as members of the Athenian cavalry—were participating in an international athletic festival, as at Olympia. Secondly, what was true for the second century B.C. may not have been the case in the fifth century.

[50] 2.13.8. Cf. Thuc. 6.94.4: thirty *hippotoxotai* were included in the reinforcements to the Athenian army at Syracuse in 414.

[51] Gomme, Andrewes, and Dover, *HCT* 4, p. 155. But this can be done by *hippeis* as well.

Chalkidians in Thrace and the Bottiaians early in the following summer (2.79.1).[52] It is not clear from Thucydides' language whether the horsemen, as in the case of the hoplites, were Athenians: *Athenaioi dischiliois hoplitais heautôn kai hippeusi diakosiois epestrateusan*. We do know that during the early stages of the operations around Poteidaia and Olynthos in 432, Athens had relied exclusively on a foreign cavalry force of 600 Macedonian horsemen under the command of Philip, brother of King Perdikkas, and Pausanias (Thuc. 1.61.4, 62.4, 63.2). But with the first (recorded and probably actual) use of cavalry transports in 430 (2.56.2), Athens could now transport Athenian horsemen to more distant fields of battle. The Athenians could thereby augment the foreign cavalry force and also avoid the complete loss of a cavalry arm in the event of a reconciliation between Perdikkas and Philip.[53]

We can at least confirm the Athenian participation in the force of 200 horsemen. A certain Menexenos I, son of Dikaiogenes I, serving as a phylarch, was killed in the subsequent engagement with the Chalkidian cavalry outside the walls of Spartolos in the spring of 429.[54] Despite Thucydides' use of *euthus*, "straightaway," it is debatable whether Hagnon's expedition set out before the new military commanders, including Menexenos, had assumed office for the year 430/29.[55] Is it not possible that Xenophon, son of Euripides, and his two co-generals accompanied Hagnon's expedition and remained to take command of the operations at Poteidaia? They are attested as gen-

[52] According to Gomme, *HCT* 2, p. 212, this should fall in late May 429.

[53] For instance, alliances with Perdikkas had not proved to be of a lasting nature, see Thuc. 1.57.1, 62.2. Could more be expected from his brother?

[54] See Is. 5.42 and *APF* 3773, pp. 145–49.

[55] Nevertheless, Gomme, *HCT* 2, p. 164, supposes that the expedition left in July, before the end of the archon year, i.e., July 23, 430 B.C. (according to B. Meritt, *Athenian Financial Documents of the Fifth Century* (Ann Arbor, Mich., 1932), 176. The bouleutic year, however, ended on July 1. First of all, the chronology is inexact, based primarily on a *terminus post quem* for the end of the Peloponnesian invasion of Attika in that year (see Thuc. 2.56.6, 2.47.2, 56.2). Secondly, it is not absolutely certain that in this period the generals and other military officers took office on Hekatombaion 1, the first day of the new archon year, rather than on the first day of the first prytany of the new Council of Five Hundred (as suggested by H. T. Wade-Gery, "Studies in the Stucture of Attic Society: II. The Laws of Kleisthenes," *CQ* 27 [1933]: 28). The two calendrical systems were not synchronized until the last decade of the fifth century B.C. (see Rhodes, *CAP*, pp. 224–29). Thus, if the military officers were inaugurated along with the councillors, i.e., on July 2, 430, then the expedition might well have had aboard the military commanders for 430/29. Even if they entered office with the archons, one cannot assume that the Athenians did not wait a few weeks until the transfer of commands was effected.

erals in the Thracian regions in the winter of 430/29 (2.70.1).[56] This same Xenophon, incidentally, was one of the three hipparchs of the 450s whose name is inscribed on the base of the equestrian monument found on the Acropolis (see Chapter Two). He, like Menexenos I, died in the Spartolid campaign (2.79.7).

In the summer of 422 Kleon persuaded the Athenians to dispatch an expedition to Thrace. The force consisted of 1,200 Athenian hoplites, 300 horsemen, a larger complement of allies, and 30 ships (5.2.1). After brief stops at Skione (where he enlisted some hoplites from the army beseiging the city) and Torone in the Chalkidike, Kleon proceeded around Mt. Athos to Amphipolis. It is not to my purpose to detail the subsequent battle of Amphipolis in which both Kleon and Brasidas fell. The Spartan force was victorious. We hear nothing in Thucydides' account about how the 300 Athenian cavalrymen fared in the battle, but one scholar has inferred that a charge of cowardice was subsequently leveled against them.[57] This suggestion has little to commend it. It is of some interest, however, that Tzetzes' scholion on Aristophanes' *Clouds* 549a records that Kleon was a *hippeus* and that he was killed in a cavalry battle (*hippomachia*) with Brasidas. The idea that Kleon may have been a cavalryman is probably correct (see later in this chapter), but the report that he participated in the battle of Amphipolis in that capacity has only Tzetzes' scholion to support it and is rather unlikely considering his command position as *strategos*. Finally, Thucydides 5.10.9 reports that while fleeing, Kleon was overtaken by a Myrkinian peltast (light-armed soldier). Surely, if Kleon had been mounted, Thucydides would have noted this amazing display of speed.[58]

The last time we hear of the Athenian cavalry involved in activities in the northern Aegean during the war occurred in the winter of 416/5. Thucydides reports that the Athenians transported some of their cavalry by sea to Methone, along with some Macedonian exiles, and that they ravaged the land of Perdikkas (6.7.3)

[56] Fornara, *Board of Generals*, p. 55, suggests that Xenophon and his colleagues "far from replacing Hagnon and Kleopompos, probably held command over the three thousand troops already stationed before Potidaea," yet Fornara is not confident enough to list them as *strategoi* for the previous year, 431/30 (p. 53). And we simply cannot say whether Menexenos was in his second consecutive year as phylarch in 430/29.

[57] J. MacInnes, "The Athenian Cavalry in the Peloponnesian War," *CR* 25 (1911): 193–95.

[58] Gomme, *HCT* 3, pp. 652–53, argues that Kleon has been falsely accused of cowardice because of Thucydides' prejudice toward him.

Sicily

We turn our attention now to the events of the Sicilian Expedition (415–413).[59] Thucydides reports that one horse-transport with thirty horsemen accompanied the mighty Athenian host to Sicily in 415 (6.43).[60] The meager size of the cavalry force is puzzling at first glance. It is clear that even during the debate in the assembly over the expedition Nikias recognized the threat of the Syracusan cavalry. He declared that the chief advantage of the Syracusans lay in their possession of an abundant supply of horses (Thuc. 6.20.4) and that a large land force would be necessary so that the allies would not be shut off from the land by the numerous Syracusan cavalry (6.62.1).[61] Even so, he never suggested that an Athenian cavalry (or any from another allied state) be sent.[62] There can be no doubt, therefore, that the size of the Athenian cavalry actually sent reflected the unquestioned belief that Egesta could not only furnish a respectable cavalry force of its own, but could guarantee cavalry contributions from other Sicilian cities. Nikias' warning was prophetic: if the Egestaian promises were not fulfilled, the Athenian efforts in Sicily were doomed to failure (6.21.1).

It is probable that the thirty horsemen were never expected to engage in a major confrontation with the Syracusan cavalry. Rather, they were sent to act as mounted scouts,[63] counterparts to the *hippês kataskopoi* of the Syracusan cavalry (6.63.6), or as messengers.

I have suggested elsewhere that another, more important, responsibility devolved upon them after the true situation in Sicily became known.[64] Thucydides reports that shortly before the first major en-

[59] Along with Kagan, *Sicilian Expedition*, see for general discussion, P. Green, *Armada from Athens* (London, 1970).

[60] Diod. 13.2.5 claims that this expedition included horse-transports, hoplites, slingers, and cavalry (the text on the last is disputed).

[61] Compare the similar danger and phrasing with respect to the expedition into Thessaly in 455 and the effect of the Thessalian cavalry's control of the land (Thuc. 1.111, see Chapter Two).

[62] Kagan, *Sicilian Expedition*, p. 242, speculates that "the oversight was more a failure of purpose than of judgment, that it resulted, at least in part, from his original disinclination for the expedition, from his hope that it would never be necessary to fight at all." This explanation is not satisfactory.

[63] As the Athenian cavalry would be used in 370 B.C., although Xenophon criticizes Iphikrates' decision to send all of the cavalry, rather than a few who would have been more than sufficient (Xen. *Hell.* 6.5.52). On mounted scouts, see also Aineas Taktikos 15.5; Onasander 6.7; and Asklepiodotos 7.1. For modern discussion, see Pritchett, *GSW* I, pp. 127–33.

[64] See my "The Athenian Cavalry in the Sicilian Expedition. Some Notes on Thucyd-

gagement of the Sicilian campaign (winter, 415/4) the generals anxiously sought a suitable landing place and a defensible position for the army near Syracuse, a place where the light-armed troops and the *ochlos* might not suffer from the attacks of the Syracusan cavalry, "since they had no horsemen," *sphisi d'ou parontôn hippeôn* (6.64.1).[65] The authors of *HCT* 4, p. 310, express surprise that the thirty Athenian horsemen were by this account absent. I propose that they were otherwise engaged—as heralds. After the Athenians discovered that they had been deceived by the Egestaians with respect to the promised money, Alkibiades counselled that heralds be sent (*epikêrykeuesthai*) to all the cities except Selinous and Syracuse and to the Sikels with the purpose of making friends from whom the Athenians might procure grain and troops (6.48). Lamachos supported Alkibiades' plan (6.50.1), so we may assume that the heralds were in fact sent on their way. I suggest that Alkibiades employed the Athenian horsemen for some of these diplomatic missions, especially to the more distant and inaccessible areas of the interior, and that they were so occupied at the time of the battle.[66] We may possess a *comparandum* in the Athenian

ides 6.43," in *Studia Pompeiana et Classica in Honor of Wilhelmina F. Jashemski*, edited by R. I. Curtis, vol. II (New Rochelle, N.Y., 1986).

[65] I am tempted to suggest that Thucydides intended the phrase *sphisi d'ou parontôn hippeôn* to mean "they had no horsemen *present*" at the time of the battle, not that they had no cavalrymen available at all (Loeb and R. Warner translations) nor that they were there but stood apart from the battle because of their few numbers—thus constituting no real and legitimate cavalry force: W. S. Ferguson, *CAH* V. 291. For possible *comparanda* to my suggested use of *pareimi*, see Thuc. 5.73.1: at Mantineia (418), the Athenians would have suffered even more "had not their horsemen been present (*parontes*) and been useful"; and earlier in the same campaign, the Argives and their allies awaited the arrival of the Athenians since "they had no horses [i.e., cavalry]," meaning "they had no cavalry present there for the battle" (*hippoi de autois ou parêsan*; but see n. 32); and perhaps more obvious, though not concerned with cavalry, Thuc. 4.94.1: at the battle of Delion (424), "but the light-armed troops . . . were neither present then nor did the city have any" (*oute tote parêsan oute egeneto têi polei*); and also Xen. *Hell.* 6.5.30 describing the invasion of Lakonia in 370 by the Thebans and their allies, "as many horsemen of the Phokians, Thessalians, or Lokrians as were present" (*hosoi Phôkeôn ê Thettalôn ê Lokrôn hippeis parêsan*). Perhaps the usage is always ambiguous: those who do not have cavalry cannot, of course, have them present on the battlefield; but those who have cavalry, may choose for one reason or another, not to have them present at a given battle.

[66] After the flight of Alkibiades in the face of impending criminal charges back at Athens, Nikias sent word to Athenian allies among the Sikels, bidding that troops be sent (6.62.5). One should not assume from this that these allies represent new friends gained for the Athenians as a result of the diplomatic missions proposed by Alkibiades. A distinction is clearly made between old friends and allies as opposed to new ones. During the debate over which strategy to execute, Nikias had suggested that the Athenians should sail by the other cities showing off the power of Athens and demonstrating their

horseman, specifically identified as a *kêryx*, who was sent to inquire as to the disposition of Pausanias' forces at Plataia in 479 (Hdt. 9.54–55). This may be simply a herald who is mounted, but Herodotus' words *sphêon hippea* seem to imply an Athenian cavalryman (see Chapter One). We might also note the role of Xenophon as an aide-de-camp to Cyrus the Younger in the ill-fated Persian campaign in 401 B.C. and his organizing of a small Greek cavalry after the defeat (Xen. *Anab.* 1.8.14–17; 3.1.4–10; 3.3.19–20).

The first major battle in the winter of 415/14 near Syracuse demonstrated that the Athenian and allied infantry was superior to that of the Syracusans, but that pursuit was made impossible by the presence of an undefeated and unopposed Syracusan cavalry. The Athenians finally acknowledged the impossibility of continuing the campaign without a respectable cavalry force, one composed of Sicilian and Athenian cavalrymen (6.71.2; Diod. 13.6.6). Passing the winter in Naxos (Sicily), the Athenians dispatched a trireme to Athens for money and cavalry, in order to be prepared for the spring campaign. They also made overtures to Carthage, Tuscany, and Egesta to send as many horses as possible (6.88.6). These actions were absolutely necessary because the Athenians never considered sending a large number of horses to Sicily. We are informed that in the spring of 414 Athens sent as reinforcements thirty *hippotoxotai* and 250 *hippeis* "with their equipment" (*meta skeuês*), but without their horses, "for it was expected that horses would be procured in Sicily" (6.94.4). The money for that purpose was to come out of the 300 talents of silver sent with the reinforcements.

It is not really surprising that Athens declined to send nearly 300 horses with their riders but chose rather to depend on mounts purchased on the spot. Thucydides remarks that this expedition was sent further than any other from home (6.31.6). But the contrast with previous expeditions was not limited to distance. Whereas Perikles transported 300 horsemen and their mounts to the northern Peloponnesos in 430 (and Hagnon the same force to Poteidaia the same summer),

zeal towards their friends and allies and then they could sail home—unless they were able to help the Leontines or win over some of the other cities (Thuc. 6.47). Although Nikias speaks of "friends and allies" somewhat abstractly here, he seems to be speaking of the same "allies" in 6.62.5 and less likely of newly won allies. Alkibiades' recommendation to make friends was specifically directed towards the "other" cities (*es tas poleis . . . tas allas* (6.48). I propose that most, if not all, of the Athenian horsemen had just begun the effort of recruiting new friends among the Sikels when the Athenian generals commanded the allies of the Athenians to fulfill their treaty obligations by sending troops in the late summer (6.62.5), and that the horsemen were still recruiting the following winter when the battle took place.

and Kleon took an equal complement by ship to Thrace in 422, the fleet was always within reach of land in case of inclement weather. The sometimes treacherous open-sea crossing of the Straits of Otranto, on the other hand, no doubt made the Athenians less inclined to risk such a large proportion of their cavalry mounts.[67] If a horse-transport sank in open waters, there was little hope of the horses surviving; triremes could rescue drowning men, but what accommodations could be made for horses, even if there existed some means of hoisting them aboard? One bad storm and nearly one-third of Athens' cavalry's mounts would be lost.

Furthermore, as I have remarked earlier in this chapter, I am inclined to believe that Athens never actively maintained many more than ten horse-transports. A larger number would have been uneconomical. We may assume that one transport normally carried thirty horses and their riders (and probably attendants)[68] from the information provided in Thucydides 6.43 and we have no evidence that Athens ever committed more than 300 cavalrymen on any major seagoing expedition during the Peloponnesian War. Certainly there must have been replacements in dry dock in case of loss, but probably only a few. Therefore, the Athenians' reluctance may be attributed not only to fear for their horses but also concern for the specially reconstructed ships which carried them.

By the summer of 414 Athens was able to field a cavalry force numbering 650.[69] Egesta contributed 300, the Sikels, Naxians, and some

[67] Perhaps a modern parallel: in the Spanish-American War (1898), there was a good deal of hesitation about shipping American cavalry mounts to the Philippines on a journey expected to last thirty to forty days; it looked as if the troopers would have to fight dismounted, unless horses could be procured from Australia. But the American horses were finally shipped, secured in specially constructed quarters below the decks (Carter, *Horses, Saddles, and Bridles*, pp. 385–401).

[68] What about the mounts, whether horses or mules, of the attendants? We know from Thuc. 7.13.2 and 7.75.5 that the Athenian cavalrymen (and hoplites) had attendants (*therapontes, akolouthoi*) or grooms (*hippokomoi*, see Xen. *Hell.* 2.4.6 and *Hipparchikos, Peri Hippikes, passim*). It is reasonable to suppose that if their masters procured mounts for them, they did so upon their arrival in Sicily. Even their masters did not come with horses on this particular expedition, so all mounts had to be procured on the spot. Generally, while attendants normally accompanied both hoplites (as at Poteidaia, Thuc. 3.17.2–3) and cavalrymen on foreign expeditions, for the last they probably did so without mounts. In Attika, during land operations close to home, some of them would have been mounted (as in Xen. *Hipparch.* 5.6). If the state was not willing to provide a *katastasis* to enable an attendant to join his master mounted, there is no reason to believe it would have been willing to transport his mount.

[69] Diod. 13.7.4 reports that the Athenians and allied cavalry totaled 800, apparently reckoning 250 from the Sikels; see Gomme, Andrewes, and Dover, *HCT* 4, p. 732.

others about 100, and the Athenians 250. They had received some horses from the Egestaians and Katanaians and purchased others to mount their own cavalrymen (6.98.1). Again the thirty Athenian horsemen (6.43) are absent from this list. Unless we assume that they had returned to Athens or had been killed in some unrecorded action the previous year, we must suppose that they were still engaged in diplomatic/recruiting duties. Perhaps some trace of their success may be found in the cavalry contributions from the Sikels, Naxians, and some others (6.98.1) who were not present in the previous campaigning season. It is also possible, of course, that by this point in the narrative Thucydides has simply forgotten to add them to the tally.

The effect of a cavalry force was immediate. A tribal division (*phylê*) of Athenian hoplites joined by all of their horsemen (*hoi hippês met'auton pantes*) routed part of the Syracusan cavalry attempting to hinder construction of a round fort. The Athenians set up a trophy to commemorate the cavalry victory (6.98.2–4). Following a string of military successes during the summer, many of the Sikels who had been hesitating joined the Athenians as allies (6.103.2).

Upon the arrival of Gylippos, the Spartan commander, and reinforcements from Himera, Selinus, and some Sikels, the tide of the campaign began to turn against the Athenians. Gylippos, effectively utilizing the Syracusan cavalry and javelin throwers, put a halt to the construction of the Athenian circumvallation wall (7.6.2–4). The Syracusan cavalry meanwhile continued to harass the Athenian foraging parties and effectively maintained control of the land (7.4.6).[70] By the end of the summer Nikias sent a desperate letter to Athens requesting reinforcements, specifically pointing to the large number of the opposing cavalry and their control of the countryside (7.11.2, 4). The letter reached Athens in the winter of 414/13. After Nikias had so clearly indicated the problems caused by the Syracusan cavalry, we might be surprised to discover that although the Athenians dispatched 80 triremes and 500 hoplites under the command of Demosthenes, no mention is made of cavalry, with or without horses.[71]

The reason that the Athenians failed to send additional cavalry seems obvious: they feared a Spartan invasion of Attika; consequently,

[70] On the Syracusan cavalry's continued dominance of the countryside and attacks on foragers and stragglers, see Thuc. 7.11.4, 13.2, 44.8, 78.7; cf. Plut. *Nik.* 19.6.

[71] Martin, *Cavaliers*, p. 438, believed that despite Thucydides' silence, the reinforcements must have included cavalry. He conjectured that the number of horsemen may have been too small to mention. Considering the fact that Thucydides does mention the one horse-transport with thirty horsemen in 415, it is very difficult to believe that he could have neglected to mention the presence of cavalrymen at this time.

a sizeable cavalry force had to be on hand to defend the homeland—a continuation of its primary responsibility from the 420s. The fragile Peace of Nikias was officially broken in the summer of 414 when Athens aided Argos against an invasion by the Lakedaimonians and their allies. Thucydides 6.105.1 (cf. 7.18.2) reports that the presence of thirty Athenian ships was viewed as the most overt violation of the peace treaty. Thus, at the same time that the Athenians were assembling the reinforcements destined for Sicily, Sparta was making preparations for a new invasion of Attika (7.18.1). At the very beginning of the spring of 413, Sparta did invade Attika, ravaged the plain of Attika, and proceeded to fortify Dekeleia (7.19.1). Thucydides tells us explicitly that the dispatch of Demosthenes with the reinforcements to Sicily occurred synchronously with the fortifying of Dekeleia (7.20.1–2).[72]

There can be little doubt that the Athenians knew of Sparta's planned invasion. This fact alone argued against the dispatch to Sicily of any more cavalrymen. And with the establishment of a permanent Spartan garrison in Attika at Dekeleia, Athens was not about to diminish its cavalry force further. Only the cavalry allowed Athens to maintain any semblance of control in the countryside. The cavalrymen sallied forth every day against Dekeleia and tried to keep guard throughout Attika (7.27.5). No small task! I have already mentioned what hardships these operations inflicted upon both horsemen and unshod mounts.

It is not my task to detail the tragic events that befell Athens in Sicily in 413—the naval defeat in the harbor, the retreat inland, the massacre of the Athenian army at the Assinaros River, and the incarceration of the survivors. Plutarch reports that the victorious Syracusan cavalry adorned their own horses, cropped the manes and tails of the captured Athenian mounts, and marched in procession into Syracuse (*Nik.* 27.6). Some Athenian horsemen did escape the carnage at the Assinaros. We can identify two of them: Kallistratos son of Empedos, the hipparch of the Athenian cavalry in Sicily, and an unidentified son of Polystratos, a simple cavalryman. According to Pausanias 7.16.3, Kallistratos charged through the enemy at the head of his cavalry, led them to safety at Katana, and then returned alone to Syracuse, only to perish attacking Syracusans who happened to be looting the deserted Athenian camp.[73] The first part of this account is corroborated by the story

[72] For this relative chronology, see also Isok. 8.84; Aischin. 2.76.

[73] This story is repeated in [Plut.] *Lives of the Ten Orators* (*Demosthenes*) 844B, but there is confusion between this Kallistratos and the homonymous fourth-century B.C. politician.

of the son of Polystratos recorded in [Lys.] 20.4, 24–25 (also see above). The son reports that after the massacre he himself made his way to Katana and using that city as a base of operations raided the enemy and acquired money to ransom Athenian prisoners. Thus it is clear that a certain proportion of the Athenian cavalry escaped to Katana led by the hipparch Kallistratos. The son of Polystratos was subsequently obliged to serve in the cavalry and infantry of Katana. This service can have lasted only a few years, since he was back in Athens at the time of the speech, 410 B.C.[74]

Asia Minor

The effects of the disaster in Sicily were felt throughout Athens' Aegean empire. During the final years of the Peloponnesian War, the Athenians directed their attention to keeping their restless subjects in obeisance. Although we have no evidence in the literary tradition that a cavalry force accompanied the Athenian fleet in its operations in the eastern and northern Aegean in 412/1, an inscription, recording casualties suffered in a naval battle believed to be connected with the Athenian victory in that year over a Peloponnesian fleet in the Hellespont (Thuc. 8.104–6),[75] lists a phylarch named Pythodoros (*IG* I^2 950, line 180). Perhaps Pythodoros was on board as a marine (*epibates*), as is attested for some hoplites after the revolt of Chios (8.24.2), but is more likely that he commanded a tribal contingent or a small cavalry force during the land-based operations against revolting states like Chios, Lesbos, and Miletos. We know at any rate that the Athenians included troop transports (*hoplitagagoi*) in the expedition to Asia Minor in 412 (8.25.1, 30.2).

From 410/9 to 408/7 there was a small but effective cavalry force operating in the eastern Aegean and Hellespontine region. Early in the summer of 409 Thrasyllos set sail for Samos with a force of hoplites, 50 ships, and 100 Athenian cavalrymen (Xen. *Hell.* 1.1.34; 1.2.1). After some skirmishes near Miletos and Kolophon and a defeat at Ephesos, Thrasyllos rejoined the rest of the Athenian army under the command of Alkibiades at Sestos. During the winter Alkibiades led an expedition against Abydos from his base at Lampsakos. Pharnabazos, the Persian satrap, arrived with a large number of cavalrymen, but was defeated and put to flight. Alkibiades pursued Pharnabazos with the

[74] We cannot guess how many men of the cavalry force succeeded in reaching home. Thuc. 8.1.2 states generally that Athens, because of the disaster, was deprived of many hoplites and horsemen.

[75] For a new study of this inscription, see D. W. Bradeen, "Athenian Casualty Lists," *Hesperia* 33 (1964): 41–42, no. 13, pl. 7.

Athenian cavalry and 120 hoplites until darkness prevented further chase (*Hell.* 1.2.16).

In the following year Alkibiades, accompanied by only a few hoplites and the cavalry, was able to intimidate Bithynian Thracians into surrendering valuables which the city of Kalchedon had deposited into their safe-keeping (*Hell.* 1.3.3). In the subsequent siege of Kalchedon, Pharnabazos, despite his infantry and numerous horsemen, was unable to assist Hippokrates, the Spartan governor of Kalchedon. A hotly contested hoplite battle took place between the troops of Thrasyllos and those of Hippokrates, until Alkibiades turned the tide of battle with a few hoplites and the cavalry (*Hell.* 1.3.6). The success of this small Athenian cavalry force against the numerically superior satrapal cavalry was no mean accomplishment.

After his triumphal return to Athens in 407, Alkibiades set sail for Andros and Samos with a force of 1,500 hoplites, 150 horsemen, and 100 ships. We have no other information on these horsemen. The last we hear of the Athenian cavalry during the Peloponnesian War comes in 406. The Athenian fleet under the command of Konon found itself blockaded in the harbor at Mytilene by a Spartan fleet. When the Athenians received word of this predicament they equipped 110 ships, manning them with all those of military age, whether slave or free man (*Hell.* 1.6.24). Xenophon adds that even many of the cavalrymen went aboard, *eisebêsan de kai tôn hippeôn polloi.* Some of the cavalrymen may even have served as rowers, although the verb might suggest their function as *epibatai.* In the autumn of 428 an Athenian force of approximately equal size was dispatched to Mytilene. It had included 1,000 Athenian hoplites, some undoubtedly doing the rowing themselves (Thuc. 3.18.3), others perhaps serving as *epibatai* (so Gomme, *HCT* 2, p. 271). The seriousness of the emergency in 406 is highlighted by the fact that in the previous expedition, the *hippeis* (and the *pentakosiomedimnoi*) had not been required to embark (3.16.1).[76] The loss

[76] This is a troubling passage. And the scholiast to Thuc. 3.16.1 leaves us wondering. After telling us that the Athenian constitution was divided into the four [Solonian] classes, he explains the passage: the *pentakosiomedimnoi* were not compelled to embark on the ships because of the very great honor (*megistên timên*) they held on account of paying so many of the taxes, *dia ta polla telein* (perhaps *eisphora*?, see discussion later), and the *hippeis*, because there was no "need for them" (*toutôn chreia*). But what does he mean by "need"? That the Athenians had sufficient numbers with the two lower classes and therefore did not need the *hippeis* and *pentakosiomedimnoi* in a quantitative sense, or for what was viewed as essentially a naval operation, there was no need for cavalry *qua* cavalry? It is not easy to decide—Gomme, *HCT* 2, p. 271, obviously thought that Thucydides' use of *hippeis* meant cavalry. I would suggest that the selection, or exemption if you will, was primarily based on the numbers needed to man the ships, so

of nearly a fourth of their fleet with almost all of their crews in the costly victory at Arginousai in 406, no doubt including a number of *hippeis*, may have increased the hostility of the cavalry towards the democratic regime and encouraged their support for the oligarchy of 404/3.

POLITICAL UNITY

Kleon and the Athenian Cavalry

In the early 420s, Kleon became involved in a bitter dispute with the Athenian cavalry.[77] The animosity between the two parties is first mentioned by Aristophanes in his *Acharnians* performed at the Lenaian festival in 426 and received notoriety in his *Knights* produced at the same festival in the following year. Unfortunately, both the chronology and the details of the conflict have to be reconstructed mainly from scholia to the Aristophanic comedies.[78]

Difficulties may have arisen when Kleon himself was a member of the cavalry. The scholion to *Knights* 225 reports that "the horsemen attacked him, since when he was one of them, he treated them badly."[79] The claim that Kleon had been a cavalryman is also voiced by Tzetzes on Aristophanes *Clouds* 549a, but, as I have noted above, his comment that Kleon was mounted during his fatal skirmish at Amphipolis in 422, is unlikely. Kleon's membership in the Athenian cavalry at some point in his career is not really particularly startling: W. R. Connor has shown that Kleon probably belonged to the *nouveaux riches* of Athens, and that the portrait of him as a lowly tanner in the plays of Aristophanes smacks of comic irony.[80] We do not know the cause of this quarrel, but it must have been serious since it apparently involved the whole corps. Perhaps his status as *neoploutos* brought abuse from a

a division by census classes seemed a convenient way of accomplishing that end, but the Athenians were also not unaware that the top two classes represented the cavalry needed to defend Attika and the class engaged in the usual liturgical duties (implied in the scholion)—in fact, there should have been at least 100 upper-class Athenians already committed to this campaign, serving as trierarchs. In any case, the upper-class bias against serving as lowly rowers would not have gone unvoiced.

[77] For basic sources, see Fornara, *Archaic Times*, no. 131.

[78] On these problems, see W. R. Connor, *Theopompus and Fifth-Century Athens* (Washington, D.C., 1968), 50–59.

[79] For comment, see G. Gilbert, *Beiträge zur innern Geschichte Athens im Zeitalter des peloponnesischen Krieges* (Leipzig, 1877), 133.

[80] *New Politicians*, p. 152 and n. 32. Connor believes that it is perfectly reasonable to accept Kleon's membership in the cavalry. See *APF* 8674, pp. 318–20 and now F. Bourriot, "La famille et le milieu social de Cléon," *Historia* 31 (1982): 404–35.

strong aristocratic element of the cavalry and this mushroomed into a wider dispute affecting the whole cavalry. We cannot be sure.

The introduction of the war-tax (*eisphora*) and Kleon's association with it may have further angered the Athenian cavalry. During the winter of 428/7 the Athenians, engaged in the siege of Mytilene, needed additional funds. According to Thucydides 3.19.1, the Athenians "then for the first time (*tote prôton*) imposed a war-tax on themselves to the amount of 200 talents." This information has stimulated a lively debate among scholars as to whether it means for the first time in the Peloponnesian War or in Athenian history.[81] What cannot be disputed is that this measure reflects a condition of extreme financial difficulty which necessitated the tapping of a *normally* unused source of revenue. There is also every reason to believe that the *eisphora* was intended to fall most heavily on the wealthy classes.[82] Its introduction or re-introduction at this time could not have been welcomed by these Athenians. In *Knights* 923–26 the Paphlagonian slave (Kleon) threatens to enroll the Sausage-seller among the wealthiest citizens in order to burden him "with war-taxes" (*tais eisphorais*). Earlier in the same play (223–24), we learn that the rich (*hoi plousioi*) fear the Paphlagonian slave.

In *Knights* Aristophanes certainly leaves us with the impression that Kleon was primarily responsible for guiding the fiscal policies of the state:

> And how, among all public men except me, could you find
> such an absolute friend?
> Didn't I, when I used in the Council to sit, hand out to
> you funds without end,
> By the art of the squeeze and the twist and the rack; and
> blackmail in varieties, too?
> Not caring how mere private citizens squealed if I just
> gave pleasure to you?[83]

[81] This is not the place to enter the debate. The reader can consult Gomme, *HCT* 2, p. 278; A. Andreades, *A History of Greek Public Finance* (Cambridge, Mass., 1933), 333; R. Thomsen, *Eisphora* (Copenhagen, 1964), 119–46; and the relevant epigraphic evidence: *IG* I^3 52, frag. b. line 17 (=I^2 92; ML 58; 434/3 B.C.); I^3 41, frag. a, line 38 (=I^2 42, 446/45 B.C.) along with Meiggs, *Athenian Empire*, p. 519; H. B. Mattingly, "Athenian Finances in the Peloponnesian War," *BCH* 92 (1968): 450–85; and G. T. Griffith, "A Note on the First Eisphora at Athens," *AJAH* 2 (1977): 3–7.

[82] It has been debated whether the *eisphora* was a progressive or a proportionate tax; I accept the conclusion of Andreades, *Greek Public Finance*, p. 340, that "we must acknowledge at any rate that in actual practice, the wealthier citizens paid much more than their proportionate share of the taxes." Thomsen, *Eisphora*, pp. 166, 252 concurs.

[83] Lines 773–776; translation by G. Murray, *Aristophanes The Knights* (London, 1956), 74–75.

We may detect a veiled reference to the *eisphora* in Aristophanes' use of "private citizens" (*tôn idiôtôn*), those whose overriding concerns were private, not public matters. Kleon might then be referring to those wealthy Athenians who did not contribute sufficiently to the democracy's efforts and therefore deserved to be taxed additionally. Even if Kleon did not himself propose the tax measure, there is no reason to doubt his support of or complicity in its imposition.[84]

Since the *eisphora* affected the richest citizens of Athens most directly, clearly all of the members of the Athenian cavalry or perhaps the supporting fathers of many young sons serving in the cavalry found themselves with an additional wartime burden. It is not hard to imagine that Kleon felt some satisfaction at "paying back" the cavalry for its earlier abuse, even if this was not the primary reason behind the introduction of the *eisphora*. As I have noted above, this measure was in part necessitated by the naval operation of 428/7 (3.16.1). Since the *hippeis* and *pentakosiomedimnoi* had been exempted from service on that occasion, it was natural to assume that they would contribute to the war effort in other ways, e.g., monetary. Thucydides at least leaves the impression from 3.17.1–3 and 3.19.1 that the naval operation was still going on when the tax measure was passed.

The well-to-do Athenians were no doubt unhappy with a measure which seemed to be directed exclusively at them. Their dissatisfaction probably spilled over into the *symposia* and the political clubs (*hetaireiai*).[85] In this form, it remained relatively amorphous and innocuous, but among the members of the Athenian cavalry, already at odds with Kleon on other matters, the dissatisfaction assumed a sharp and hostile character.

Already hated by the cavalry and many of the well-to-do, Kleon made the unforgivable mistake of getting caught accepting bribes from certain tribute-paying islanders. In *Acharnians* 5–8, Aristophanes has his character Dikaiopolis delightfully recount that the cavalry forced Kleon to vomit up five talents; "I love the horsemen for this deed," he continues, "one most worthy for Greece." According to the scholion to this passage, citing Theopompos as his source,[86] Kleon was fined

[84] Gomme, *HCT* 2, p. 279, is overly concerned by Thucydides' failure to mention Kleon's name in connection with the measure. See Thomsen, *Eisphora*, pp. 168–70, for examination of the evidence connecting Kleon to the introduction of the *eisphora*; he argues that none of the evidence proves that Kleon was the originator, but he must have warmly supported the measure.

[85] The standard work on political clubs is F. Sartori, *Le eterie nella vita politica ateniese del VI e IV secolo a.c.* (Rome, 1957); see also Connor, *New Politicans*, p. 197 and n. 83.

[86] *FGrH* 115 F 94; for discussion see Connor, *Theopompus*, pp. 53–59.

five talents because he had insulted (*hybrizein*) the cavalry. The scholion goes on to record that the islanders had paid Kleon five talents in the hope that he might persuade the Athenians to lower their tribute (*eisphora*). The cavalry, discovering this, spoke against him, and demanded he return it.

One problem, as Connor rightly points out, is that the wording of the scholion might seem to imply a "formal heliastic trial."[87] This interpretation would seem to be supported by the scholiast's use of *ezêmiôthê*, "he was fined."[88] In spite of this, Connor is probably correct when he suggests that "the words of the fragment itself need mean no more than that the knights spoke against Cleon's proposal to lower the islanders' tribute and demanded that he return the five talents he allegedly exacted."[89] That Kleon accepted a bribe amounting to five talents seems certain from the combination of *Acharnians* 6 and the scholion to this verse quoting Theopompos' words "he received five talents." If Kleon had really been fined five talents for accepting a bribe, in point of fact he would have suffered no penalty at all since he would only be relinquishing that amount which belonged to him illegally anyway, i.e., the five talents from the islanders. Furthermore, if Kleon had really been forced to submit to a formal trial, it would be hard to account for his apparently undiminished popularity throughout the 420s.[90]

In addition, I suggest that the scholiast's explanation "on account of insulting the horsemen" probably encompasses both Kleon's earlier troubles with the cavalry and the affair of the bribe. The direct force of *hybrizein* closely mirrors the words "he treated them badly" in the scholion to *Knights* 225. The hounding of Kleon because of the bribe is nicely explained by some outrage (unspecified) perpetrated by Kleon against the cavalry at a time when he was a member of that body and aggravated by the imposition of the *eisphora* in 428/7. The issue of *eisphora* carried a bitter twist. It had been imposed because there was a desperate need for additional revenue to wage war. To discover that the man held responsible for its introduction had accepted bribes, or

[87] *Theopompus*, p. 57.

[88] The same legalistic term is repeated in the Second Hypothesis to the *Knights*; in Tzetzes on the *Clouds* 549a; and in Bios XI of Aristophanes (lines 22–24).

[89] *Theopompus*, p. 57.

[90] Professor Ed Carawan has kindly allowed me to see a typescript of his forthcoming article, "Eisangelia and Euthyna: The Trials of Miltiades, Themistocles, and Cimon," in *GRBS*, vol. 28, where he suggests that the case against Kleon was initiated as a *probolê*, a proposal for investigation of wrongdoing before the Athenian assembly, which apparently never made it to trial.

at least was accused of that indiscretion, for the purpose of reducing the amount of *eisphora* from the islanders and consequently making the collection of domestic revenue more necessary must have incensed the well-to-do Athenians and especially the cavalry corps, an established enemy.

It strikes me as deliberate that immediately following Paphlagon's threat to enroll the Sausage-seller in the richest classes and thus force him to pay *eisphorai* (*Knights* 923–26), the Sausage-seller describes a plan for getting a talent out of his rival (929–40): An assembly is meeting to discuss the tribute quota required from the Milesians.[91] When Kleon steps forth, eager to earn the talent with which he has presumably been bribed by the Milesians to hold down the assessment increases or even to effect a decrease, in his gluttony he will gobble down a free pan of cuttle-fish and in his haste to finish the meal and get to the task of earning the talent bribe, will choke. When this happens the assembly, i.e., the people of Athens will get the talent instead.

What I suggest is that Aristophanes deliberately juxtaposed the two passages, knowing that the more perceptive members of his audience would detect the comic hypocrisy of Kleon's imposing extraordinary taxes on the citizenry at the same time as he was filling his own coffers with bribes from those states desiring to have their quotas lightened.[92]

Kleon's problems with the cavalry were not limited to this. We possess a perplexing scholion to *Knights* 226, which cites Theopompos again as the source.[93] Aristophanes has Demosthenes tell the Sausage-seller that the latter may find allies in his struggle against the Paphlagonian slave in the 1,000 good cavalrymen "who hate him," *misountes auton*. The scholiast comments:

> Theopompos says in Book Ten of his *Philippics* that the horsemen hated him. For, having been treated in an abusive manner and provoked by the horsemen, he "attacked the constitution" (*epetethê têi politeiai*) and continually worked evil towards them. For he accused them of having deserted the battlefield. For this reason, therefore, Aristophanes has made mention of them.

[91] Suggested by B. B. Rogers in the Loeb edition, p. 217. Kleon is also accused of accepting a forty-mina bribe from the Mytilenaians in Ar. *Knights* 830–35; see Gomme, *HCT* 2, pp. 331–32.

[92] Aristophanes expounds on the theme of corruption by Kleon in *Knights* 1211–25: Paphlagon's little storage chest reveals all the things rightly belonging to Demos which Kleon has hidden away for his own benefit.

[93] See *FGrH* 115 F 93.

I believe that this scholion reveals two areas of dispute between Kleon and the Athenian cavalry.

(1) Having had a bad time of it with the cavalry, Kleon attacked the constitution. The exact meaning of the last phrase is disputed. Connor has argued that *epetethê têi politeiai* means that Kleon "applied himself to holding political power" and sees "the fragment as a reference to Cleon's decision to win the greatest political power in the state and thereby teach the knights a lesson."[94] Though reasonable, this interpretation is not entirely convincing.

Ch. Fornara has ingeniously suggested that the words *têi politeiai* have been mistakenly substituted for the words *têi katastasei*.[95] Since the word *katastasis*, like *politeia*, can mean "constitution," the scholiast inserted what he believed to be a synonymous and more common word for that found in the text of Theopompos. But if, he argues, we suppose that the *têi katastasei* referred not to a political constitution, but to the system of state loans granted to the cavalrymen, then the more natural meaning of the passive form *epetethê*, i.e., "he attacked," leaves no ambiguity. Thus, according to Fornara, Kleon "attacked the *katastasis*," or more properly, "the system of *katastasis*." Although Fornara's theory requires a textual emendation, it is very attractive.

Small wonder that "the cavalrymen hated him." And is this not just another attack on the wealthiest citizens, specifically against a group who had been responsible for forcing Kleon to disgorge a five-talent bribe? Since the cavalry came from wealthier Athenian families, Kleon no doubt argued that they should forfeit their *katastaseis* (although they had to be paid back upon retirement) and—as was the case with the hoplites—foot the bill for that item appropriate to their military function, i.e., their horses.

(2) Connor assumes that the charge of military desertion (*lipostratia*) was but one example of the evil actions continually devised by Kleon against the knights, as stated in the scholion.[96] But might it not be possible that the explanatory conjunction, "for" (*gar*) introducing the sentence on desertion explains the entire preceding sentence and not just the second clause? The sequence of events would then go something like this: (a) Kleon accuses the cavalry of desertion, (b) the cavalry, either indignant at the false charge or stung by the truth of it, quite naturally attacks him in return, to which abuse he responds with

[94] See Connor, *Theopompus*, p. 52.

[95] "Cleon's Attack Against the Cavalry," *CQ* 23 (1973): 24.

[96] *Theopompus*, p. 52.

(c) an attack on the state-loan system of *katastasis* along with other unfriendly actions.

It has been suggested that the charge of *lipostratia* mentioned by the scholiast was connected with the campaigning season of 427.[97] Thucydides 3.26.3 records that the Spartan invasion of 427 destroyed vast stretches of Attika and proved more grievous to the Athenians than any other invasion except the second, yet does not mention any Athenian cavalry response. A charge of cowardice could have arisen from the heightened frustration of the Athenians at the progress of the war, the effects of the plague, and now this terrible destruction of Attika and the belief that the Athenian cavalry was not doing enough to curtail or prevent it. Anyone with an axe to grind could easily elevate this perception to a charge of desertion, or more properly, a dereliction of duty. Still, the silence of Thucydides is hardly decisive, especially in light of the assumed routine cavalry activity during most years of the war.

Furthermore, just because Kleon made the charge does not make it true; he also rails against the cavalry—falsely according to Aristophanes—accusing them all of being conspirators, *xynômotai*.[98] It is still too early to be accusing the cavalry of seeking the overthrow of the democracy. Their dissatisfaction is directed not so much at the democracy, but at its spokesman. Later, there will be good reason to call their loyalty into question. In any event, Kleon's accusations of desertion and conspiracy made the epithet which Aristophanes coined for him most appropriate: *taraxippostraton* (Ar. *Knights* 247), translated by B. B. Rogers as "troubler of our knightly train."

Although the chronology of events is arguable at almost every point and the particulars provided by the scholia are not secure, I tentatively offer the following reconstruction of Kleon's deteriorating relationship with the cavalry in the 420s. When Kleon had been a cavalryman, he had already experienced a falling out with the cavalry. Subsequently, he was credited with playing a supportive role in the imposition of the *eisphora* (428/7). Kleon was caught accepting a bribe and was forced to "cough it up." Kleon charged the cavalry with *lipostratia* (427?) which evoked the angry and abusive response from the cavalrymen.

[97] Gilbert, *Beitrage*, pp. 134–36. Connor considers this hypothesis "plausible" (p. 52). Busolt, *Griechische Geschichte*, vol. 3, pt. 2, p. 994, n. 6, supposes that the charge of *lipostratia* and the disgorging of the five talents were directly connected, i.e., Kleon, to punish their cowardice, tried to withhold five talents due the cavalry [perhaps, the *sitos* allowance?], but the cavalry forced him to give it up. Gomme, *HCT* 2, p. 290, calls this "an ingenious and possible explanation for Aristophanes' jesting."

[98] Ar. *Knights* 626–30; cf. 236, 257, 452, 476–77.

Kleon responded by attacking the most fundamental element of the new Athenian cavalry, the system of *katastasis*, assuming that Fornara is correct.[99]

In addition, I believe that the time during which Kleon served in the cavalry ought to be separated not only from the time he is accused of accepting a bribe (Theopompos F 94),[100] but also from the time he may have attacked the *katastasis* and charged the cavalry with desertion (Theopompos F 93). It is not likely that Kleon, heavily committed to policy-making, continued to belong to the cavalry after 428/7 (despite Tzetzes' remarks on Ar. *Clouds* 549a), nor immediately prior, if in fact he was close to his mid-forties or older at the time.[101] The thousand good and noble young horsemen of Athens hated and attacked one outside their number.[102] One point at least stands secure: the conflict between Kleon and the cavalry must have been serious and timely to have warranted the composition of a play entitled *The Knights*, which by its clear reference to the cavalry success at Solygeia appears to refute Kleon's earlier charge of *lipostratia*.

The Athenian Cavalry and the Revolution of 411 B.C.[103]

With the occupation of Dekeleia and the continuous devastation of the land, the Athenian cavalry not only suffered the hardships of constant duty in Attika, but many of its members saw their family estates being

[99] Of course, Connor has interpreted Theopompos F 93 differently both in content and sequence. For a more simplified chronology, see his *Theopompus*, p. 53.

[100] However, on the basis of the scholion to *Knights* 627 (repeated by the *Suda* s.v. *hippeis*) which claims that "it was permitted for the cavalrymen to impose a penalty," Connor, *New Politicians*, p. 152, n. 32, suggests that the cavalry may have forced Kleon to give back the five-talent bribe by means of the disciplinary power implied in words, "to impose a penalty" (*epitiman*). This presupposes that Kleon was still officially a member of the cavalry at the time of the dispute. I am disinclined to agree with this suggestion for the following reasons: (1) the entire line reads *exên de autois epitiman kai koman* (Dindorf edition and the *Suda*), yet the most recent edition omits the key word *epitiman* (Koster, 1969), thus there is a question about the reading; (2) *epitiman* need not mean "to impose a penalty"; it can mean "to show honor to"—this makes more sense in the context of the entire phrase, "to show honor to" and "to wear their hair long"(*koman*), particularly since it immediately follows the scholiast's reference to the involvement of the *hippeis* in sacrifices called *hippades* which they conduct in the ceremonial processions. The context favors an honorific, social explanation, not a judicial one; and (3) Connor himself repeats Tzetzes on *Clouds* 549a to the effect that "he [Kleon] was once a member of that well-to-do long-haired group of Athenians, the Knights" (p. 152)—in other words, an event sometime in the past.

[101] See *APF* 8674, p. 319: "Kleon is unlikely to have been born after 470."

[102] See Ar. *Knights* 225–26, 242–54.

[103] For relevant sources and discussion on the Revolution of 411, see now Andrewes and Dover, *HCT* 5, pp. 184–256.

systematically destroyed.[104] Many blamed the democracy for these reverses, not to mention the ill-fated Sicilian Expedition, and became receptive to the conspiratorial designs of aristocratic factions.

It is natural to assume that the cavalry, as a body, was fairly unified in its support of the Council of Four Hundred and also of its successor, the government of the Five Thousand.[105] These oligarchies limited political power to those who were best qualified to serve the state "with their wealth and bodily vigor" (*tois te chrêmasi kai tois sômasin*, Thuc. 8.65.3; *AP* 29.5; cf. Xen. *Hell.* 2.3.48) echoing, in less exclusive terms of course, the prerequisites for cavalry duty (*tous dynatôtatous kai chrêmasi kai sômasi*, Xen. *Hipparch.* 1.9). We should not, however, naively suppose that the cavalry was in complete agreement about the deposition of the democracy in 411 B.C.: the younger members, representing the major component of the corps, were probably on the whole more enthusiastic about political change than the older, more mature members who had a good deal more to lose, e.g., estates, family, social standing, than their younger colleagues should they fail. It will be argued that the oligarchic governments' most avid supporters were the younger members of the cavalry, even though a great number of these would not have been old enough to hold political office. Presumably, they would have been voting members of the oligarchic assembly.

We would like to know the exact relationship between the cavalry and the two oligarchic governments. Obviously, no Athenian could have continued to serve in the cavalry and also have been a member of the Four Hundred, as this body was the natural successor to the Council of Five Hundred and by that fact exempted its members from military service. If the Four Hundred had actually published its secret membership list of the Five Thousand, we may suppose that most of the *hippeis* would have been included, barring questionable political loyalties, personal grudges, or other factors about which we can only guess. The assembly which met on the Pnyx and deposed the Four Hundred after four months of rule, and then turned the government over to the Five Thousand, decreed that all those who could provide hoplite armor (*hopla parechontai*) were to be included (Thuc. 8.97.1; *AP* 33.1). This franchise stands in contrast to the Five Thousand whose names were promised by the Four Hundred and would not have

[104] See footnote 4 above. Busolt-Swoboda, *GS* 2, p. 839, n. 1, suggest that the property devastation accompanying the seizure of Dekeleia led to the setting aside of the traditional Solonian census classes and the adoption of a classification based on the possession of sufficient wealth to provide hoplite armor.

[105] Hignett, *HAC*, p. 272.

included all the hoplites, judging by Thucydides' remarks at 8.92.11. The "new" Five Thousand, whatever its actual numbers,[106] would presumably have included all the *hippeis* since they were in a sense always hoplites.

Let us take a closer look at the events of that year. When the Athenian fleet stationed at Samos elected to remain democratic (including even the oligarchs among them: Thuc. 8.75.2), a restlessness spread among the people, even extending to the oligarchy's own supporters. In response, the oligarchic Council of Four Hundred commissioned Athenian hoplites to construct a fortified mole, the Eëtioneia, across the harbor at the Peiraeus (Thuc. 8.90.1). However, Theramenes, a moderate oligarch, along with his followers, began to clamor that the Eëtioneia was not meant to keep out the democratic fleet, but was really intended to admit the army and fleet of the Lakedaimonians to whom the Four Hundred were then making overtures of peace. In this atmosphere of suspicion, Aristokrates, a taxiarch commanding his tribal contingent in the building of the mole, arrested Alexikles, a general of the oligarchic faction and one very well-disposed towards the political clubs (8.92.4). The hoplites overwhelmingly supported the taxiarch's actions (8.92.4). The Council of Four Hundred prepared to punish this act of rebellion with arms; at the same time, they sought to place the blame on Theramenes. Theramenes, however, convinced them to let him prove his loyalty by personally seeing to the rescue of Alexikles. With him went Aristarchos, one of the generals of the oligarchy (8.98.1) and some "young men of the cavalry," *tôn hippeôn neaniskoi* (8.92.6). The size of the rescue force must have been considerable, to judge from the tumult generated among the citizens of Athens at their departure and the consternation among the hoplites at the Peiraeus when they spied the oncoming procession (8.92.7–8). Thucydides tells us that while Theramenes feigned anger at the hoplites, Aristarchos and those opposed to the populace were incensed (8.92.9). In the end Theramenes agreed that the fort should be demolished and the hoplites, aided by many people from the Peiraeus, proceeded to tear it down.

In this account we get the clear contradistinction between the hoplites in revolt and the oligarchic Council of Four Hundred which the

[106] For a critical appraisal of the influential thesis of G.E.M. de Ste. Croix ("The Constitution of the Five Thousand," *Historia* 5 [1956]: 1–23) that the Five Thousand excluded the thetes only from political office, not from the assembly or the *dikasteria*, see P. J. Rhodes, "The 5000 in the Athenian Revolution of 411 B.C.," *JHS* 92 (1972): 115–27 (repeated in his *CAP*, pp. 382–84, 411–12) and Andrewes and Dover, *HCT* 5, pp. 323–29.

cavalry supported. We learn later that the hoplites held an assembly in the theater of Dionysos in Munychia (Peiraeus) and then proceeded to the city. Representatives of the Four Hundred met with them and promised to constitute the government of the Five Thousand, news which quieted the whole body of hoplites (8.93.3).

The solidarity with which the younger members of the cavalry supported the Four Hundred may be confirmed at several other places in Thucydides' narrative of the Revolution of 411 B.C. It has been suggested,[107] and I think correctly, that some of the 120 young men (*neaniskoi*) who accompanied and assisted the Four Hundred in removing the regular members of the Council of Five Hundred from the *bouleuterion* (8.69.4) may have belonged to the cavalry. Furthermore, "some of the younger men" (*tines tôn neôterôn*) who had earlier assassinated Androkles (and others), a popular leader and the one most responsible for the banishment of Alkibiades (8.65.2), must surely be included among the 120 *neaniskoi*. It makes good sense to associate the *tines tôn neôterôn*, the *neaniskoi*, and the *tôn hippeôn neaniskoi* with those conservative political clubs which were particularly supportive of the Four Hundred.[108]

A question remains. What punishment did the *hippeis* receive from the restored democracy for their support of the Four Hundred? Andokides (1.75) reports that the soldiers (*stratiôtai*) who had remained in Athens during the rule of the Four Hundred were forbidden to speak in the assembly or to become members of the newly restored Council of Five Hundred. The prohibition was not repealed until 405/4 (And. 1.73, 77–79). Was the term *stratiôtai* meant to include *hippeis*? It is hard to believe that the most avid of supporters, the young *hippeis*, would not have been held accountable. We already know that the younger son of Polystratos was serving in the cavalry in the period after the fall of the Four Hundred (410/9) and even fought against them as exiles ([Lys.] 20.28) at the very time his father was being prosecuted as a former member of the Four Hundred (although apparently serving for only eight days: [Lys.] 20.13–16). But there is no evidence one way or the other to determine if the son of Polystratos was subjected to partial disenfranchisement while in service. Technically, he

[107] Martin, *Cavaliers*, p. 472; Lintott, *Violence*, pp. 140, 142, 145.

[108] W. J. McCoy, "Theramenes, Thrasybulus, and the Athenian Moderates" (unpubl. diss., Yale University, 1970), 98, n. 35, prefers to disassociate the *neaniskoi* from the *tines tôn neôterôn*. For the impact of aristocratic youth on Athenian politics, with ramifications for our understanding of the events of 411 B.C., see W. G. Forrest, "An Athenian Generation Gap," *YClS* 24 (1975): 37–52.

might not have been enrolled in the cavalry until after the fall of the Four Hundred and thus was not culpable. Again, we do not know.

At any rate, we must assume that (a) those who committed serious crimes, like the young assassins (Thuc. 8.65.2), whether *hippeis* or not, were duly punished as criminals; (b) those *hippeis* who had supported the Four Hundred suffered the loss of full citizenship (i.e., assembly and council); and (c) those *hippeis* (some of the older ones?) who managed to avoid being linked too closely with the Four Hundred were absolved. Now for many of the *hippeis*, the penalty really meant only the loss of *isêgoria* in the democratic assembly, since they had not yet attained the minimum age for admission to the Council of Five Hundred. And for the youngest, the bouleutic clause never became operative because it was repealed five years later while they were still in their twenties. The exclusion from bouleutic service on this occasion may have offered a legal precedent for the penalty (though denied by the young Mantitheos: Lys. 16.8) imposed upon the *hippeis* by the restored democracy after 403 as punishment for their support of the Thirty Tyrants in 404/3. More on this in the next chapter.

If we assume that the *hippeis* serving in 411 lost the rights of full citizenship between 410 and 405, we can understand more easily the overwhelming support of the cavalry for the Thirty Tyrants. Some of these horsemen, particularly those described as young in 411 and who had not fled with the Four Hundred, would still be serving in 404/3 and still, I suggest, nursing a grudge against the democracy despite the repeal of the penalty in the preceding year. Joined by returning exiles and invigorated by a new young generation of *hippeis* in 404, even more sympathetic to oligarchic ideas because of the democracy's perceived role in the defeat and humiliation of Athens, the *hippeis* as a body were more than ready to ensure the success of the oligarchy.

Conclusion

Even before the Peloponnesian War, the Athenian *hippeis* had shared common interests of wealth and social class, and no doubt a generally conservative political outlook, though some may have described themselves as Periklean-style democrats. Highly visible, they participated as a body in festivals and processions and some, perhaps many, were identifiable by their long hair.[109] But it was during the Peloponnesian

[109] See *Knights* 580, 1121; Schol. to 627; the *Suda* s.v. *hippeis*; Ar. *Lysistrata* 561 ("long-haired phylarch"); and perhaps Ar. *Clouds* 14. For other references to the "aristocratic look" without, necessarily, cavalry associations, see Ar. *Clouds* 545, *passim*;

War that the Athenian cavalry affirmed its sense of self-worth and importance in that one area which had found only modest acclaim in the past—on the battlefield. The cavalry's unique obligations of year-round service in the defense of Attika and its active participation abroad, with some notable successes, e.g., Solygeia and Mantineia, forged a new and vigorous *esprit de corps*. Some may have found the post-Periklean democracy troubling (cf. Thuc. 2.65.5–10; *AP* 28.1) and began to frequent the oligarchic political clubs in the 420s. The confrontation with Kleon served to fuel the fires of antidemocratic sentiment among many of the cavalrymen and made possible a political solidarity which mirrored this new *esprit de corps*. The disaster of the Sicilian Expedition and the occupation of Dekeleia by the Spartans from 413 to 404 B.C. only confirmed to many upper-class Athenians, from whom the cavalry was recruited, that the democracy was the cause of Athens' woes. This set the stage for the Revolution of 411 in which we have detected signs of cavalry support. The repercussions of that support created a new tension between the democracy and the *hippeis*, although there is no evidence that they fought any less bravely because of it. The imposition of additional *eisphorai* between 410 and 405 B.C.[110] could not have been well received by the cavalrymen, any more than the first *eisphora* associated with Kleon had been to their fathers and relatives twenty years before. Whatever good service the cavalry had provided in the great war with Sparta was soon to be lost in the events of 404/3.

Wasps 1066–70. For brief discussion, see W. Donlan, *The Aristocratic Ideal in Ancient Greece* (Lawrence, Kans., 1980), 160–62.

[110] See Thomsen, *Eisphora*, pp. 175–77.

CHAPTER FOUR

The Year of the Thirty Tyrants

"... a boon to the democracy if these Athenian horsemen went abroad and perished."
(Xen. *Hell.* 3.1.4)

In the following narrative I will trace the role of the Athenian cavalry in the events of the year of the Thirty Tyrants.[1] I will argue that the enthusiastic support of the cavalry for the two oligarchic regimes in Athens in 404/3 B.C. initiated a decline of that body in the fourth century B.C.

When Lysander destroyed the Athenian fleet at Aigospotami in 405 B.C., the Athenian cause was lost. While Lysander removed the last vestige of Athenian control from the Aegean, the Athenians prepared themselves for the expected siege. Agis marched out from the Spartan stronghold at Dekeleia, camped in the region of the Academy, and effectively besieged Athens from the landward side. Lysander arrived and blockaded the Peiraeus with 150 ships. After a time the Athenians were forced to sue for peace. Along with other conditions, they agreed to destroy the Long Walls and those of the Peiraeus, surrender all but twelve of their ships, and allow their exiles to return (*Hell.* 2.2.20).

In the autumn of 404 an oligarchy was established in Athens. Thirty men were chosen to create a constitution based on the ancestral laws (*Hell.* 2.3.2.). The Thirty appointed a new Council of Five Hundred and other magistrates at their own discretion, but refused to carry out the mission for which they had been selected—to frame a new constitution.

The power of the Thirty remained unchallenged so long as they were content to bring to trial certain sycophants who had been odious to the aristocrats (*tois kalois k'agathois*),[2] but when they began to prosecute anyone who was hostile to the new order or presumed to be, whether aristocrat, metic, or commoner, a schism formed between the moder-

[1] For the most recent work on the Thirty Tyrants, see P. Krentz, *The Thirty at Athens,* (Ithaca, N.Y., 1982) with bibliography. The classic work by P. Cloché, *La restauration démocratique à Athènes en 403 avant J.-C.* (Paris, 1915; repr. Rome, 1968) is still valuable. For a reasonable chronology of events with a view to the conflicting sources, see G. Adeleye, "Studies in the Oligarchy of the Thirty" (unpubl. diss., Princeton University, 1971), 167–70.

[2] Xen. *Hell.* 2.3.12; *AP* 35.3; Diod. 14.4.2; cf. Lys. 12.5; for a summary account of the events of 404/3 with no mention of the cavalry, see Justin 5.8–10.

ate oligarchic members of the Thirty, led by Theramenes, and the extreme oligarchic majority led by Kritias. Fearing that Theramenes would gain the support of the citizens and precipitate the overthrow of the Thirty as he had done to the Council of Four Hundred in 411, Kritias decided to enroll 3,000 Athenians to share in the new government. Theramenes, however, argued that a body of 3,000 would still exclude good and noble men (*kalous kai agathous*) from political participation.[3] Furthermore, the fact that a man belonged to the upper classes was no certain guarantee that he identified his interests with oligarchy or with the Thirty in particular.[4] For example, Nikeratos, the son of the famous general and politician Nikias, was accused of being favorably disposed to the democracy and was killed by the Thirty. Lysias (18.6) claims that "neither his birth nor his property (*ousia*) nor his age would have been thought to disqualify him from being included in the new government."[5] Moreover, *AP* 37.1 reports that the 3,000 would not include those who had participated in the demolition of the fort at Eëtioneia or any who had acted in opposition to the Four Hundred in 411. If this statement can be believed, it would have excluded many of the hoplites and theoretically the whole body of Five Thousand! Nevertheless, despite Theramenes' arguments, the Three Thousand were enrolled and the rest of the populace disarmed.

After the populace had been rendered harmless, a reign of terror commenced in Athens. The Thirty Tyrants and their supporters exiled or executed many Athenians either because of personal enmity or simply to acquire their estates. They likewise targeted metics[6] as a means of financing Lakedaimonian guardsmen stationed at Athens. When Theramenes protested these outrages, Kritias and his comrades decided to remove the political gadfly once and for all. The deed was accomplished in a manner reminiscent of the abolition of the Council of Four Hundred in 411. Kritias commissioned the boldest young men among his supporters (*neaniskois hoi edokoun autois thrasytatoi*; *Hell.* 2.3.23) to arm themselves with daggers and to be present at the council meeting in which Theramenes was to be accused. As with the

[3] Xen. *Hell.* 2.3.18–21; *AP* 36.1–2.

[4] For evidence that the upper classes were not monolithic on the issue of oligarchy and that some of them suffered under the Thirty, see Xen. *Hell.* 2.3.14, 19–21; *AP* 35.4; Diod. 14.4.4, 14.5.5–7; and Isok. 8.133; 15.318; 16.40; 18.35; 21.2–3, 11–12.

[5] For the family and its troubles with the Thirty, see *APF* 10808, pp. 404–5.

[6] Xen. *Hell.* 2.3.21 (30 metics); Diod. 14.5.6 (60 metics), though there is apparently some disagreement on the chronology; on closer inspection, however, it is easy to argue that the singling out of rich metics was proposed before the trial and execution of Theramenes, but only carried out after his death—so Adeleye, "Oligarchy of the Thirty," p. 168, n. 33.

revolution of 411 B.C., we may suspect that some of these young men were probably also members of the cavalry.

When Theramenes ably defended himself[7] and received a round of applause from the councillors, Kritias acted quickly to circumvent the normal judicial process. He removed Theramenes from the legally protected body of the Three Thousand and thereby cleared the way for a direct vote of death by the Thirty. To silence protestation from the councillors at this judicial chicanery, Kritias ordered the young men to stand in clear view with their daggers in hand (*Hell.* 2.3.50–51). This action intimidated the council, and Theramenes was led away to his death.

Once Theramenes had been removed, the Thirty accelerated their purge of the populace. Those Athenians who were not listed among the Three Thousand were forbidden to enter the city, and their lands confiscated.[8] Refugees from Athens and the Peiraeus poured into Megara and Thebes,[9] their former enemies. At this point Xenophon, who narrates these events, reports that Thrasyboulos and a band of seventy exiles set out from Thebes and captured Phyle, a strong fortress in the Parnes range protected by precipitous slopes around most of its perimeter. The Thirty quickly marched out from Athens, "accompanied by the Three Thousand and by the cavalry," *syn te tois trischiliois kai syn tois hippeusi* (*Hell.* 2.4.2).

"Certain young men" (*tines tôn neôn*) eagerly tried to attack Phyle but accomplished nothing; some were even wounded (*Hell.* 2.4.2). This is not surprising. The terrain of the region is extremely rugged, and it is very likely that these "young men" were members of the cavalry. Phyle was (and is) no place for horses. The Thirty and their army were forced to retire to Athens because of a heavy snowstorm (cf. Diod. 14.32.2), but they dispatched most of the Lakonian guardsmen from the city and two tribes of the cavalry, *tôn hippeôn duo phylas*, to protect the farms below Phyle. In the early morning hours Thrasybou-

[7] At Xen. *Hell.* 2.3.48, Theramenes makes an interesting remark: he has always viewed, both now and in the past, that government is best which has the support of those most capable of benefiting it, "either with horses or with shields," i.e., with cavalry and hoplites. Certainly, he must be referring to his support of the Five Thousand in 411 and his recommendation for a larger franchise than the Three Thousand in 404 B.C.

[8] Xen. *Hell.* 2.4.21: For their own personal gain, the Thirty have nearly killed more Athenians in eight months than in ten years of the Peloponnesian War. Isok. 4.113 claims that more men were put to death without trial in three months than had been tried during the whole period of her empire! Isok. 7.67 gives the number killed at 1,500 (as does *AP* 35.4), and the refugees at 5,000; Isok. 20.11; Diod. 14.32.1; Nepos *Thrasybulus* 1.5; Lys. 16.5.

[9] Xen. *Hell.* 2.4.1; Diod. 14.5.7–6.1–3; 14.32.1.

los and a much augmented force attacked the camp at a time when grooms (*hippokomoi*) were noisily rubbing down their masters' mounts. The Athenian camp was taken by surprise and put to flight with a loss of 120 hoplites and 3 horsemen, one of whom, Nikostratos, was called "the beautiful."[10] By the time a relief column of cavalry arrived from Athens, Thrasyboulos had made it safely back to Phyle.

In light of these setbacks, the Thirty decided to appropriate Eleusis as a future haven of refuge. By a subterfuge Kritias and the Thirty, aided by the Athenian cavalry, succeeded in arresting a certain number of Eleusinians who were deemed politically troublesome. Lysimachos, one of the hipparchs, then transported the prisoners to Athens and turned them over to the police officials called the Eleven for incarceration. The next day, Xenophon informs us, the Thirty summoned to the Odeion[11] *tous en tôi katalogôi hoplitas kai tous allous hippeas* (2.4.9). Martin interpreted this passage to mean that "they assembled at the Odeon the hoplites inscribed on the list and the other citizens [likewise on the list] who were knights" (p. 476). He took this passage as proof that the Athenian cavalry was included in the body of the Three Thousand, despite the fact that Xenophon's words at 2.4.2, "with the Three Thousand and with the horsemen," seem to imply that they were not.[12]

Grote, on the other hand, had suggested that Xenophon intended the words *tous allous* in 2.4.9 to refer to "the rest" of the *hippeis* who had not participated in the recent actions at Eleusis.[13] This interpretation is appealing. From the comments made by Kritias at the gathering, it is clear that his plan was to win a unanimous verdict of death for the imprisoned Eleusinians and thereby to implicate all those assembled in the Odeion in the crime (cf. Diod. 14.32.4).

In his defense of the theory that the Athenian cavalry was included

[10] *Hell.* 2.4.6; cf. Diod. 14.33.1. If Xenophon himself was a young member of the cavalry in that year (certainly explaining his absence from Athens in subsequent years), then we might speculate that he had had more than a passing acquaintance with the "fair Nikostratos," thus explaining the personal note.

[11] The Odeion of Perikles was located to the east of the theater of Dionysos. This building figures prominently in the activities of the cavalry later: Xen. *Hell.* 2.4.24. In 415 B.C. it was used as a meeting place for the conspirators implicated in the mutilation of the Herms (And. 1.38).

[12] *Cavaliers*, pp. 472–76. Martin's interpretation has been accepted by Cloché, *Restauration*, pp. 7–9; Krentz, *Thirty*, p. 51; and Rhodes, *CAP*, p. 458. Other relevant passages in Xenophon (2.3.20–21, 51; 2.4.23) are enigmatic.

[13] *History of Greece*, 12 vols. (New York, 1858) 8:267; and implied by A. Lintott, *Violence, Civil Strife, and Revolution in the Classical City 750–338* B.C. (London, 1982), 166. For criticism, see Cloché, *Restauration*, pp. 8–9.

in the Three Thousand, Martin pointed out that it would be difficult to explain the ambiguous place of the cavalry in a constitutional system delineated by "those on the register" and "those not on the register" (pp. 474–75). This is a legitimate point. And if we assume that the cavalry was to be included in the Five Thousand in 411, it would appear inconsistent to suggest otherwise for 404/3. But consider the chronology: The Four Hundred never published their list of Five Thousand, yet the cavalry seems to have supported the oligarchy without written guarantees. And we have no reason to suppose that the cavalry did not support the Thirty Tyrants—before the publication of the list of Three Thousand. For example, in his speech *Against Agoratos*, Lysias reports that during the negotiations with the Spartans following Aigospotami, the oligarchic conspirators targeted the democratic leaders and the *strategoi* and taxiarchs as obstacles to their plan to subvert the democracy (13.7–18, 30–32, 62–63). It is hard to resist the implication that by their absence from this list, the cavalry commanders, the hipparchs and phylarchs, were firmly in the camp of the conspirators. This can also be applied to the cavalry as a whole. Furthermore, *AP* 36.2 claims that the Thirty kept changing the list of names—does that sound like a tactic the Thirty would have applied to its staunchest supporters? I think not. The Athenian cavalry was granted a special status in the oligarchy.

On this point, it is not inappropriate to recall the basic nature of the force. It was a unique military establishment of specific membership (quite unlike the hoplites), molded by the recent traditions of service in the Peloponnesian War, exhibiting a camaraderie characteristic of a military corps, and financially supported by the state, whether oligarchic or democratic. Dominated by well-to-do young men, it had acquired definite sympathies for oligarchic rule, but it tended to obey the ruling government, however questionable or unconstitutional. The young horsemen of Athens followed orders and obeyed their commanders, whether democratic or oligarchic. But if the opportunity presented itself, they were inclined to favor oligarchies, thus explaining their support of the Four Hundred and the Thirty.

I would argue that in 404 B.C. the Athenian cavalry was thought to be politically safe and militarily distinct; it did not have to be officially registered. This situation changed only when the Thirty turned against fellow aristocrats and executed Theramenes, a moderate oligarch. At that point some of the cavalrymen, perhaps the older members, may have conceived doubts about the new order. The assembly at the Odeion was intended to enforce the loyalty of all the cavalrymen and implicate them in the actions of their more eager comrades.

Of course, it is possible that the summons to the Odeion of "the other horsemen" was dictated by logistical considerations. We have left the impression that a number of *hippeis* had somehow avoided accompanying the Thirty to Eleusis and that the oligarchs were concerned about their loyalty, but it may simply be that those absent horsemen had been ordered to remain in Athens to protect the interests of the Thirty or that they continued to patrol and protect the farms around the city as they had been instructed to do before (*Hell.* 2.4.4). Generally, it seems unlikely that the Thirty would have taken all of the horsemen to Eleusis, regardless of Xenophon's nonpartitive references to "the horsemen" (*Hell.* 2.4.8).[14]

None of these considerations, however, diminishes the primary reason for the summons—the Thirty's need to associate as many of their followers as possible in their crimes. And considering how smoothly the horsemen transferred their allegiance to the Ten who replaced the Thirty says much about their disaffection with the Thirty. In any case, it does become clear that as the acts perpetrated by the Thirty and their successors became more desperate, many Athenians began to devise ways of avoiding full culpability.[15]

We should not imagine, however, that the doubts which the cavalrymen may have had were directed at the existence of an oligarchy at Athens. Rather, it was the high-handed way the oligarchy was being administered that probably caused discontent among the members of the cavalry. By this time the cavalry must have removed from its ranks the last of its few genuinely prodemocratic sympathizers through execution or flight. An interesting case is that of our young aristocrat Mantitheos, who was probably enrolled and later probably served briefly as a cavalryman in the year 404/3 despite his protestations to the contrary. Mantitheos claimed that his father sent him abroad to reside at the court of Satyros on the Pontos in 405, before the disaster at Aigospotami, and that he returned only five days before Thrasyboulos led his followers from Phyle to the Peiraeus in the spring of 403 (see below).[16] Nevertheless, the fact that Mantitheos returned while the Thirty were still very much in power convinces me that he or his father

[14] Cloché, *Restauration*, p. 8, is too dogmatic on this point.

[15] For instance, among the Three Thousand after the battle at the Peiraeus with Thrasyboulos: Xen. *Hell* 2.4.23. See also Isok. 18.17: the Thirty commanded some people to commit criminal actions to implicate them. The best known example is when Socrates and four others were ordered to arrest Leon of Salamis; the Thirty gave many such orders (Pl. *Apology* 32 C–D).

[16] Lys. 16.4.

had become reconciled to the rule of the Thirty and that Mantitheos was returning to fulfill cavalry duties.

In the spring of 403 Thrasyboulos led an army numbering 1,000 to the Munychia at the Peiraeus. The Thirty responded with a force of the Lakonian guardsmen, the horsemen, and the hoplites (*Hell.* 2.4.10). A battle subsequently took place between the oligarchic and democratic forces. Xenophon's failure to ascribe any role to the cavalry during this battle may be explained by the fact that Kritias and his army had to attack uphill (*Hell.* 2.4.15; cf. Diod. 14.33.3), and thus lost the advantage of cavalry. When Kritias was killed and the supporters of the Thirty defeated, the oligarchic partisans voted to depose the Thirty and to elect in their place the Ten, one from each tribe. The remaining members of the Thirty retired to Eleusis, but the cavalry, or at least the vast majority, accepted the constitutional legitimacy of the new oligarchic regime and remained in Athens. In fact the cavalry became the enforcer of the new order. Both cavalry commanders and the corps assisted the Ten in keeping watch over the citizenry (cf. Diod. 14.33.5) and guarding against those enemies outside. Xenophon tells us that the cavalrymen were quartered in the Odeion together with their mounts and their shields. At night they patrolled the walls as hoplites, ever watchful for treachery, and during the day they operated outside the walls as cavalrymen (*Hell.* 2.4.24).

It is quite significant that while Thrasyboulos could muster a force comprising "many hoplites," he could only mount about seventy horsemen (*Hell.* 2.4.25). We may suppose that some of these seventy horsemen were recruited from among those who would have been serving in the Athenian cavalry had they or their families not sought safety from the Thirty in exile and from those who had defected from the Ten. The small number of cavalrymen in Thrasyboulos' army says much to confirm that the Athenian cavalry overwhelmingly supported oligarchic rule.[17]

At the same time, we must recognize that the Athenian cavalry was composed of hundreds of individuals, a mixture of older men who had served the Four Hundred in 411 and of younger men who genuinely believed that oligarchy was the best form of government, of whom most shared a common political goal, but who did not always agree on the methods by which that ideal could be best actualized. One thing

[17] Just as the *hippeis* of Orchomenos, 300 in number, were induced to support the efforts of refugees from Thebes to install an aristocratic government in Thebes; in this case, however, the ringleaders backed out at the last moment and betrayed their hapless Orchomenian allies (Diod. 15.79.3–5). For discussion, see J. Buckler, *The Theban Hegemony 371-362* B.C. (Cambridge, Mass., 1980), 182–84.

seems clear: they, like many others, had become disaffected with the rule of the Thirty. This explains why the cavalry did not protest the demise of the Thirty, and why it was subsequently assigned such great responsibility by the government of the Ten. In fact, Xenophon reports that only the cavalry ventured outside the walls; moreover, the horsemen succeeded in capturing foraging parties and inflicting some damage on the main body of the troops of the Peiraeus (*Hell.* 2.4.26).

Even under the rule of the Ten, while individual atrocities could create dissension among the cavalrymen, their loyalty to the oligarchy was secure. One illustrative incident occurred during a cavalry attack on one of the foraging parties. A cavalry force led by the hipparch Lysimachos captured some Athenians who had ventured out to secure provisions at their farms in the deme of Aixone. Lysimachos had the prisoners summarily put to death, despite their pleas for mercy and the fact that "many of the horsemen were sorely distressed" by this action (*Hell.* 2.4.26). Understandably, the entire Athenian cavalry was blamed for this deed, not just Lysimachos and his supporters among the cavalry force. In retaliation, the democrats at the Peiraeus executed a captured cavalryman, Kallistratos of the tribe Leontis.

It may be this incident or others like it that prompted Aristotle to write that, outside of Kallibios and his Peloponnesian garrison, the staunchest supporters of the Ten were "some of the horsemen" (*eniôn tôn en tois hippeusi*). For some of these stood out among the citizens as particularly eager to keep the democratic faction at the Peiraeus from returning to power (*AP* 38.2). Obviously, some of the *hippeis* believed that their actions under the oligarchic regimes were sure to lead to harsh reprisals under a restored democracy. Perhaps young Xenophon was one of those "many" horsemen who objected to the execution of the foragers; perhaps it is simply *apologia* and a plea not to judge all *hippeis* by the actions of some. We cannot know.

At about this time, both the Thirty, ensconced at Eleusis, and also the Ten at Athens, appealed to the Spartans to dispatch an army to put down the democratic faction at the Peiraeus. The Lakedaimonian-mercenary army was commanded by the Spartan king Pausanias and by the victorious general of the Peloponnesian War, Lysander. During a fiercely contested, but indecisive, battle, three tribes of Athenian cavalrymen led a successful charge against a portion of the army from the Peiraeus.[18] One can imagine the reaction among the democrats; not

[18] Xen. *Hell.* 2.4.31. See *IG* II² 11678; *SEG* XXVII.23. F. Willemsen, "Zu den Lakedämoniergräbern im Kerameikos," *AM* 92 (1977): 135–40, argues that Lakrates, the Olympic victor, mentioned at *Hell.* 2.4.33, was probably not a Lakedaimonian, but one of these Athenian *hippeis*; see Pritchett, *GSW* 4, pp. 133–34.

only had the cavalry disgraced itself by supporting the oligarchy, but now it had actually joined ranks with the enemy which had crushed their empire and robbed them of their freedom a scant year before.

After the battle Pausanias effected a reconciliation between the oligarchic and democratic factions. This reconciliation was accompanied by a general amnesty for the warring parties.[19] Only the Thirty, the Eleven, and the Ten who had ruled in the Peiraeus were excluded.[20]

To declare an amnesty, however, is quite a bit easier than to abide by it. *AP* 40.1 reports that many of those who had supported the Thirty lived in fear of reprisals from the democratic victors. In an earlier passage Aristotle states that any Athenian who wished to leave Athens and take refuge in Eleusis could do so if he registered himself within ten days after the peace and departed within twenty days.[21] In typically human fashion the hopeful refugees waited until the last moment before registering in order to decrease the chances of reprisal from enemies. When Archinos, an Athenian statesman, realized just how many Athenians intended to flee to Eleusis, he shortened the remaining registration time, forcing many of them to remain in the city involuntarily. Although Archinos appears to have taken this measure out of a genuine concern for the welfare of Athens, some of the returning exiles were not quite so magnanimous and began to violate the amnesty with complaints against the supporters of the oligarchy. Aristotle reports that Archinos brought one of these violators into court and had him executed as a warning to others that the amnesty had to be maintained completely (40.2). After this execution, Aristotle claims, no one remembered past injuries again. That this was not the case with respect to the Athenian cavalry, we shall observe shortly.

We do not know how many of the cavalry found their way to Eleusis, but the majority appear to have stayed in Athens. Some cavalrymen probably left Attika after 403, perhaps Xenophon the historian and Lykios the son of Polystratos, both of whom participated in the march of the Ten Thousand from Cunaxa in 401. We cannot prove that they

[19] On the amnesty, see *Hell.* 2.4.38, 42–43; the Lysianic corpus, *passim*; Isok. orations 16, 21; *AP* 39.6–40.4; see also Diod. 14.33.6; Nepos 8.13.1–3; Justin 5.10. For discussion, see A. P. Dorjahn, *Political Forgiveness in Old Athens: The Amnesty of 403 B.C.* (Evanston, Ill., 1946) and T. C. Loening, "The Reconciliation Agreement of 403/402 B.C. in Athens: Its Content and Application" (unpubl. diss., Brown University, 1981), 160–63 (on cavalry).

[20] *Hell.* 2.4.38; cf. *AP* 39.6: the Ten who had replaced the Thirty in the city were also blacklisted. But Aristotle adds that all of these oligarchs could be granted amnesty if they were willing to render an accounting of their actions and successfully passed this examination; for discussion, see Rhodes, *CAP*, pp. 468–72.

[21] *AP* 39.1–4; see also Xen. *Hell.* 2.4.38.

were cavalrymen in 404/3, but the fact that they were experienced horsemen in the Asian campaign lends credence to the possibility. It would not be surprising if they had left Athens because of the hostility of the Athenians towards the cavalry—it cannot have been a friendly environment in which to stay. And if Lykios can be identified as the youngest son serving in the cavalry in 410 (see Chapter Three), then he still could have been a member of the cavalry in 404/3.

The oligarchic faction at Eleusis held out until 401/0, when the Thirty began to hire mercenaries, presumably for an attempt to eject the restored democracy from Athens. The democracy responded with their whole force (*pandêmei*), the oligarchic generals were treacherously murdered at a conference with the democrats, and the rest of the Eleusinian faction was persuaded to rejoin the democracy. So ended the civil war in Athens.

Impact of the Year of the Thirty Tyrants

As I have noted, the Athenian cavalry had not been excluded from the amnesty; of course, a *hippeus*, like any citizen, found guilty of intentional homicide or malicious wounding by his own hand could be prosecuted (*AP* 39.5; Lys. 16.8). But the cavalry, in contrast to the seemingly less culpable Three Thousand, demanded special attention. The restored democracy immediately attacked the financial foundations of the cavalry—not to abolish the organization—but to sting it for its misplaced loyalty.

In the Lysianic oration, *In Defense of Mantitheos*, we learn that at his *dokimasia*[22] for admission to the Council of Five Hundred Mantitheos was charged with having served as a cavalryman under the Thirty. His accuser had apparently claimed that Mantitheos' name could be found on the cavalry list, the *sandion* (6). Mantitheos charged that the list was unreliable: it did not include some who admitted they had served that year, yet did include others who had not been in Athens at the time—as Mantitheos claimed in his own case. It is not hard to imagine that many cavalrymen removed their names from the public list and added the names of enemies when the end of the oligarchy seemed imminent. In addition, Mantitheos argued that when the democracy had been restored, the people ordered the phylarchs to determine who had actually served in the cavalry and then to force them to

[22] For the use of the *dokimasia* for political purposes in the early fourth century, see J. Roberts, *Accountability in Athenian Government* (Madison, Wis., 1982), 20–21 and G. Adeleye, "The Purpose of the *Dokimasia*," *GRBS* 24 (1983), 295–306.

return their state loans (*katastaseis*).[23] Mantitheos declared that no one could prove that his name was placed on the phylarch's list, nor that he had been brought before the *syndikoi*,[24] nor that he had been forced to return his *katastasis* (7). In view of our understanding of the institution of *katastasis*, we can now appreciate the severity of this measure, a measure which, if it did not legally violate the terms of the amnesty, certainly violated the spirit of the amnesty and the reconciliation. The Athenian cavalry was placed in a unique position. The cavalrymen who had served in 404/3 were not judged as culpable as the oligarchs themselves, but they were not allowed to escape unpunished, as the Three Thousand appear to have been. All the more reason to differentiate the cavalry from the Three Thousand.

We can eliminate the possibility that this measure was designed simply to force all of the cavalrymen into an early retirement by demanding back the loan which the state had provided them upon entrance into the corps, and which they were expected to repay upon leaving the corps. In *Hellenika* 3.1.4 Xenophon provides the proof that some of the cavalrymen who had served under the tyrants were still serving in 400/399. The Lakedaimonians had decided to send an expedition to Asia Minor to relieve the Ionian cities harassed by the Persian satrap Tissaphernes. Thibron, selected as harmost (military governor) and commander of the expedition,[25] requested that the Athenians contribute 300 cavalrymen, whom he offered to pay himself. Xenophon comments parenthetically that the Athenians dispatched some of those who had served under the Thirty, thinking it to be a boon to the democracy if these cavalrymen should go abroad and die there. There is no further mention of these 300 Athenian cavalrymen.

Xenophon's remark is particularly striking because it reveals, as late as 400/399, just how much the Athenians still hated the cavalrymen who had served in 404/3. The continued presence of the oligarchs at Eleusis until 401/0 could only have served as a constant reminder of

[23] 16.6. I think we can assume that the phylarchs who had served under the Thirty are meant here. Lysias informs us that the Athenian democracy stipulated that if the phylarchs did not produce the list of those having received the *katastaseis*, they would be fined (7). This may mean that the phylarchs would have to bear the loss of the *katastaseis* themselves, as W.R.M. Lamb translates in the Loeb edition, or simply that they would be fined some amount. Either way the stipulation reflects the restored democracy's hostility towards the phylarchs of 404/3.

[24] These officers were involved in claims to property confiscated under the Thirty, see R. J. Bonner and G. Smith, *The Administration of Justice from Homer to Aristotle*, 2 vols. (Chicago, 1938) 2:38; and A. W. Harrison, *The Law of Athens: Procedure* (Oxford, 1971), 34–35.

[25] For this campaign, see also Diod. 14.36.1–3; Isok. 4.144.

the troubled times but a few years past. The measure meant that although a cavalryman had to return his *katastasis*, he could not retire and would have to produce a horse at his own expense. If his mount were injured or killed, he would have to purchase a replacement with his own money, a significant financial burden.

Although more problematical, there are some grounds for the belief that the grain allowance for the cavalry, the *sitos*, also came under fire in this period. Evidence exists which indicates that not only the guilty cavalrymen were affected, but presumably also the recently enrolled and therefore blameless cavalrymen. Mutilated papyrus fragments of an oration by Lysias entitled *Against Theozotides*, dated to 403/2,[26] informs us that the *misthophoria* for the Athenian cavalryman had been reduced from one drachma (per day) to four obols; the Athenian assembly had taken this action upon a motion of Theozotides (restored but secure). The *hippotoxotai*, on the other hand, were slated to gain six additional obols.[27] Gernet and Bizos suggest that although the measure against the cavalrymen may have been prompted by the sad state of Athenian finances at that time, it could also be explained by the disfavor which the cavalry corps had brought on itself.[28] The second possibility seems the stronger; if the state finances were shaky, why then increase the *misthophoria* of the *hippotoxotai*?[29]

This measure may have been repealed at a later date. Demosthenes recommends that Athens spend one drachma a day for each of 200 cavalrymen on a proposed expedition to the North around 351 (4.28). Kroll, however, has argued that the word *misthophoria* on the papyrus pertains to the introduction of a peacetime *sitos* allowance amounting to four obols, and that one drachma represented the wartime allowance throughout the fourth century.[30] This would square with Demosthenes, who clearly refers to a war-time situation. Troubling, however, are the words on the papyrus, *peri polem[ou]*, "concerning war" and

[26] First published by B. P. Grenfell and A. S. Hunt, *The Hibeh Papyri: Part* 1 (London, 1906), 49–55; see the Budé edition, ed. L. Gernet and M. Bizos, frag. 6. pp. 257–59 and brief discussion by Pritchett, *GSW* 1, pp. 21–23. For discussion, bibliography, and date of the decree, consult R. S. Stroud, "Greek Inscriptions, Theozotides, and the Athenian Orphans," *Hesperia* 40 (1971): 280(297)–301; see also for brief remarks, Lintott, *Violence*, p. 176; M. H. Hansen, "Nomos and Psephisma in Fourth-Century Athens," *GRBS* 19 (1978): 320, n. 18 (*SEG* XXVIII.46); and his "Did the Athenian *Ecclesia* Legislate after 403/2?" *GRBS* 20 (1979): 32, 37.

[27] See Appendix D.

[28] Lysias, Budé edition, pp. 234–36; and so Loening, "Reconciliation," p. 174, n. 79: the net savings *per diem* would amount to 800 obols.

[29] So Stroud, "Theozotides," p. 298 and Krentz, *Thirty*, pp. 116–17.

[30] "Archive," p. 98, n. 36.

peri phylakês, "concerning guard" (over the land?, cf. *IG* II² 791, lines 16–17, 30–31). These expressions should reflect a new general policy, that is, with regard to garrison/patrol duty and to future outbreaks of war, the state subsidy is now reduced to four obols.[31] Altering Kroll's suggestion, then, this might mean that the cavalrymen were to be paid from this time on only the four-obol peacetime rate, whether Athens was at war or not. If this is correct, then we must concede that the measure was indeed repealed sometime before 351 B.C. (unless Demosthenes is proposing a return to the former one-drachma rate because of the unusual demands of the expedition, see Chapter Five). We will have occasion to address this issue again because of its implications for the size of the cavalry in the fourth century; in any event, the fragmentary state of the papyrus makes any conclusion in this regard only tentative.

Stroud has convincingly dated the Lysianic papyrus to 403/2. He connects column I of the papyrus referring to orphans with an inscription found in the Athenian Agora in 1970 which mentions Theozotides and pertains to the children of Athenians who died during the oligarchy while aiding the cause of the democracy.[32] The two texts reveal a striking contrast: Theozotides directed his attention to the welfare of the children whose fathers were lost in the struggle with the Thirty, and also directed his attack against the cavalry—some of whom must have been responsible for making those very children orphans, e.g., in the incident of the foragers of Aixone—by reducing their state maintenance allowance. This measure then represented a second attack on the finances of the Athenian cavalry.

Whether one argues that the reduction in the cavalry *sitos* allowance applies to peace- or wartime conditions, it ought to reflect the democracy's hostility toward the cavalry at this time. This attitude fed on the knowledge that not only had the cavalrymen remaining in Athens escaped punishment, but that some had fled to Eleusis where they were outside the injunctions involving *katastasis* and *sitos*. When the refugees at Eleusis joined the restored democracy in 401/0, the process of

[31] R. Renehan, *Studies in Greek Texts* (*Hypomnemata*, 43, Göttingen, 1976), 69–74, has analyzed the uses of *peri tou polemou* and *peri polemou* and has concluded that the anarthrous expression should mean "concerning war," i.e., in a generic sense. Grenfell and Hunt, *Hibeh Papyrus*, p. 54 argue that the *misthophorein* pertains to wartime pay over and above the *sitos* allowance. Pritchett, *GSW* 1, p. 23, has shown, however, that the *misthos* for soldiers in the fifth century "must have been primarily for rations and could be called trophe or sitos."

[32] Stroud, "Theozotides," pp. 285–87.

searching out those who had been cavalrymen must have resumed and surely rejuvenated some old hatreds. Xenophon's remarks about the hoped-for destruction of the 300 cavalrymen in the following year testifies to the continued strength of that hostility.

The reputation of the cavalry was further diminished by the actions of individual citizens. One such case involved the son of the famous Alkibiades, also named Alkibiades, who refused to serve in the infantry on the expedition against the Spartans at the Haliartos River in Boiotia in 395.[33] The events are described in the two Lysianic orations (14, 15), *Against Alkibiades.* It appears that young Alkibiades, in his twenties, [34] had not been enrolled in the cavalry that year and consequently had not passed the requisite scrutiny (*dokimasia*) before the Council of Five Hundred (14.8).[35] In fact Lysias tells us that Alkibiades had never been enrolled in the cavalry (14.10).[36] In any event this expedition was quite unusual in that some of the cavalrymen were expected to serve as hoplites even though they had ordinarily fulfilled their military service as cavalrymen (14.10).

It has been suggested that since Boiotia and Athens were allies at this time and the former possessed an excellent cavalry, there was greater need for Athenian hoplites than Athenian cavalry against a hoplite-dominated Spartan army.[37] Perhaps, but it is hard to believe that Athens could not have mustered a sufficient hoplite force without injecting cavalrymen into the hoplite ranks. Moreover, in the following year at

[33] On this trial, see also Isok. 16.46; W. Donlan, *The Aristocratic Ideal in Ancient Greece* (Lawrence, Kans., 1980), 175.

[34] *APF* 600, p. 19: "he must have been born in 417 or 416."

[35] Moreover, Alkibiades' accuser cites a law (*nomos*) that purports to say that anyone who serves in the cavalry without first having passed the *dokimasia* (see 14.10,22) loses his citizenship and has his property confiscated (14.9). It would be interesting to know how recently this law was passed and if its intent was more pointedly specific than simply to punish desertion or refusal to perform military service.

[36] Despite Alkibiades' claim that he had inherited little of what remained of his father's wealth (Isok. 16.46; Lys. 14.31), it seems safe to say that young Alkibiades should have ordinarily been a likely recruit for the cavalry (see Lys. 14.27 and *APF* 600, p. 21), yet he had not served. Luck? Or perhaps a case where the young man has avoided cavalry service previously because of influence with the cavalry commanders, who still attached some importance to his namesake, his ancestry, and his tragic circumstances? Or perhaps in anticipation of rejection by the Council of Five Hundred *because* of his namesake? It can work both ways. The charge of not being qualified (14.10) surely cannot have meant that he lacked sufficient wealth to qualify. That he was not well disposed towards the "majority," i.e., the democracy, may be true (14.10), despite having been expelled by the Thirty (Isok. 16.46; Lys. 14.17?).

[37] As implied by Anderson, *AGH*, p. 13.

Korinth, Athens committed 600 horsemen (!) against the Lakedaimonians, even though the Boiotians were there with 800.[38] We know from remarks made by the young aristocrat Mantitheos, who also participated in the Haliartos campaign, that it was generally felt that the cavalry was assured of safety, whereas the infantry anticipated real danger.[39] I suggest that the Athenian democracy wanted to insure that the cavalry did not escape the dangers present for the rest of the citizenry. The fact that some of those men had probably served under the Thirty may have encouraged that decision.

Lest it be thought that young Alkibiades was an exception, Mantitheos declared that during the campaign others had also joined the cavalry ranks without first having been approved in the customary *dokimasia*, contrary to the law, *adokimastôn para ton nomon*—further confirmation that a *hippotrophos* need not be a *hippeus*.[40] We may assume that Alkibiades was not the only one prosecuted for deserting the hoplite ranks to join the cavalry and each publicized case must have contributed to the cavalry's loss of prestige.

The orations *Against Alkibiades* also reveal distinct signs of irregularity at the command level. It was expected that the generals would intercede in the case on Alkibiades' behalf—no doubt as a personal favor to a fellow aristocrat—with the claim that they had given him approval to join the ranks of the mounted archers, the *hippotoxotai* (15.5–6).[41] Yet, Lysias provides the somewhat curious information that it was not in the power of the commanders to enroll a man in the infantry if he was a *dokimastos hippeus* nor to transfer a hoplite into the cavalry who had not undergone the requisite *dokimasia* (15.7). But the former is exactly what was done in the Haliartos campaign. The intent, as the Loeb edition hints with its translation "at their pleasure," must be that this was a measure that required (or assumed) a collective policy decision by the state or possibly approval from the Council of Five Hundred; it was not considered one of the prerogatives of the commanders—if we may believe Lysias. But there must have been complicity on someone's part, perhaps initially merely "overlooking" (*perieôrôn*) Alkibiades' illegal status as *hippotoxotês*; later the generals obligingly testify that Alkibiades entered the cavalry on their orders

[38] Xen. *Hell.* 4.2.17. If one wanted to debate this point, one could argue that since the Orchomenians dispatched no cavalry in this campaign, it was felt that a larger contingent from Athens was required. But I doubt it.

[39] Lys. 16.13; cf. 14.7.

[40] Lys. 16.13; cf. 15.8.

[41] Interestingly enough, this is the first time we learn that Alkibiades had joined the *hippotoxotai*, and not the *hippeis per se*.

(15.6). Complicity of this sort is apparently not unique in Athenian history: in Aristophanes' *Knights*, the rejuvenated Demos declares that no longer will an Athenian enrolled as a hoplite be able to change that status "by means of well-placed connections" (*kata spoudas*, lines 1369–71).

Although the suit took place after the Haliartos campaign, we cannot be sure how long Alkibiades maintained his illegal status. But some of the cavalry officers finally did take action against him. [Lysias] 15.5 comments that if the generals had in fact officially ordered him to be enrolled in the cavalry, then they should have cited Pamphilos because he had deprived the city of a horseman by taking away his horse (*aphairôn ton hippon*). They also should have attacked the phylarch because, having expelled Alkibiades from the tribal contingent, he had caused disorder (*akyron*) in the ranks. I interpret this passage to mean that the phylarch expelled Alkibiades from the tribal ranks because of his dubious status and then notified the hipparch Pamphilos. Pamphilos then literally "took away his horse," not his own horse nor one purchased with a *katastasis* loan (being *adokimastos*), but one which the state apparently supplied directly to a *hippotoxotês*.[42] To have gotten as far as he did presumes well-placed connections or else a high degree of administrative carelessness.

The irregularity of service in the Haliartos campaign, coupled with the widespread belief that cavalry service at this time offered little danger, reinforced the antipathy of the general populace towards their cavalry. Whatever good service the Athenian cavalry had performed in the Peloponnesian War was rapidly being overshadowed by the events during and after 404/3. It did not matter that the majority of the cavalrymen assigned to hoplite duty in this campaign probably executed their duty without question, or that the cavalrymen who served in a cavalry capacity did likewise; the stain created by the case of Alkibiades and others only seemed to confirm the belief that something was amiss with the cavalry and its personnel.

For many of the younger cavalrymen who had not seen service under the Thirty, the inherited guilt must have been particularly galling. And if, as everyone believed, there was no risk involved, how then could there be glory?

In the campaigns of 394 at Korinth and Koroneia, however, the Athenian cavalry may have regained some measure of the respect which had characterized its service in the Peloponnesian War. In July 394 a Spartan army met the combined force of Athenians, Boiotians,

[42] See Appendix D.

Argives, Korinthians, and Euboians near Korinth.[43] The Athenians contributed 600 cavalrymen, the largest cavalry force ever sent outside Attika (excluding the Megarid and possibly Boiotian campaigns in the 420s).[44] The Lakedaimonians, however, were victorious. A month later, Agesilaos met the allied forces, including Athenian horsemen (*IG* II² 5222), at Koroneia in Boiotia.[45] Xenophon reports that the horsemen on either side were about equal in number (*Hell.* 4.3.15), but we are not told what that number was. Although Agesilaos claimed victory (4.3.21), he failed to accomplish his military objective and was compelled to retreat from Koroneia.

We are very fortunate to have two public monuments commemorating the battles of Korinth and Koroneia and a third one celebrating an act of personal bravery at Korinth. The first is an ornately sculptured anthemion (Figure 11a) found in the Kerameikos. Its epistyle contains an inscription[46] recording the names of one phylarch, a certain Antiphanes, and ten cavalrymen who died at Korinth: Melesias, Onetorides, Lysitheos, Pandios, Nikomachos, Theangelos, Phanes, Demoklees, Dexileos, and Endelos; and at Koroneia, only one, Neokleides. It had been suggested that this monument lists the fallen horsemen of a single tribe, Akamantis; it is known that Dexileos, one of the cavalrymen honored, belonged to that tribe.[47] Tod (p. 19) comments that this suggestion is supported by the relatively small number of the dead in relation to the 600 Athenian cavalrymen known to have been present at Korinth,[48] and by the fact that the tribe Akamantis is believed to

[43] The most trustworthy account of the battle can be found in Xen. *Hell.* 4.2.9–23. For other sources and an insightful analysis of the topography of the battlefield, see Pritchett, *Studies* 2, pp. 73–84 and also B. S. Strauss, *Athens After the Peloponnesian War: Class, Faction, and Policy 403–386* B.C. (Ithaca, N.Y., 1987), 122–25. For recent studies of this period, see Ch. Hamilton, *Sparta's Bitter Victories: Politics and Diplomacy in the Corinthian War* (Ithaca, N.Y., 1979) and P. Funke, *Homonoia und Arche: Athen und die griechische Staatenwelt vom Ende des peloponnesischen Krieges bis zum Konigs Frieden (404/3–387/6 v. Chr.)* (Wiesbaden, 1980; *Historia Einzelschriften*, 37).

[44] Diod. 14.82.10 reports that the total cavalry force of the confederates numbered only 500, whereas Xen. *Hell.* 4.2.16–17 gives the total at 1,550.

[45] For topography, see Pritchett, *Studies* 2, pp. 85–95.

[46] *IG* II² 5222; for text and commentary, see Tod, *GHI* 2, no. 104, pp. 18–20. For the relief, see Ch. Clairmont, *Patrios Nomos: Public Burial in Athens during the Fifth and Fourth Centuries* B.C. (Oxford, 1983), no. 68b, pp. 212–14; cf. S. Kaempf-Dimitriadou, "Ein attisches Staatsgrabmal des 4. Jahrhunderts v. Chr.," *Antike Kunst* (1986): 23–36.

[47] So Martin, *Cavaliers*, p. 416; see Pritchett, GSW 4, p. 213.

[48] Pritchett, *Studies* 2, p. 83, apparently disagrees, arguing that the cavalry "were apparently scarcely engaged," during the battle, and that this should account for the small number of casualties. This is an assumption based primarily on Xenophon's silence about the role of the cavalry in the battle. Also, Strauss, *Athens After the Peloponnesian*

have suffered greater losses than any other tribe in the battle.[49] One thing is certain: if we are dealing with a monument to one tribe, then it must be that of Akamantis. If we are not, questions must inevitably arise. It is troublesome that the name of the tribe is nowhere mentioned on the stone, but then we do not have the whole monument; the anthemion was designed to rest on another element, probably a relief.[50] On the other hand, in proportion to the size of the force, the number of casualties does seem very small; if cowardice explains it, then the splendid monument doubles both as a panegyric for those horsemen who fought and died and at the same time a public condemnation for those who performed their duty less than courageously. Whichever interpretation one favors, it is hard to resist the idea that this monument says as much about 404/3 as it does about 394. These men are being honored for their bravery under the restored democracy.

The private dedication linked to the battle at Korinth concerns the same Dexileos mentioned in *IG* II² 5222. A spectacular relief shows the young Dexileos mounted upon a rearing horse poised in the act of hurling his javelin at the fallen enemy hoplite (see Figure 12).[51] Below the relief is an inscription which reads:

> Dexileos, son of Lysanias of the deme
> Thorikos, was born in the archonship of
> Teisander and died in the archonship of
> Euboulides at Korinth, one of five horsemen.[52]

Scholars have been puzzled by the words "one of five horsemen." Tod (p. 20) dismissed the suggestion that Dexileos was "one of the five horsemen who lost their lives at Corinth," because we know that at

---

War, pp. 123–24, is inclined to think that the casualties are not limited to one tribe and that the small number may reflect reluctance on the part of the cavalry to fully engage the enemy—see reference to charge of cowardice brought by Thrasyboulos, though specifically directed at the hoplites: Lys. 16.15–16.

[49] Lys. 16.15. This is not certain, being based on the assumption that the Orthoboulos mentioned in Lys. 16.13 was a phylarch and not a hipparch. If Orthoboulos was a hipparch, then the fact that he belonged to the tribe Akamantis has no value in determining Mantitheos' tribal affiliation. For discussion, see *APF* 9667, pp. 364–65. Moreover, it is clear from Lysias 16.16–17 that Mantitheos was a hoplite and therefore he should be referring to hoplite casualties. The cavalry was usually stationed on the wings as a separate force, at some undetermined distance from their infantry comrades.

[50] So Hölscher, *Historienbilder*, p. 108.

[51] For discussion of the relief, see Hölscher, *Historienbilder*, pp. 102; Clairmont, *Patrios Nomos*, no. 68a, pp. 219–21; and Kaempf-Dimitriadou, "Staatsgrabmal," pp. 31–34.

[52] *IG* II² 6217; for text and commentary, see Tod, *GHI* 2, no. 105 and Martin, *Cavaliers*, pp. 415–17.

least eleven did (see above). It has been suggested that the words refer to a privileged rank in the squadron, or perhaps to an elite group of "four or five of the best men and horses from each division (*taxis*)" who, Xenophon advises, should remain behind the rest of the force to fall unexpectedly upon the enemy as he turns to charge (*Hipparch.* 8.25).[53] There are questions about this as well. Here, as in so many other cases, we cannot be sure whether Xenophon is reporting standard military practice or recommending the introduction of something new. We are not even sure what a *taxis* of the Athenian cavalry is supposed to number.[54]

It is natural to assume that Dexileos fought and died with four members of his own tribe. Consequently, if we were to suppose that *IG* II² 5222 catalogued the total cavalry casualties at Korinth and Koroneia, then we would be forced to acknowledge that Dexileos' tribe, Akamantis, accounted for nearly half of the cavalry casualties of the two campaigns. This is not impossible. The tribe Akamantis may have been stationed at a point where the fighting was most intense.[55] We know that the Athenian cavalry had been dispatched by tribal units in the past, e.g., in 404/3 two tribes were sent to protect farms near Phyle (Xen. *Hell.* 2.4.4) and three tribes accompanied King Pausanias against the party in the Peiraeus (Xen. *Hell.* 2.4.31). And in the preparations for the Lamian War (323–322), the assembly resolved that three tribes were to guard Attika, the other seven to be prepared to campaign abroad (Diod. 18.10.3), but the passage seems to imply that hoplites are meant, though cavalry cannot be excluded.

In the end our uncertainty on these points permits two possibilities: either Athens sent only six tribes of horsemen to Korinth, i.e., 6 X 100–600 horsemen, or it dispatched a smaller proportion, i.e., 60, from each of the ten tribes, perhaps selected by lot (cf. Dem.21.133: Euboian campaign, 349/8). Only one thing can be safely concluded: *IG* II² 6217, in conjunction with II² 5222, confirms that the tribe Akamantis suffered at least half of the cavalry casualties at Korinth and possibly all of them, though this last now strikes me as less likely.

Scholars have also marveled that the inscription records the dates of Dexileos' birth and death, information unique in Attic epitaphs (Tod,

[53] Tod, *GHI* 2, no. 105, pp. 20–21.

[54] On its application to the infantry, see J. K. Anderson, *Military Theory and Practice in the Age of Xenophon* (Berkeley, 1970), 97.

[55] Perhaps analogous, Plut. *Aristeides* 19.2 reports that when the Greeks stormed the Persian camp at Plataia in 479 B.C., fifty-two Athenians perished, all members of the one tribe, Aiantis; according to the Atthidographer Kleidemos, this tribe fought the most courageously.

p. 20). The archon dates prove that Dexileos was born in 414/3 and died in 394/3. Colin Edmonson has offered an ingenious explanation for this exceptional epitaph.[56] He suggests that Dexileos' parents wanted to announce publicly that their young son, only twenty years old, had not only died bravely for the democracy in cavalry service but was also too young to have participated in the cavalry and its disgrace in 404/3. The exact dating stood as a testament to that fact.

This type of defense is paralleled in two Lysianic orations, *On the Scrutiny of Euandros* (Oration 26) and *Against Theomnestos* (Orations 10 and 11). In the first, the accuser boasts that he had undergone the scrutiny for manhood (in the eighteenth year) *after* the oligarchy in Athens and, furthermore, that his father had died in Sicily long before 404/3 (21). In the second case, the speaker claims that he was only thirteen years old—at a age when he knew nothing of oligarchies—when his father was executed by the Thirty (10.4, 31–32; 11.2, 12). And in two works by Isokrates, the same sort of argument appears: in 16.46, Alkibiades the Younger emphasizes the fact that he was but a *pais*, a child, in 404/3 and indeed was expelled from Athens by the Thirty; and in 20.11, the accuser has to admit that Lochites, a young aristocrat, was too young to have participated in the oligarchy of the Thirty Tyrants and has to fall back on character assassination to link him with the oligarchs. Being a minor at the time of the Thirty, therefore, removed a person from complicity with the oligarchs. Such is likely to have been the intent of the Dexileos epitaph.

Both *IG* II² 5222 and *IG* II² 6217 represent an effort to refurbish the image of the Athenian cavalry, to publicize the bravery and the loyalty of at least some of its members to the democracy. But they may also bear witness to the cowardice of the rest of the cavalry—in which case the ambivalent feelings towards the cavalry would have certainly persisted. When these monuments were dedicated, many of the cavalrymen who had served under the Thirty had since retired from service. At Korinth and Koroneia, therefore, the personnel of the cavalry consisted of a majority of younger men who had not served in the cavalry under the Thirty and who deeply resented the stigma which continued to cling to the corps.

To these two monuments from the Kerameikos cemetery, one public and one private, we may add yet a third, the official casualty list of the campaigns at Korinth and Koroneia. Of this monument only the top

[56] In a lecture, "*IG* II² 6217: A Footnote to the Restoration of the Athenian Democracy," delivered at the American Philological Association/Archaeological Institute of America meetings in Washington, D.C., 1975, which Professor Edmonson kindly allowed me to read in typescript in Athens in 1977. Strauss, *Athens After the Peloponnesian War*, p. 124 and n. 12, also accepts Edmonson's explanation.

right-hand corner exists, including a battle relief and the column headings of the last six Athenian tribes, Akamantis to Antiochos (see Figure 11b). An inscription makes it certain that the monument (which Pausanias probably saw, 1.29.11) pertains to the war losses. The battle relief[57] is of some interest to our discussion. We see a cavalryman in *chlamys* and *petasos* mounted on a rearing horse and ready to hurl a javelin at two hoplites, one standing in a protective posture over his fallen comrade. There can be no doubt that the Athenians were symbolized by the victorious cavalryman, not by the defeated hoplites.[58] We possess only the right-hand panel of the monument; perhaps Athenians were depicted as hoplites in the panel to the left, atop the lost casualty columns of the first four Athenian tribes. In any event, the Athenian state chose to depict a victorious Athenian cavalryman on an official casualty list of the two campaigns.

However we may interpret the exact intent of these monuments, the general Athenian hostility continued unabated in the law courts and at the customary scrutinies of individual cavalrymen before they assumed public office. If it became known that a man seeking office had been a cavalryman in 404/3, he became open to attacks by his personal enemies or opportunistic rivals.

The case of the young aristocrat Mantitheos illustrates this point. He was accused of having been a cavalryman in 404/3. We have already noted that Mantitheos denied having served, though he was probably not telling the whole truth (see above). Moreover, Mantitheos claimed that even if he had served, he could still pass the scrutiny by simply demonstrating that he had caused no citizen any harm (8).[59] Mantitheos goes on to say that many of those cavalrymen were now serving on the council, and many others had subsequently been elected to the *strategia* and hipparchy.

For the moment, let us assume that it was possible for a cavalryman who had served under the Thirty to become a member of the council. This information at first glance would appear to undermine the argument that individual cavalrymen still encountered hostility. This is not the case. First of all, the fact that Mantitheos vigorously denied his membership in the cavalry in 404/3 indicates a genuine respect for the seriousness of that charge. Secondly, the fact that a person who admitted his membership still had to prove that he had never been personally responsible for a crime against a fellow citizen serves to accentuate the

[57] For discussion of the relief, see Hölscher, *Historienbilder*, pp. 105–7, pl. 9.1; and Tod, *GHI* 2, no. 104, p. 19.

[58] So Hölscher, *Historienbilder*, p. 106.

[59] For similar defense, see Isok. 18.16–18.

rather unusual status of the cavalry under the terms of the amnesty. The difficulty of proving one's innocence during the confused events of 404/3 probably convinced most of those "many" who had been cavalrymen then, and were councillors now, that the easiest course was simply to deny ever having served, as Mantitheos did. Finally, I think that Lysias exaggerates the number of those who elected to go on to public office or high military commands. To be sure, there must have been those who were willing to risk the scrutiny and the chance that personal enemies might expose them, but I think Lysias magnifies their numbers to benefit his client. And it is a clever rhetorical ploy. Who of the council or of the current military commanders would have been foolish enough to stand and identify himself as a former cavalryman under the Thirty?

As late as 382 B.C., the date of the Lysianic speech *On the Scrutiny of Euandros*, the ghost still haunted those in cavalry service under the Thirty. At his scrutiny for the archonship, Euandros[60] was accused of having been a supporter of the oligarchs in 404/3. To make matters worse, he was identified as a member of the oligarchs' most avid supporters, the cavalry(2):

> If he were now being scrutinized for admission to the Council of Five Hundred and his name had been inscribed on the tablets (*en tais sanisin*) as one who had served in the cavalry under the Thirty, you would reject him, even without an accuser. But now, when it appears that not only was he a member of that cavalry and the [oligarchic?] council, but also to have caused harm to the populace, you would be exhibiting strange behavior if you did not treat him in the same way. (10)

This passage offers compelling reason to be cautious about statements found in the rhetorical literature of the fourth century B.C. Lysias' client Mantitheos has just stated that many former cavalrymen had successfully passed the scrutiny and were serving in the Council of Five Hundred. Yet here another client of Lysias claims that if a man's name was found to be inscribed on the cavalry lists for the year 404/3, his admission to the Council would be automatically rejected. It is tempting to accept this last statement as fact, as D. M. MacDowell has done,[61] because it would further highlight in explicit legal terms the

[60] See *APF* 5267, pp. 187–88.

[61] *The Law in Classical Athens* (Ithaca, N.Y., 1978), 75, no doubt encouraged by the precedent of barring the stratiotic supporters of the Council of Four Hundred in 410 from bouleutic and assembly participation (And. 1.75 and above).

extent of the hostility directed against the oligarchic cavalry. But this information cannot be accepted without clarification.

First of all, in addition to membership in the cavalry of the Thirty, Euandros is also charged with belonging to the oligarchic council and perpetrating criminal acts against the people (cf. 26.8, 12, 18). One wonders whether Euandros could have been both a *hippeus* and a member of that council in the same year, a conflict of public duties not apparently permitted under the democracy. But considering the irregularities in that year, we cannot be certain. In any event, if he could prove to the councillors that he had not been a *hippeus* in 404/3 (and Mantitheos claims that the cavalry lists for that year were wildly inaccurate, 16.6), then the prohibition could still be valid, only now not applicable to Euandros. And we know that Euandros did pass the scrutiny and assumed the archonship for 382/1 (Diod. 15.20.1).

Secondly, Lysias employs a contrary-to-fact imperfect conditional clause, that is, "if he were now being scrutinized," etc., then "you would reject him, even without an accuser." Even though the apodosis should imply the existence of an official decree to that effect, perhaps the real sense is that "you, as a body, have the legal authority to reject him and *would*, if you exercised good judgment, do so." This interpretation can be culled from Lysias' reference in the preceding passage (26.9) to an apparent amendment to the law on scrutinies passed sometime between 403/2 and 389 B.C.[62] which targeted oligarchic partisans by permitting the Council of Five Hundred or the jury-courts to reject a candidate for office "even without an accuser" (*katêgoros*; see *AP* 55.4). The Athenian cavalry, as the most avid of the "oligarchic partisans," merited specific reference, particularly in light of Euandros' alleged participation.

Thirdly, if membership in the cavalry of 404/3 barred one from admission to the Council of Five Hundred, logically it should also have barred one from the archonship. And Lysias' client himself emphasizes that to permit Euandros access to a single important magistracy, which in turn would reward him with lifetime membership in the Areopagos, would be more dangerous than membership in the council, wherein he would be neutralized by sheer numbers (26.11). Therefore, by reverse reasoning, the fact that the archonship was not automatically barred to Euandros as a stated member of the cavalry of the Thirty (otherwise the accuser would have cited a specific law to that effect) makes it unbelievable that the Council of Five Hundred was.

[62] See Hansen, "Athenian Ecclesia," pp. 36–37 and Adeleye, "Purpose of the *Dokimasia*," 304. I am indebted to Professor Hansen for drawing my attention to this point.

We are left with two options. Since Euandros did pass the scrutiny, either he successfully proved that he had not been a horseman in 404/3 or the *automatic* prohibition against admission to the Council of Five Hundred did not exist. Taken together with the claims by Mantitheos, the second option is to be preferred. No law or decree forbade a *hippeus* who had served under the Thirty from admission to the Council of Five Hundred any more than to the archonship. Euandros probably was guilty of having supported the oligarchy, and as a member of the liturgical classes (26.3–5) would have been a likely candidate for cavalry service, but his alleged crimes were not serious enough to warrant special prosecution, let alone rejection at his *dokimasia.*

To conclude this section. Although individual horsemen who had served under the Thirty were protected by the amnesty, the cavalry as an institution did not get off quite so lightly. The unanimity of their support for the oligarchic regimes;[63] the atrocities committed by some of the more zealous horsemen; and their willingness to obey the commands of a foreign king against their own countrymen, all of these combined to evoke an intense hatred towards the cavalry which persisted at least a generation after the event. And it must have particularly galled the democrats to think that these well-to-do young men had favored oligarchy so readily and enthusiastically over the democracy which had in fact made possible, by the grants of *katastasis* and *sitos*, the very existence of the Classical Athenian cavalry. So, the restored democracy took immediate measures to vent its displeasure—it attacked these two cornerstones of the organization. As for legal prohibitions, although it appears that these cavalrymen were not legally barred from political office, it is probable that many chose to forego a political career out of fear that at every *dokimasia* their personal enemies would trot out their past association with the Thirty for public condemnation and character assassination. And if Hansen is correct, the Council of Five Hundred could reject their candidacy on its own initiative and authority. Each such public display opened old wounds and ultimately violated the very spirit of the amnesty.

Military Decline of the Athenian Cavalry

In 387/6 the Greek states became subject to the King's Peace which ended major hostilities, and by that fact, military activity for the Athenian cavalry in mainland Greece until at least 379. In this year a small

[63] Especially striking when one considers that the Athenian upper classes were divided on the issue of oligarchy; see footnote 4 above.

group of Theban conspirators, in exile in Athens since 383 when the Spartans seized the acropolis of Thebes, the Kadmeia, returned to their home city and began an insurrection to remove the Lakedaimonian garrison. A force of 5,000 hoplites and 500 horsemen, presumably a citizen army, under the command of two Athenian generals, one perhaps being Demophon, assisted the Theban conspirators in their successful expulsion of the enemy garrison. Then the Spartan king Kleombrotos arrived with an army in Boiotia, and the Athenians, fearing that war would soon be brought to Attika, put their two generals on trial, executing one and sentencing the other to exile when he failed to appear.[64] But when Sphrodrias, the Lakedaimonian governor at Thespiai in Boiotia, was "bribed" by certain Thebans (Xen. *Hell.* 5.4.20) into invading Attika in 379/8 with the plan of capturing the Peiraeus, actually reaching Thria in Attika before giving up the venture as hopeless, the Athenians were incensed and demanded that Sphrodrias be punished by the authorities in Sparta. When instead he was acquitted, the Athenians declared the truce between them as invalid, levied 20,000 (?) hoplites, 500 cavalry, and 200 ships, and took common cause with Thebes.[65]

In 378 Agesilaos invaded Boiotia. Diodoros 15.32.2 reports that Athens sent 5,000 foot soldiers (*pezoi*) and 200 horsemen. Later in the narrative (15.32.5), Chabrias the Athenian is introduced as the commander of the mercenary troops (*misthophoroi*). Are we to assume

[64] Xen. *Hell.* 5.4.9–12, 19; Diod. 15.26.1–4. The two basic sources disagree on particulars: Diodoros (agreeing with Deinarchos 1.39) explicitly states that the Athenians voted to send the expeditionary force under one general, Demophon, but Xenophon leaves the impression that the *two* Athenian generals, waiting at the Theban border, more or less acted on their own authority when they took their army to Thebes. But the army could not have been there unless there had been an approved levy. Perhaps this is a case similar to that in 323 B.C., when word arrived that Alexander the Great was dead; the Athenians gave secret instructions to Leosthenes to recruit 8,000 mercenaries in preparation for open revolt, but he should act as if he were doing it on his own authority. Thus, if the rumors of Alexander's death proved false, the Athenians could disclaim any knowledge of, or responsibility for, Leosthenes' provocative activity (Diod. 18.9.2).

[65] Xen. *Hell.* 5.4.20–35; Diod. 15.29.5–30.1–2. See also *IG* II² 43 (= Tod, *GHI* 2, no. 123) for the so-called charter of the Second Athenian League of 378. Lines 24–25 clearly indicate that Athens and Thebes were allies prior to the decree. For discussion of this very important document (standing prominently in the foyer of the Epigraphic Museum in Athens), see C. D. Hamilton, "Isocrates, *IG* II² 43, Greek Propaganda, and Imperialism," *Traditio* 36 (1980): 83–109 and J. Cargill, *The Second Athenian League: Empire or Free Alliance?* (Berkeley, 1981), who takes a revisionist position. For the suggestion that the Chalkotheke on the Acropolis was used, or even first constructed, as an arsenal in the 370s reflecting "revived imperial aspirations," see L. La Follette, "The Chalkotheke on the Athenian Akropolis," *Hesperia* 55 (1986): 75–87.

that the Athenian force was composed entirely of mercenaries, or only a part, or that the citizen contingent was commanded by some unnamed officer? Demosthenes claims that Chabrias commanded Athenian citizens on this occasion (20.71).[66] The simplest solution is to assume that it was a combined force, as Demosthenes (4.24) has described the forces operating around Korinth in the late 390s—mercenaries and citizens under the commands of Polystratos, Iphikrates, and Chabrias. With this in mind, there is no reason that we should identify the 200 horsemen as mercenaries; after all, the Athenians had "established" their state cavalry to be prepared for occasions such as this. We should consider them members of the Athenian cavalry.

War between Athens (with her new Naval League) and Sparta continued on and off through the 370s, but we hear nothing of Athenian cavalry during this period. Though technically still an ally of Thebes as late as 371,[67] Athens did not participate in the momentous battle at Leuktra in Boiotia where the Thebans under Epameinondas defeated the hitherto invincible Spartan army. Xenophon describes the cavalry of the Thebans as "well-trained" because of the recent war with the Orchomenians and Thespians.[68]

Between Leuktra (371) and Mantineia (362), the Athenian cavalry was deployed fairly regularly in the operations in the Peloponnesos. In late winter 370/69, responding to appeals from his allies in the region, Epameinondas and his army invaded the Peloponnesos, penetrated Lakonia itself, and operated almost at will.[69] Sparta appealed to Athens and in the early months of 369 the Athenians dispatched Iphikrates with an Athenian army (Diod. 15.63.2: 12,000 young men) to deal

[66] Pritchett, *GSW* 2, pp. 73–74 is inclined to accept Demosthenes' testimony; but H. W. Parke, *Greek Mercenary Soldiers from Earliest Times to the Battle of Ipsus* (Oxford, 1933), 76–77 views them as mercenary peltasts, both native and foreign. Xenophon omits any reference to Athenians in his narrative of the battle (*Hell.* 5.4.35–41).

[67] Athens and Thebes had had a falling out in 371, before the battle of Leuktra, when Athens had succeeded in negotiating a peace with Sparta, the terms of which were disputed by Thebes (*Hell.* 6.3.1–20).

[68] *Hell.* 6.4.10; on the other hand, Xenophon ridicules the Spartan cavalry for its incredibly inefficient organization: the richest (*plousiotatoi*) raise the horses, but do not ride them into battle. When a levy is called, an appointed trooper presents himself, receives his horse and arms (*hopla*) at that time, and takes the field. Worse still, according to Xenophon, is the fact that horsemen are the least physically fit and least ambitious (*hoi tois sômasin adynatôtatoi kai hêkista philotimoi*; 6.4.11). Such a system must reflect the Spartan disinterest in, even disdain for, cavalry in contrast to the pride of its military forces, the hoplites.

[69] On the campaigns of the Thebans in this period, see Buckler, *Theban Hegemony*, pp. 70–219.

with Epameinondas. Iphikrates chose to allow Epameinondas to pass through the Korinthia unopposed on his return to Boiotia.

Xenophon digresses in his narrative to relate an incident connected with the passage of the Theban army out of the Peloponnesos. The historian censures Iphikrates for displaying poor judgment, first of all because whereas he guarded the pass at Mt. Oneion,[70] he left unguarded the best route past Kenchreiai and thus failed to prevent Epameinondas' withdrawal.[71] Secondly, when he sought information as to the proximity of the Theban army near Mt. Oneion, Iphikrates sent out all of his horsemen, both Athenian and Korinthian—whereas, according to Xenophon, only a few would have been sufficient (cf. *Hipparch.* 7.6, 8.10–14) and they would have returned in relative safety. But instead, the scouting party was apparently attacked by the Thebans and it lost twenty men, such a large force being unable to maneuver in the difficult terrain.

In the following summer (369), Athens assigned to Chabrias and a force of cavalry and hoplites the task of blocking the passes at Korinth.[72] But again, Epameinondas succeeded in entering the Peloponnesos. On his return, he occupied the plain around Korinth. According to Xenophon, the Athenian and Korinthian horsemen declined to approach the Theban army when they saw the strength and numbers of it.[73] We must assume that Athens had dispatched at least several hundred horsemen to aid Sparta in view of the reputation of the Theban cavalry. This trepidation on the part of the Athenian cavalry, then, stands in sharp contrast to the bravery and efficiency displayed by a small Syracusan cavalry numbering only fifty men which had also come to help Sparta (7.1.20). Xenophon goes on to recount that the Syracusan horsemen spread themselves out and rode along the hoplite ranks of the enemy. Then they charged and hurled javelins at the soldiers. When the infantry pushed forward to stop them, they quickly retreated, then whirled about and attacked again. If they were pursued any distance, the Syracusan horsemen kept just beyond their reach and then harassed the enemy as they retired to their lines (7.1.21). It is not

[70] On the topography of the Korinthia, see R. S. Stroud, "An Ancient Fort on Mt. Oneion," *Hesperia* 40 (1971): 127–45.

[71] Buckler, *Theban Hegemony* p. 89, however, calls Iphikrates' strategy "humane and sensible, inasmuch as his raw levy stood no chance of victory in open combat with Epameinondas' veterans." His detailed response to Xenophon's criticism of Iphikrates is reasonable (p. 295, n. 30).

[72] *Hell.* 7.1.13–15; Diod. 15.68.1–3.

[73] *Hell.* 7.1.20; Diod. 15.68.4. Diod. 15.68.1 records 7,000 infantrymen, and 600 horsemen from Boiotia; we are not told the numbers of their allies, the Arkadians, Argives, and Eleians.

hard to imagine that in recording this incident Xenophon recalled the equally effective force of fifty Greek horsemen who repelled the attacks by 1,000 Persian horsemen during the retreat of the Ten Thousand from Cunaxa in 401/0 (*Anab.* 3.4.2–5). The Athenian cavalrymen at Korinth certainly earned no *kudos* in their second encounter with Epameinondas, although Diodoros 15.69.1–4 reports that Chabrias and the Athenians did force the Thebans to break off their attack on Korinth and return home. Perhaps the Athenian cavalrymen acquitted themselves better in the later stages of the campaign.

The Athenian cavalry may have performed some good service in the following year (368) when the Argives and Arkadians invaded Phleious. In this encounter the Phleiasian horsemen and their picked troops, aided by a number of Athenian cavalrymen who were present (*syn tois parousi tôn Athenaiôn hippeusi*) defeated the invading force (*Hell.* 7.2.10). But we should remember that the Argives were not known for their cavalry (they apparently had none as late as 418, see above) and neither were the Arkadians. Since the Phleiasians were fighting on their home territory, their own cavalry may have carried the real brunt of the fighting on that day. We cannot be sure.

In 366 a strange series of events brought Athens into an alliance with the Arkadians. This alliance, however, led the Korinthians to view the Athenians with some suspicion and resulted in the disbanding of the Athenian garrisons around Korinth and the termination of all Athenian military presence in the area. The Korinthians were now compelled to hire mercenary troops, both infantry and cavalry, to replace the Athenians. But Athens still had a commitment in the Peloponnesos: according to the terms of the new treaty, the Athenians were to dispatch their cavalry to assist the Arkadians if their country were ever attacked (*Hell.* 7.4.4–6). In compliance, Athens sent 400 horsemen to aid the Arkadians and the Pisatans who had forcibly replaced the Eleians as administrators of the Olympic Games for 364. Although the Arkadians were supported by these horsemen and 2,000 Argive hoplites, the Eleians routed the usurpers (*Hell.* 7.4.29–31). There can be little doubt that the Eleian horsemen, who had played a major role in the defeat of the Arkadians in the previous year's fighting at the sanctuary (7.4.14), successfully neutralized the Athenian cavalry.

In none of the incidents related above do we detect the military efficacy which the Athenian cavalry had displayed during the Peloponnesian War and perhaps again in 394. I contend that the Athenian cavalry was summoned not so much because of its military reputation but because it constituted the third largest cavalry in mainland Greece, inferior in numbers only to those of Thessaly and Boiotia. Athens could field a force of 400 to 600 cavalrymen in the Peloponnesos, yet still maintain an adequate cavalry at home as a precaution against a possible Theban attack on the city. The Peloponnesian states simply could not match the cavalries of Boiotia and Thessaly without Athenian assistance. And as undistinguished as the Athenian cavalry may have become by the 360s, it was still significantly better than the Spartan cavalry which Xenophon describes as "very poor" (*ponêrotaton*) at the time of Leuktra (6.4.10).

Several days before the main engagement at Mantineia in 362, which led to the fall of the Theban hegemony, a cavalry battle took place between Athenian cavalrymen on the one side and Theban and Thessalian horsemen on the other. At the very moment when the Athenian horsemen arrived at Mantineia, a body of Theban and Thessalian cavalry sent by Epameinondas from his headquarters at Tegea began to attack some Mantineians caught by surprise outside their walls. Xenophon reports that the Mantineians begged them to intercede on their behalf, and even though the Athenian cavalrymen had not breakfasted and their horses had not been fed, they sallied forth to the rescue. Xenophon rhetorically asks, "Who would not admire the excellence (*aretê*) of these men?" Moreover, they were attacking a larger force and one composed of the best horsemen of Greece (*Hell.* 7.5.16). In spite of this and the fact that the Athenian cavalry had just engaged hostile forces near Korinth en route to Mantineia, it forthrightly charged the enemy "eager to restore its ancestral glory" (*erôntes anasôsasthai tên patrôian doxan*, 7.5.16). This is a telling remark on Xenophon's part. To restore something requires first the loss of it.

For a brief brilliant moment the Athenian cavalry regained the respect which we first observed in the Peloponnesian War.[74] In the engagement at Mantineia they prevented the enemy cavalry from killing the Mantineians and from stealing their cattle. In the process the Athenians lost some brave men (7.5.17). Xenophon had a personal reason for such high praise: one of the Athenian cavalry casualties was Gryllos, his own son; also killed was Kephisodoros of Marathon, the hipparch.[75] Unfortunately, the sources have inserted this cavalry engagement into the actual battle of Mantineia which occurred a short time later, and even credited Gryllos with having been responsible for the death of Epameinondas.[76] In any case, this cavalry action inspired a

[74] See the importance attached to it by Aristides, *Panathen.* 312 (Behr).

[75] Paus. 8.9.5, 10; 8.11.6; 9.15.3; Diog. Laert. (Xenophon) 2.5.4; Plut. *Moralia* 346 D–F, 350 A.

[76] Paus. 9.15.3. Anderson, *Military Theory and Practice*, p. 224, prefers to think that Gryllos was killed in the main battle.

funerary monument to be erected at Mantineia in Gryllos' honor (Paus. 8.9.5), and a painting by Euphranor depicting Gryllos in the battle was commissioned to be placed in the Stoa of Zeus in Athens (Paus. 1.3.4.).[77]

In the actual battle of Mantineia,[78] the Athenian cavalry played a much reduced role, according to Xenophon's account (*Hell.* 7.5.24–25). Epameinondas successfully neutralized the Athenian cavalry stationed on the left wing of the allied army by positioning his own horsemen and some hoplites on some hills opposite them. This move was intended to instill the fear in the Athenian horsemen that if they moved to aid the infantry, the Theban cavalry and hoplite force would attack them from the rear (7.5.24). This strategy was working until Epameinondas was wounded and his victorious troops gave up the pursuit and retired to their own lines. In the resulting confusion some of the infantry accompanying the Theban cavalry and peltasts encountered the undefeated left wing of the Athenians and were slain. We may suppose by this time the Athenian cavalry had become involved.

Diodoros, however, claims that a fierce cavalry battle took place at the battle of Mantineia on the flanks (15.85.2–8). The Athenian cavalry suffered defeat at the hands of the Theban cavalry, "not because of a lack of excellence of their mounts or because of the lack of courage or horsemanship on the part of the cavalrymen," but because of their inferiority in light-armed troops (15.85.4). Under an unceasing hail of missiles the Athenian cavalry finally fled the field. But in its flight, it did not break ranks and it managed to slay the Euboians and certain mercenaries stationed on the nearby hills. At this point the Theban cavalry apparently directed its efforts at the Athenian hoplites, and the latter were only saved by the timely arrival of the Eleian cavalry, not the Athenian, which is credited with saving the left wing of the allied army.

Most agree that Diodoros' account, as it stands, is untrustworthy.[79] One might speculate that the confusion evident in the painting by Euphranor depicting Gryllos facing Epameinondas and also the story which claimed that Gryllos had actually killed Epameinondas both originated as early as Diodoros' source. Perhaps there was an intentional effort to associate the cavalry skirmish closely with the main battle of Mantineia in order to downplay the negligible role of the

[77] For discussion, see Hölscher, *Historienbilder*, pp. 116–19.

[78] See Pritchett, *Studies* 2, p. 71. For topographical considerations of Mantineia as the site of several important battles, see pp. 37–72; for the battle of Mantineia in 362, see pp. 70–72 and Buckler, *Theban Hegemony*, pp. 205–19.

[79] See Buckler, *Theban Hegemony*, p. 316, n. 57.

Athenian cavalry in the actual battle and to allow the Athenians to claim a genuine victory at Mantineia through its cavalry. In Xenophon's account, the Athenian cavalry on the left wing was neutralized by the positioning of Theban horsemen and hoplites on some hills overlooking them; concern about these troops kept the Athenian horsemen out of the main action. This may indicate Epameinondas' respect for their fighting capabilities, particularly after the results of the earlier skirmish—Epameinondas obviously attacked the right wing of the allied army.

Ironically, the failure of the Athenian cavalry to contribute more to the outcome of the battle may have resulted in charges of "lack of courage" or "poor horsemanship" which found their way into Diodoros' (Ephoros'?) account. I am very inclined to think that Diodoros' remarks about the Athenian cavalry were intended as an apologetic explanation to deflect these charges. They seem too tendentious not to have some basis in historical fact, or at least to reflect contemporary perceptions. And his efforts to applaud their victory over the Euboians and certain mercenaries (15.85.6; 87.3) smacks of special pleading. How can these troops possibly compare with the Theban cavalry? In Xenophon's account, it was only after Epameinondas fell in the fighting and his army withdrew from the attack that the Athenians, presumably both cavalry and hoplites, encountered in their area some careless Theban *hamippoi* (infantry mixed with cavalry) and peltasts; most of these they destroyed (*Hell.* 7.5.24–25).

It is obvious that Xenophon directed special attention to the cavalry skirmish outside the walls of Mantineia because his son had perished in it. But it also seems clear that this skirmish attracted the interest of Xenophon and the Athenians—enough to merit a prominently displayed painted document of it—because it was the first time since the Peloponnesian War that the Athenian cavalry had acquitted itself with genuine courage against the best cavalries of Greece. Of course we are at the mercy of our sources. Perhaps the military history of the Athenian cavalry recorded other such notable moments, but Xenophon's words "to restore their ancestral glory" argue against it. The heroic cavalry action in 362 shone like a diamond in the rough, and the Athenians clasped it tightly as a memento of an earlier, more illustrious, age.

Concluding Remarks

What, then, can we suggest about the Athenian cavalry between 404/3 and 362?

In spite of the amnesty of 403, the Athenian democracy, as well as individual citizens, refused to forgive and forget the part played by the cavalry in the service of the oligarchs in 404/3. The stigma associated with this service endured well into the 380s. In response, some cavalrymen must have withdrawn from the political arena because of their unremitting antidemocratic sentiments and out of apprehension for the official scrutiny required of every high public office. I suggest that when these men returned to private life (a phenomenon already evident in the Peloponnesian War),[80] shunned the political process, and avoided the high profile and popularly elected military offices—at the very least the hipparchy, too easy a target (*contra* Lys. 16.8)—they set an example for their sons, many reaching maturity in the 380s and 370s, and engendered in them a disinterest in, or worse, a hostility towards the service which had proved such a bane to their fathers. At every *dokimasia* and lawsuit, the crimes of 404/3 could be used to discredit the sons, much as young Alkibiades inherited the sins of his controversial father.[81] As a result, cavalry enrollment of sons and relatives of identifiable cavalrymen of 404/3 probably declined. The young men from wealthy hippotrophic families did not relinquish their love for horses, but this predilection was channeled into agonistic victories and sleek racehorses, not the hardy war-horse appropriate to cavalry service. In a sense, the young men simply resumed the horsey interests of their class. Furthermore, some cavalrymen of 404/3, perhaps Xenophon himself and Lykios, son of Polystratos, probably left Athens after the fall of the oligarchs in an attempt to escape the hostility at home. The process of self-imposed exile further diminished the pool of experienced cavalrymen.

With the loss of those Athenian families bred in the tradition of horsemanship and cavalry service, at least since the mid-fifth century, the hipparchs and phylarchs were increasingly forced to recruit from among the lower ranks of the well-to-do, from those citizens, less able, perhaps, to use bribery to gain exemption from cavalry service. Most of these recruits could not be considered *hippotrophoi* nor did they concern themselves at all with horses. In spite of the financial support of *katastasis* and *sitos*, cavalry service was still the business of the leisured classes, in particular, of the otherwise idle sons of the wealthy.

[80] See Connor, *New Politicians*, Chapter 4.

[81] See Isok. *On the Team of Horses*, a case involving one of the racing teams with which his father had won a victory in the Olympic Games of 416, now (ca. 397 B.C.) being brought against Alkibiades the Younger. Particularly, see sections 2–3, where it is clear that his father's political career has been used by the plaintiff for character assassination.

The grudging service performed by less-than-enthusiastic recruits from moderately well-to-do families did little to enhance the military prestige of the cavalry. It was not that these men could not be trained to be good horsemen (some no doubt were), but rather that many probably devoted less attention to the purchasing of good cavalry mounts, to the proper care of them,[82] and to the acquisition of the requisite equestrian skills (recall Diodoros' remarks).

These internal developments and the ambiguous performance of the cavalry in the military activity of the first half of the fourth century led Xenophon to conclude that the military reputation of the Athenian cavalry had declined in recent decades, i.e., from his own time. This conclusion, perhaps unfair, nonetheless may have encouraged him to compose his *Hipparchikos* and *Peri Hippikes* as a corrective. And we should not forget that his own sons now belonged to that organization.

To be fair, we must be careful not to suggest that the Athenian cavalry had lost all honor and respect. Throughout this period (first half of the fourth century), the cavalry officers and even at times the *hippeis* of Athens were called upon to act as witnesses or oath-swearers to treaties with foreign states—no small honor.[83] But this was pretty much a *pro forma* procedure,[84] and probably should be viewed as a recognition of two factors: (1) defensive alliances presumed a military capability to back them up—thus the emphasis on the military as guarantors, and (2) all of the military and civic officials listed served full-year terms (unlike the hoplites)—thus perhaps inspiring greater confi-

[82] See Anderson, *AGH*, p. 139.

[83] For example, by the *hippeis*: *IG* II² 16, line 11 (=Tod, *GHI* 2, no. 103; *SEG* XXX.55; treaty with Eretria, 394 B.C.); by the hipparchs: *IG* II² 41, line 10 (=Tod, *GHI* 2, no. 121; *SEG* XXI.227; treaty with Byzantion, 378 B.C.); *IG* II² 42, line 10 (=Tod, *GHI* 2, no. 122; *SEG* XXI.231; treaty with Methymna, 377 B.C.); by the *hippeis*: *IG* II² 96, lines 16–17 (=Tod, *GHI* 2, no. 126; *SEG* XXI.234; treaty with Kerkyra, Akarnania, and Kephallenia, 375 B.C.); the two hipparchs as witnesses: *IG* II² 102, lines 17–19 (=Tod, *GHI* 2, no. 129; *SEG* XXI.236; treaty with Amyntas, 375–373 B.C.); *IG* II² 104, line 3 (=Tod, *GHI* 2, no. 134; *SEG* XXI.238; treaty with Leukas, 368 B.C.); by the hipparchs: *IG* II² 105, line 34 (=Tod, *GHI* 2, no. 136; treaty with Dionysios of Syracuse, 367 B.C.); by the hipparchs, phylarchs, and *hippeis* (restored): *IG* II² 112, lines 39–40 (=Tod, *GHI* 2, no. 144; *SEG* XXIX.90; treaty with Arkadia, Achaia, Elis, and Phleious, 362/1 B.C.); by hipparchs and *hippeis*: *IG* II² 116, lines 15–16 (=Tod, *GHI* 2, no. 147; *SEG* XXI.243; treaty with the Thessalians, 361/60 B.C.). In the third century the hipparchs (restored) and phylarchs are still involved in oath-swearing, e.g., the famous Chremonidean decree of 268 B.C.: *IG* II² 687, lines 53–54.

[84] Just as Eretria had its generals, hipparchs and *hippeis* act as witnesses to the treaty with Athens in 341/40: *IG* II² 230.II.10–11. See W. Wallace, "The Demes of Eretria," *Hesperia* 16 (1947): 113–46, esp. 144–45. By autopsy I can confirm Wallace's reading of the final OI of what must be restored as the heading STRATEGO*I* in frag. b, line 3.

dence in the durative power of the treaty. It is also possible that this selection insured that the treaty was supported by some of the most distinguished individuals in Athens, adding dignity to the treaty.

Other forces were also at work to make cavalry service seem less appealing and even unnecessary in the fourth century B.C. First of all, cavalry service was in a real sense a liturgy, a financial burden on the upper classes of Athens. And it has been observed that in the fourth century there was a tendency among the wealthier Athenians to want to avoid both political and liturgical obligations.[85] Secondly, Xenophon informs us that after 391, the combatants in the Korinthian War waged their campaigns primarily with mercenaries, particularly peltasts, not with citizen levies. This trend is well documented throughout the rest of the fourth century [86] and, with respect to Athens, was decried by such orators as Isokrates (e.g., 8.44–48: now [355 B.C.] even with hoplites, not just peltasts) and Demosthenes (*Philippics* and *Olynthiacs*). This reflected an increasing reluctance on the part of the Athenian citizenry to perform military service, particularly to distant battlefields. A certain proportion of those eligible for cavalry service must have shared this attitude, that much more so because of the unique year-round and long-term obligation involved. The splendor of the festival cavalcade has to be weighed against the demands of wartime readiness and the not inconsiderable financial burden.

[85] See Isok. 7.35; 15.159–66; for the ample evidence from Isaios, see R. F. Wevers, *Isaeus. Chronology, Prosopography, and Social History* (The Hague, 1969), 121: "the picture of society thus seen is one where the wealthy are trying to escape their financial and political responsibilities..."; and G.E.M. de Ste. Croix, "Demosthenes' *timêma* and the Athenian *eisphora* in the 4th century," *C&M* 14 (1953): 34. For the case of Demosthenes' father, see *APF* 3597, pp. 128–31.

[86] See Parke, *Mercenary Soldiers* and Pritchett, *GSW* 2, pp. 34–125.

CHAPTER FIVE

The Athenian Cavalry in the Age of Philip of Macedon

"To restore their ancestral glory."
(Xen. *Hell.* 7.5.16)

IN HIS treatise *Poroi*, Xenophon speaks of the benefits of peace following the Social War (357–355 B.C.). He explains that with the money not squandered in war with recalcitrant allies, the Athenians will be able to celebrate their festivals with more splendor, restore the temples, repair the walls and docks, and restore to the priests, councillors, magistrates, and "cavalrymen their ancestral honors," *hippeusi ta patria apodôsomen* (6.1). Although it is not absolutely clear what is implied by *ta patria*,[1] Xenophon believes that there has been a decline in the state and its cavalry and that it is linked with the financial health of the commonwealth. It is not uncommon in the fourth century literature to find nostalgic references and apologetics about the late, great Athenian empire of the preceding century. And it is true that the Second Athenian League (378–338 B.C.) was but a pale reflection by comparison. But we should not dismiss these remarks about "restoration" simply because they smack of nostalgia; in particular, the reference to the cavalry is too specific and too close to Xenophon's own interests.

It is appropriate at this point to return to the question of the *sitos* allowance. Xenophon (*Hipparch.* 1.19) reports that the state spent forty talents a year on its cavalry so that it would be fully prepared to respond to any military emergency throughout the year. The hoplites, as we have noted, were levied for specific seasonal campaigns. In 403/2 the state reduced the grain allowance from one drachma to four obols (Lys. *Against Theozotides*, frag. 6, Budé), yet Demosthenes in his *First Philippic* (351 B.C.) recommends an expeditionary force to harass Philip of Macedon which is to include cavalrymen each paid at thirty drachmas a month, or one drachma *per diem*. The traditional view[2] argues that Xenophon uses the one drachma rate in his calculation of

[1] See Ph. Gauthier, *Un commentaire historique des Poroi de Xenophon* (Geneva, 1976), 218; R. Sealey, "Athens after the Social War," *JHS* 75 (1955): 74–81.

[2] See Martin, *Cavaliers*, pp. 351–53 and now P. Brun, *Eisphora-Syntaxis-Stratiotika: Recherches sur les finances militaires d'Athènes au IVe siècle av. J-C.* (Paris, 1983), 150 (arguing for an Athenian cavalry numbering a little over 660).

forty talents. Mathematically this leads to the conclusion that by the 360s the Athenian cavalry had fallen to around 650 men, i.e., 1 drachma X 360 days X 650 = about 40 talents. It follows, then, that the four-obol rate mentioned in the Lysianic fragment must have been rescinded sometime (unrecorded) between 403/2 and the composition of the *Hipparchikos* in the 360s.[3] On the other hand, Kroll prefers to view the four obols as the peacetime *sitos* allowance and that one drachma continued as the wartime allowance throughout. He comments:

> Since 40 talents can also be divided into payments of 4 obols daily to a fully enrolled cavalry of 1,000 and since 4 obols seems to have been the actual *sitos* stipend, Xenophon is apparently giving the conventional maximum figure that assumes a cavalry at ideal strength, whether the cavalry was at full strength at the time he was writing or not.[4]

But as I have noted earlier, there is no indication in the Lysianic fragment that it was to be considered only a peacetime rate and the reference to war, *peri polemou*, in the passage casts a good deal of doubt on that conclusion. While conceding the problematic state of the papyrus, nonetheless, I am inclined to my earlier suggestion that the reduction was to apply to both war (*peri polemou*) and peacetime (*peri phylakês?*) conditions, and represented a form of punishment for the cavalry. And Xenophon makes no distinction between peace- and wartime conditions; after all, the state is paying for *war-ready* horsemen. But Kroll is quite correct when he says that Xenophon assumes the canonical size of 1,000 men (see *Hipparch.* 1.2; 1.9–10; and especially 9.3) when he calculates a forty-talent expenditure on *sitos*. M. H. Hansen had suggested that the four-obol rate was restored to one drachma between 355 and 351.[5] He interpreted the reference to *ta patria* in Xenophon's *Poroi* 6.1 as a recommendation to restore the old one drachma rate and that the Athenians had done just that in this period. But Professor Hansen has now reconsidered his suggestion in conversation, leaning now to the position that Demosthenes is recommending an extraordinary expedition, thereby justifying a higher subsistence rate. In the end, we cannot be certain that the four-obol rate was ever

[3] J. K. Anderson, *Xenophon* (London, 1974), 184, believes this treatise "clearly must have been written after Xenophon's recall from exile (369?)."

[4] "Archive," p. 98, n. 36.

[5] "Perquisites for Magistrates in Fourth-Century Athens," *C&M* 32 (1980) 109, n. 6.

restored to one drachma for ordinary (or peacetime?) cavalry service in the fourth century.

Having stated this, I do think that there is sufficient evidence to suggest that Athens was indeed experiencing difficulties in meeting its cavalry quota and that its numbers probably regularly fell below the 1,000 level. Xenophon seems to be quite concerned about recruitment; he emphasizes it at the beginning of his *Hipparchikos*: "unless the hipparch recruits additional men, the force will continually diminish" (1.2). At 1.9–12, he speaks of the options, i.e., legal and personal, that the hipparch has at his disposal to secure additional recruits, both wealthy and less wealthy, and arguments to employ to win over their fathers, e.g., cavalry service will put an end to the purchasing of expensive racing horses by their young sons.[6] As they stand, these passages could be dismissed as routine facts of recruitment that every hipparch had to confront during his term of office. But that is not the case, as is best illustrated by *Hipparchikos* 9.3–7.

In this passage Xenophon suggests that the 1,000-man cavalry could be more quickly raised and prove less onerous to the citizens, if a force of 200 mercenary horsemen were established. There can be no doubt that Xenophon is proposing a cavalry composed of 800 citizens and 200 foreigners (*xenoi*).[7] He believes that the presence of these mercenaries will increase discipline and create an eager competition (*philotimoteron*) among the horsemen to excel in manly virtue (*andragathia*). Xenophon cites by way of example the Lakedaimonian cavalry which he claims first gained its repute with the incorporation of foreign horsemen. Here he is clearly referring to the highly effective cavalry created by Agesilaos in Anatolia in the 390s (*Hell.* 3.4.15; *Ages.* 9.3–4), not the native cavalry of Sparta that he disparages at Leuktra (*Hell.* 6.4.10). Furthermore, Xenophon observes that in other cities the mercenary horsemen are held in high regard.

Now to pay for the cost of the horses for these mercenaries, Xenophon suggests that the money could be secured from those who vehemently resist serving in the cavalry, as well as from those current cavalrymen who might wish to pay a fee (*argyrion*; surely not be equated with the *katastasis* loan which they would have to repay regardless) to gain release.[8] In addition, money could be procured from two cate-

[6] On the theme of the decadence of contemporary youth, see also, Isok. 7.48–49; 15.285–87; Athenaios 532 D; Plato *Tht.* 173 C–D.

[7] So H. W. Parke, *Greek Mercenary Soldiers from Earliest Times to the Battle of Ipsus* (Oxford, 1933), 147, n. 3.

[8] The passage is corrupt, but it seems to allow both of the following possibilities: (a) those not yet serving in the cavalry but eligible, and (b) those already serving.

gories of Athenians who would not ordinarily be eligible or required to serve in the cavalry: (1) wealthy men who were physically unfit, and (2) orphans, heirs to large estates, but not yet, because of their minority, in independent control of them. In this case, the money would legally come from their guardians. Xenophon even suggests that some of the resident aliens, the metics, would be eager to join the cavalry, an idea which he repeats in his *Poroi* (2.5).

When discussing this passage in an article,[9] I suggested that Xenophon's use of *kathistêmi* as the "establishing" verb for the 200 mercenary horsemen implied the same system of *katastasis* as enjoyed by the Athenian citizens. I have reconsidered that position—those men earning a living as mercenaries would not usually have been financially capable of repaying a large state loan for the purchase of their mounts nor would their term of service have been long, stable, or continuous enough to justify such a system. It is a type of *katastasis*, but one in which the state would supply the horses, as it presumably did for the *hippotoxotai* and later their replacements, the *prodromoi* (see Appendix D).

The situation with respect to the metics is more complicated—again Xenophon uses the "establishing" verb *kathistêmi*. Certainly, in many respects the metics were quite like the citizens; they were expected to serve in the hoplite armies in defense of Attika, and some of them were quite wealthy and spent many years in Athens, like the family of Lysias. Therefore, they could have been incorporated into the cavalry system quite easily, except that they were prohibited from owning land—a legal point that Xenophon must have assumed could be waived to meet this recruitment crisis.

I now believe that Xenophon was differentiating between the mercenaries and the metics—perhaps hinted at by the sequence of the passage: (a) mercenaries, (b) money for them, and (c) metics in the financial system of support. That is why money had to be sought from various, usually untapped, sources, i.e., a penalty fee from eligible recruits and existing cavalrymen, and a special tax on physically unfit rich and orphans with wealthy estates, because the state would not see a financial return on its investments for cavalry mounts which it must supply to keep the mercenaries in service. On distant fields of battle, Xenophon probably assumed the state would supply a money payment for the mercenary to buy a serviceable replacement mount. But for the metics, the usual *katastasis* would have been sufficient. This analysis reaffirms what I concluded in my article (pp. 311–12)—that the verb

[9] "The Three Hundred Athenian Cavalrymen," *Phoenix* 36 (1982): 310–11.

kathistêmi is a multipurpose verb which can be used generally (in a military or nonmilitary context) or specifically with reference to the unique Athenian system of cavalry loans, *katastasis*. The "mercenary katastasis" is really but a variation on the same theme modified to fit the reality of mercenary personnel and service.

These are rather significant recommendations which attest serious difficulties in the organization of the Athenian cavalry by the 360s. Who is more qualified to detect weaknesses in the system than the old horseman himself and whose sons served in the Athenian cavalry at the time? There is, however, no indication that Athens adopted any of these recommendations, certainly not that pertaining to the metics, which he was still urging in his *Poroi* in 355. Perhaps the state did respond by restoring the one-drachma *sitos* allowance between 355 and 351, but, as we have noted earlier, there is no certainty of that based on Demosthenes.

In addition, it may have been only now that the *katastasis* reached the maximum level of 1,200 drachmas. There is no doubt that the *katastasis* had existed from the fifth century, but there is no evidence that the amount was 1,200 drachmas until the third century B.C.[10] *Contra* Kroll (see my Chapter Two), it seems unlikely that the state would have provided a loan capable of purchasing a fine racehorse,[11] particularly in light of Xenophon's comments about sons of wealthy families preoccupied with buying expensive racing horses (*Hipparch* 1.12). It seems more likely that prior to this period the state had chosen a lesser but reasonable sum for a serviceable cavalry mount. So, in order to make cavalry service more appealing at this time (350s–340s or even post-Chaironeia (338), see below) the state may have not only restored the traditional *sitos* allowance but also raised the *katastasis* to 1,200 drachmas. Of course, when I make this suggestion—and it is only that—I realize that it will necessarily affect the relationship that I have proposed between the 1,200 drachmas, the annual depreciation of the horses, and the average length of cavalry service. But inasmuch as we are dealing with so many variables anyway, the effect may still be pretty much the same; if anything, it will decrease the estimated length of service by a few years. On such things, certainty escapes us.

[10] In fact, of the tablets from the fourth century, the highest appraisal is 700 drachmas: Kroll, "Archive," p. 89.

[11] E.g., Ar. *Clouds* 21–23. In 400 B.C., Xenophon's cavalry mount did sell for approximately 1,200 drachmas (*Anab.* 7.8.6), but as I have noted earlier this horse was hardly a trooper's typical mount.

Demosthenes on the Cavalry: The Case of Meidias

Even if Athens chose not to "establish" a special permanent mercenary cavalry, another famous Athenian would urge his fellow citizens to organize a force somewhat like what Xenophon had proposed. In his *First Philippic* Demosthenes reveals much to us about contemporary attitudes towards military service and about the now clearly established use of mercenary armies. Between 358 and 338 the Greek world had to respond to the expansionist policies of Philip of Macedon.[12] Demosthenes realized that Philip was no ordinary dynast with limited abilities, ambition, and resources. Alone, Athens dare not risk all on the fortunes of a single, pitched battle (4.23). Demosthenes really offers two proposals: (a) to prepare and hold in readiness a citizen force composed of fifty triremes and sufficient horse-transports to convey half of the Athenian cavalry (!) for the purpose of making Philip think twice about attacking Thermopylai, the Chersonese, Olynthos, or elsewhere for fear of a swift response from Athens (4.16–18); and (b) a standing army composed of citizens and mercenaries to harass Philip throughout the year, even in winter, from its bases in the northern Aegean, Lemnos, Thasos, and Skiathos.[13]

He does not pursue the first proposal which he knows will not gain approval. His real hope rests with the second. For the more modest standing army, Demosthenes recommends that it be 2,000 strong, at least 500 being Athenian citizens who will serve in rotation for limited periods of time. The rest will be mercenaries. As for the cavalry, it should number 200, at least 50 being Athenians who will serve on the same rotating basis as the infantry. Horse-transports will of course be necessary. Finally, a small fleet of ten fast triremes will be required to match Philip's naval arm (4.21–22). Demosthenes is adamant about including some citizens (25 percent) in this force—this will insure a measure of control over the mercenaries; he decries the current practice of relying exclusively on mercenaries (4.24). He even censures the Athenians for not using their military officers in their proper function (4.26–27): except for the one sent off to war, what are the *strategoi*, taxiarchs, hipparchs, and phylarchs ordinarily doing?—arranging festival processions. The Athenians are busily electing military officers for the marketplace, not for war, even to the extent of allowing a foreigner

[12] For the classic study of this period, see A. Schaefer, *Demosthenes und seine Zeit*, 3 vols. (Leipzig, 1885–87). For more recent works, see J. R. Ellis, *Philip II and Macedonian Imperialism* (London, 1976) and G. Cawkwell, *Philip of Macedon* (London, 1978).

[13] 4.19–29, 31–32; for discussion, see Parke, *Mercenary Soldiers*, p. 147.

to command their cavalry at home, like Menelaos, a Macedonian, while their one active hipparch sails off to assume duties on Lemnos (see Appendix B).

Demosthenes' comments deserve restating. First, the admission that mercenaries are a dominant element in fourth-century warfare; secondly, that one cannot realistically expect the Athenian cavalrymen to serve for longer terms than the hoplites, even though the cavalry was being paid on a full-time, war-ready basis—an acknowledgement that the cavalry was no more eager to perform military service abroad than were the hoplites and that the reliance on mercenary cavalry was perfectly acceptable; and thirdly, the startling fact that not only are the cavalry officers not really engaged in military duties (no doubt exaggerated), but that the cavalry in Attika was being commanded by a mercenary hipparch!

What has happened to the Athenian cavalry? If one had only the case of the hipparch Meidias on which to judge, one could well appreciate why the Athenians turned to a mercenary hipparch to command their cavalry. Unfortunately, most of our information about Meidias comes from a vitriolic and undoubtedly biased account contained in Demosthenes' *Against Meidias* which does not appear to have been actually delivered. But if only a fraction of Demosthenes' charges are true, the hipparchy of Meidias in 349/8[14] was a disgrace.

In the early months of 348, the general Phokion took a small force to Euboia to aid a nominal Athenian ally, Plutarch of Eretria, faced with a rebellion.[15] When it was reported that Phokion was beseiged at Tamynai, the Council of Five Hundred passed a preliminary resolution (*probouleuma*) to dispatch the rest of the cavalry (*exienai tous hypoloipous hippeas*; D.21.162) to extract Phokion from his difficulty. At

[14] The date of Meidias' hipparchy is not absolutely secure, but I hold it to be associated with the Euboian campaign of 349/8. In this view I follow K. Fiehn, s.v. "Meidias," *RE* 15.1 (1931): 335; J. M. Carter, "Athens, Euboea, and Olynthus," *Historia* 20 (1971): 422–23; and J. K. Davies, *Wealth and the Power of Wealth in Classical Athens* (New York, 1981), 155, although in his *APF* 9726, p. 387, he implies he was but a *hippeus*. For the view that Meidias was a simple *hippeus* in 349/8, see Schaefer, *Demosthenes und seine Zeit* 2.87–88; Martin, *Cavaliers*, pp. 391–92; and H. W. Parke, "Athens and Euboea 349–348," *JHS* 49 (1929): 248. It must not be forgotten that Meidias is "over 50" (*APF* 9726, p. 387) by this date—acceptable for a hipparch who will see limited action, but for a *hippeus*? As I hinted earlier, part of Meidias' reluctance to serve may be related to his age, something Demosthenes would never concede.

[15] Plut. *Phok.* 12–13; for Euboian affairs in the 340s, see P. A. Brunt, "Euboia in the time of Philip II," *CQ* 63 (1969): 245–65; G. L. Cawkwell, "Euboea in the late 340s," *Phoenix* 32 (1978): 42–67; and Ellis, *Philip II*, pp. 98, 153, 162–66, 169–70. For the action at Tamynai, see Beloch, *GG* 3.1, pp. 494–95. Aischines served at Tamynai as a member of the select troops (*epilektoi* 2.169).

the next meeting of the assembly, Meidias cleverly offered to equip a ship, knowing that this voluntary gesture would excuse him from participating in the campaign as one of the hipparchs of the cavalry. But as the debate progressed, it was decided that additional horsemen would not be needed after all. Having promised a ship, Meidias could not renege on his offer, but rather than serve in person he sent a metic, the Egyptian Pamphilos, in his place (21.163).

Demosthenes then details troubles that arose between Meidias and the cavalry stationed at Argura.[16] Apparently, a cavalry force had been sent to Argura and Meidias had gone with them as their commanding officer (21.132,164,167). But when Phokion requested this cavalry take its turn of service, presumably in the continued military activities around the island, Meidias quit his post as hipparch and hastened to assume personal command of the ship that he had volunteered (21.164–65). Meidias, having returned to Athens, proceeded to attack the cavalrymen sent to Argura and charged that their dispatch to Euboia had proved to be a disgrace (*oneidos*) to the city (21.132,197). According to Demosthenes, Meidias, while at Argura, had generally heaped scorn on the cavalry by his ostentatious display of wealth and apparently haughty and abusive behavior towards them (21.132–34). They in turn reproached him for this and no doubt his perceived desertion; this led directly to his attack on them back in Athens,[17] from which he did not relent even after the same cavalry had since departed on another expedition, this time to Olynthos (21.197). Meidias even attacked his co-hipparch, Kratinos.[18] There was apparently nothing blatantly illegal in Meidias' actions, but it is clear that he was manipulating the system in order to avoid active cavalry service—in Demosthenes' words, a hipparch for the marketplace, not for war.

The relative chronology of events on Euboia in 348 is not secure; in my reconstruction I have assumed that the dispatch of cavalry to Argura occurred subsequent to the battle at Tamynai, as Demosthenes' account seems to suggest. There would have been no apparent need to

[16] There is uncertainty about the exact location of Argura: N.G.L. Hammond, *Atlas of the Greek and Roman World in Antiquity* (Park Ridge, N.J., 1981), map 12 dE, places it on the west coast of Euboia, approximately 10 km north of Chalkis. But D. Knoepfler, "Argoura. Un toponyme eubéen dans la Midienne de Demosthène," *BCH* 105 (1981): 289–329, identifies Argura as the site of Lefkandi, again on the west coast, but to the south of Chalkis.

[17] See Martin, *Cavaliers*, p. 391.

[18] The scholion to Dem. 21.132 identifies Kratinos as Meidias' co-hipparch (*synhipparchos*; Dindorf 557,14) or simply as one of the hipparchs (Dindorf 558,6). Unfortunately, the scholion also mentions him as "general of the infantry" (Dindorf 558,6), but the evidence favors him as the fellow hipparch of Meidias. See also Davies, *Wealth*, p. 155.

send cavalry to secure(?) the position at Argura until the Tamynai crisis—at which time Meidias was still in Athens. By Demosthenes' choice of words, "there did not now seem a need for the help from the cavalry," it appears that the crisis at Tamynai ended quickly, thereby negating the need for the rest of the cavalry. Only afterwards, in the context of continued campaigning on the island, does a cavalry commanded by Meidias appear at Argura, no doubt sent as reinforcements to hold a strategic position near Chalkis (Dem. 21.133 says the horsemen drew lots for the expedition). Phokion's request for cavalry from Argura does not hint at an emergency, rather, the usual rotation of cavalry personnel (*epi tên diadochên*, 21.164), perhaps along the lines of Demosthenes' recommendation for his standing army against Philip in 351. Besides, if there were cavalrymen and a hipparch already stationed at Argura, it is hardly likely that Phokion would have felt the need to appeal to Athens nor could Meidias have quit his post at Argura to assume command of his ship unless his ship were already in active service, a fact that must postdate Tamynai.

The cavalry stationed at Argura, then, could not have participated in the battle of Tamynai. But those cavalrymen of Athens who did, did not acquit themselves well during the battle; in fact, Phokion even had to contend with a number of surly and undisciplined troops (infantry) who began deserting. (Plut. *Phok.* 12). While Phokion and the rest of the army were offering the morning sacrifices, the Athenian cavalry, observing their ally Plutarch and his mercenaries rushing forward to engage the enemy, followed suit in disorder and confusion. The foremost of the horsemen were routed and the rest of the cavalry scattered. Only after the Athenian hoplites had won the day and a horseman, Kleophanes (perhaps a phylarch?),[19] had called back the cavalrymen from their flight did the Athenian horsemen provide some good service in the mopping up operations (Plut. *Phok.* 13.2–3). The Athenian success at Tamynai was short-lived; within months Athens had lost all of Euboia except for Karystos on the southern tip of the island.[20]

Meidias probably recalled the actions of the cavalry at Tamynai in the worst possible light. Demosthenes claims that Meidias, despite the fact that he was the cavalry's commander, ruined the cavalry by introducing laws (*nomoi*) which he later disowned.[21] The scholion to this passage (Dindorf 570.21) reports that these laws had been passed with

[19] See Davies, *Wealth*, p. 153, arguing cogently that he appears to be commanding the cavalry at Tamynai, but he cannot be one of the hipparchs, since Meidias and Kratinos have already been identified as such for 349/8.

[20] See Ellis, *Philip II*, p. 97.

[21] Dem. 21.173; for brief discussion, see Martin, *Cavaliers*, pp. 381–82.

the intention of abolishing the cavalry because it had disgraced the city. These last words may refer to the undisciplined action of the cavalry at Tamynai and to the confused troubles at Argura. The scholiast speculates that Meidias' motive may have been entirely personal: "he probably did this to avoid military service." Demosthenes charges that Meidias, "unable to gain release from cavalry service by any other means" (*oudena gar tropon allon en tois hippeusin auton strateias atelê poiêsai dynamenos*), introduced a new "two-percent cavalry tax" (*hippikê pentêkostê*). The *Lexicon* of Liddell-Scott-Jones s.v. *pentêkostos*, suggests that this tax constituted a 2-percent duty—presumably on one's estate—as a substitute for actual cavalry service.[22]

Perhaps Meidias was implementing Xenophon's proposal that certain cavalrymen might wish to buy their way out of the cavalry (*Hipparch.* 9.5). By Demosthenes' use of *strateia* "campaign," it might be argued that Meidias was casting about for ways of avoiding military campaigns with this new tax measure, but since he had already achieved that goal legally by his voluntary trierarchy, it ought to mean that he was trying to get out of the cavalry altogether, a duty he was obligated to fulfill for at least the one-year tenure of his hipparchy. On the other hand, Meidias may have found himself in a trap of his own design—if he sent a substitute to captain his ship, he may then have been liable for cavalry service abroad; but if he assumed personal command of his ship (and he finally did: 21.167–68), he burdened himself with another duty he obviously was trying to avoid. Perhaps, then, this new 2-percent cavalry tax was designed to release him from all potentially risky military duties—a ploy to escape being assigned to some place like Argura again?

Demosthenes further charges that Meidias had not even purchased a horse—and he being a hipparch!—and thus had to borrow a horse from Philomelos of Paiania for the requisite processions. Every cavalryman can testify to this fact (21.174). It is not surprising, then, to learn that he could not sit on a horse in the processions in the Agora (21.171). Of course, Demosthenes is exaggerating—for one thing even he admits that Meidias rode a horse at Argura (21.133)—but we cannot but conclude that Meidias did not have a tradition of horse-

[22] Both W. W. Goodwin, in his commentary on Demosthenes *Against Meidias* (New York, 1969; repr. of Cambridge ed., 1906), 96, and J. H. Vince in the Loeb edition (*Demosthenes* III, p. 115) translate (and interpret) *hippikê pentêkostê* as "cavalry-collectorship." Goodwin explains that Meidias, according to Demosthenes, sought to get "a collector's profits by pretending to serve in the cavalry." But Demosthenes makes it clear that Meidias' hope was to gain release from cavalry service—it is not a matter of "pretending" to belong to the cavalry; he did belong in fact.

manship, at least not of cavalry service, even though he ought to be considered a *hippotrophos*, owning a pair of Sicyonian greys for chariot use.[23]

What a state the hipparchy and the cavalry were in this year! Xenophon had urged the hipparch to be a model of excellence for his men.[24] How could Meidias instill respect and impose discipline if he could not ride well himself? Meidias had been elected to the hipparchy on the basis of wealth, influence (no doubt as a supporter of Euboulos, a leading politician at this time), and popularity, not because of a respectable record of cavalry service.

Why did Meidias seek the office? He assumed, as Demosthenes laments in his *First Philippic* (26), that his duties would be primarily festive and nonmilitary. Meidias sought the hipparchy as an honorific prize, hoping against serious involvement in actual combat; after all, mercenary captains and mercenary armies were at hand to be sent to some distant crisis.

The principal trouble spot in the north at this time was Olynthos, in the Chalkidike, which Philip plotted to bring under his suzerainty. In late summer or early autumn 349, Athens had sent a mercenary force of 2,000 peltasts and 30 triremes under the command of Chares to support Olynthos, not yet under siege.[25] No more is heard of this force. In the spring of 348 Philip began an offensive into the Chalkidike; the Olynthians appealed to their ally a second time and Athens ordered Charidemos, on duty in the Hellespont, to Olynthos with a force of 18 triremes, 4,000 peltast mercenaries, and 150 horsemen.[26] Surely these horsemen were mercenaries—Philochoros says that in the subsequent and third appeal, the Olynthians specifically requested that Athens send a citizen force, not a mercenary one (*boêtheia xenikê*). It was only when Olynthos was actually besieged (summer 348) and the Olynthians appealed a third time that the Athenians dispatched a force under Chares which included 2,000 citizen hoplites and 300 Athenian cavalrymen conveyed in horse-transports—and Meidias was not among them, though to be fair, his hipparchy may have just ended or was to shortly.[27] The transporting by ships of a third of the Athenian cavalry

[23] Dem. 21.158. Cf. the young *hippotrophos* Phainippos, who sold his war-horse and took to riding around in his chariot (Ps.-Dem. 42.24; see Chapter Two).

[24] *Hipparch., passim*; *Oik.* 21.6–7; *Mem.* 3.3.1–15.

[25] Dem. 3.7; 21.161; *FGrH* 328 Philochoros F 49.

[26] *FGrH* 328 F 50; see Carter, "Athens, Euboea," pp. 424–29.

[27] *FGrH* 328 F 51; Dem. 21.197 leaves the impression that Meidias still belonged to the cavalry, but since he does not press the point, we may conclude that Meidias had legally severed his ties with the cavalry by this time. Dem. 19.266 claims that Athens supported

(at ideal strength) to such a distant theatre of war was a notable event in the fourth century. Even during the Peloponnesian War Athens had transported no larger force than this—it bespeaks the seriousness of Philip's threat, particularly the respect held for the size and quality of his cavalry which even the respectable Olynthian cavalry could not match.[28]

Perhaps we are being too hard on Meidias—he might have been dragooned by the people into undertaking a military office in which he had no interest. But, Demosthenes comments, "Is a popular election being held? Meidias of the deme Anagyrous is nominated" (21.200; cf.21.15). In any case, we must fault the Athenian *demos* for very poor judgment—a reflection, perhaps, of how little importance was now attached to electing competent cavalry commanders, as Demosthenes charges? As it turned out, the year 349/8 was not a typical year—much to Meidias' chagrin. Even if he was a friend of Plutarch of Eretria (21.110,200) and had originally supported the Euboian intervention—which Demosthenes opposed (9.57)—he could not have possibly anticipated the escalating rebellion on the island and the disastrous consequences. The sting of blame for the outcome may also explain why he chose the cavalry as the scapegoat. Because of his incompetence and hostility, the Athenian cavalry became embroiled in a bitter conflict which at worst threatened to terminate its existence and at best tarnished its reputation. The cavalry action at Tamynai could not have helped in this regard. To compound the cavalry's problems, Kratinos, who had supported the cause of the cavalry at Argura, and whom Meidias probably harangued on this account, later became willing to speak on Meidias' behalf (21.132).

It is clear that the altercation between Demosthenes and Meidias which had prompted the lawsuit recorded in *Against Meidias* was settled out of court (Aischin. 3.52). More to the point, it is obvious that Meidias was never prosecuted for any of the charges relating to his cavalry service—a fact that could only be interpreted as confirming Meidias' case against the cavalry. As for the new "cavalry" laws, which Meidias later disowned (21.173), presumably after his hippar-

Olynthos with 50 triremes, 4,000 citizens, and 10,000 mercenaries. These figures are generally viewed as exaggerated: Parke, *Mercenary Soldiers*, p. 148, n. 6 and Ellis, *Philip II*, p. 264, n. 37. See also Carter, "Athens, Euboea," pp. 418–29; on Chares, see Pritchett, *GSW* 2, pp. 77–85.

[28] According to Dem. 19.266, the Olynthian cavalry numbered only 1,000. On the treachery of the two Olynthian hipparchs, Lasthenes and Euthykrates, who surrendered 500 of their cavalry to Philip, see Dem. 8.40, 9.56–68; 19.265,342, Diod. 16.53.2–3; Hypereides F 76 (*Oxford Classical Texts*; Loeb B 19.1); Ellis, *Philip II*, p. 99.

chy, we hear nothing further. But if some sort of exemption from cavalry service (*hippikê pentêkostê*) remained on the books, it can only have hurt the cavalry and encouraged a greater interest in the more dependable mercenary cavalry.

Battle of Chaironeia, 338 B.C.

The time had not yet come, however, when Athens would not risk a citizen army, if the circumstances were judged serious enough. As Athens, allied to Phokis, had committed a citizen force of 5,000 hoplites and 400 horsemen to hold Thermopylai against Philip in 353 B.C.,[29] so it would again at the decisive battle of Chaironeia in 338. But even here, as he admitted in the recommendations in his *First Philippic*, Demosthenes saw the use of mercenaries as inevitable.

The circumstances were these: Philip had been invited by the Amphictyones into Phokis to deal with one of their own, Amphissa, on a case of impiety. But when he diverted from his route and took Elateia situated along the northern road into Boiotia, great commotion arose in Athens prompted by fears that Philip intended to invade Attika.[30] Demosthenes convinced the Athenians to negotiate a treaty with Thebes, then an ally of Macedon, and saw to the raising of a mercenary army of 15,000 infantry and 2,000 horsemen from Euboia, Achaia, Korinth, Thebes, Megara, Leukas, Kerkyra, and Akarnania.[31] This army, under the command of Chares the Athenian and Proxenos the Theban, was assigned to protect Amphissa. The allied army held the passes into Boiotia. Philip defeated the mercenary army; the allies withdrew from their defensive positions into Boiotia to await Philip. No more is heard of the mercenaries.

The Athenians, "with their whole citizen army" (*pandêmei*: Diod. 16.85.2), and the allies engaged Philip at Chaironeia in August 338. Even though the Athenian cavalry is nowhere mentioned in the battle (and our accounts are exceedingly brief), we may be sure that a substantial number of them were present: (1) in the initial commotion surrounding the taking of Elateia, Demosthenes recommended that the Athenians send out all their men of military age and the cavalry under the command of a hipparch to Eleusis to show that Athens was mili-

[29] See Diod. 16.37.3–38.2; Justin 8.2; Dem. 4.17, 41; 18.32; 19.84.

[30] Diod. 16.84.2, 5; Dem. 18.169–77; Plut. *Dem.* 18.1. For discussion of events leading up to Chaironeia, see Ellis, *Philip II*, pp. 186–97.

[31] Dem. 18.237. Aischin. 3.97 claims that the actual numbers were 10,000 infantry and 1,000 horse, charging that Demosthenes pocketed the money allotted for mercenary troops who never served; for discussion, see Parke, *Mercenary Soldiers*, pp. 152–54.

tarily prepared and would not be intimidated by Philip's action or threat of invasion (Dem. 18.177, 184)—you do not make such a display unless you intend to use some of those horsemen in actual combat (though probably not all, as mistakenly implied in *pandêmei*); and (2) Philip had with him a cavalry numbering no less than 2,000 and also 30,000 infantry, a total said to outnumber the allied army.[32] Some of the Athenian cavalrymen were probably stationed in Attika;[33] after all, Athens never committed its entire cavalry to any campaign. The battle was hotly contested, but in the end Philip proved victorious, much of the glory going to his eighteen-year-old son, Alexander, who probably commanded the Companion Cavalry.[34] The Athenian defeat was costly—1,000 dead and no less than 2,000 captured (16.86.5).

The Athenians fled back to Athens and began to prepare for the worst—resolutions were passed to strengthen the garrison outposts, trenches were to be dug before the walls, and money secured for additional fortifications.[35] Demosthenes secured for himself the commission of collecting supplies and winning over new allies. This task took him out of Athens—perhaps wisely, considering the disaster of his foreign policy. Upon his return, he was appointed his tribe's representative to the "City-Fortification" Board.[36] But Philip chose to be generous—it is said through the agency of the orator Demades, son of Demeas of Paiania, a successful *hippotrophos*.[37] He released the Athenian prisoners without asking for ransom and sent envoys to negotiate

[32] Diod. 16.85.5–6 (but cf. Justin 9.3.9). There are problems with the exact size of the allied army at Chaironeia, but I think that Ellis, *Philip II*, pp. 293–94, n. 64, has been misled by Plutarch's comment at *Dem.* 17.3. When Plutarch mentions 15,000 infantry and 2,000 horse, surely he is referring to the mercenaries, and we have no idea how many of the survivors may have joined the main forces at Chaironeia after their defeat at Amphissa.

[33] In the Lamian War (323–322), Athens contributed 5,000 citizens as infantry and 500 cavalry, along with 2,000 mercenaries (Diod. 18.11.3); the assembly decreed that three tribes guard Attika, while seven be prepared to campaign abroad (18.10.2).

[34] Diod. 16.86.1–4. P. Rahe, "The Annihilation of the Sacred Band at Chaeronea," *AJA* 85 (1981): 84–87, however, credits the Macedonian phalanx equipped with the *sarissa*, (a long pike) not the cavalry, as playing the decisive role in the battle.

[35] Dem. 18.248; Lyk. *Against Leokrates* 44.

[36] Dem. 18.113, 248, 299; Aischin. 3.17, 27, 31.

[37] On the anecdote involving Philip and Demades' rebuke of the king's behavior following Chaironeia, see Diod. 16.87.1–3. On his status as a *hippotrophos*, see *APF* 3263 C, p. 101; a victory at the Olympic Games, possibly in 328, though the information comes from the *Suda* D 415; on this Davies remarks, "If the notice is accurate and not a later invention, Demades thus became one of the last Athenians to engage successfully in competitive horse-breeding." See also M. H. Hansen, "Rhetores and Strategoi in Fourth-Century Athens," *GRBS* 24 (1983): 163 and Kyle, *AAA*, Appendix B, A13.

a peace treaty.[38] This ended the war, but not the justifiable concern for preparedness—the Athenians had dealt with Philip now for twenty years and earlier peace treaties had not endured, most notably the Peace of Philokrates (346).

Impact of Chaironeia

The impact of Chaironeia was not limited to strengthening the city's defenses; the fleet was increased to 400 triremes, new ships constructed, and arms and armor collected.[39] And the defeat brought home to the Athenians the need to reform its military training practices. It is now that some scholars posit the introduction, or at least reform, of the *ephebeia*, the two-year military training program of young men, eighteen to twenty years old, described in the *Athenaion Politeia* 42.2–5.[40]

Aristotle informs us that one member, over forty years old, is elected from each tribe by the people to serve as "supervisor" (*sôphronistês*) over his tribe's ephebes, and one other is elected from all the Athenians to be "director" (*kosmêtês*) over them all. The ephebes spend the first

[38] Diod. 16.87.3. But Thebes was forced to receive a Macedonian garrison.

[39] For the concern for Athens' military strength in this period, the so-called Lykourgan age, see O. Reinmuth, "The Spirit of Athens after Chaeronea," in *Acta* of the Fifth Epigraphic Congress at Cambridge, England, 1967 (Oxford, 1971), 47–51; F. Mitchel, "Lykourgan Athens," *Semple Lectures* II (Cincinnati, 1973), 199; S. Humphreys, "Lycurgus of Butadae: An Athenian Aristocrat," in *Craft of the Ancient Historian*, 204–8; and C. Mossé, *Athens in Decline 404–86* B.C., trans. by J. Stewart (London and Boston, 1973), 81–84. See also C. J. Schwenk, *Athens in the Age of Alexander. The Dated Laws and Decrees of the "Lykourgan Era,"* 338–322 B.C. (Amsterdam, 1985).

[40] See O. Reinmuth, *The Ephebic Inscriptions of the Fourth Century* B.C. (Leiden, 1971; *Mnemosyne Suppl.* 14) 123–38, for arguments for revivification not introduction. But the inscription offered by Reinmuth (no. 1, pp. 1–4; 361 B.C.) to confirm that a formal ephebic organization existed before the first inscription that clearly attests it (Reinmuth, *Ephebic Inscriptions*, no. 2, *IG* II² 1156, 334/3 B.C.) has been disputed by F. W. Mitchel, "The So-called Earliest Ephebic Inscription," *ZPE* 19 (1975): 233–43 and E. Wheeler, "Hoplomachia and Greek Dances in Arms," *GRBS* 23 (1982): 223–33. For discussion, point by point, of the information contained in the *Athenaion Politeia*, see Rhodes, *CAP*, pp. 502–10, who, incidentally, favors a reorganization in the 330s; but he also believes that the thetes of Athens were not required to undergo this ephebic training—particularly as the state awarded the ephebe a spear and shield, the arms of a hoplite. For the opposite view, i.e., that it included every Athenian citizen, see E. Ruschenbusch, "Die soziale Herkunft der Epheben um 330," *ZPE* 35 (1979): 173–76. On the question of the exact age of the ephebes, either seventeen or eighteen, see M. Golden, "Demosthenes and the Age of Majority at Athens," *Phoenix* 33 (1979): 25–38, with earlier literature. For the history of the *ephebeia* see Ch. Pélékidis, *Histoire de l'éphébie attique des origines à 31 av. J.-C.* (Paris, 1962).

year in garrisons at the Peiraeus being trained by instructors in the use of heavy arms, the bow, the javelin, and the catapult. Each *sôphronistês* receives one drachma for himself for rations (*trophê*), and for his ephebic charges, four obols, but he takes the ration money for his tribe and buys all the provisions for the common mess. After a year of training, the ephebes drill in formation before the Athenian assembly and each receives a shield and spear from the state. In the second year, they serve on patrols (*peripolousi*) and do garrison duty around Attika (cf. Aischin. 2.167). They are identified by the *chlamydes* they wear (recalling the young horsemen on vase-paintings and on the Panathenaic frieze). And to insure that they serve a full and uninterrupted term, the ephebes are granted tax-exempt status and cannot sue or be sued by anyone. Only in cases involving the inheritance of his patrimony (*klêros*), the marriage of an heiress, or an inherited priesthood can an ephebe be excused from duty. When they have completed their two-year stint, the ephebes join the citizen body as full participating members. The nature of the acceptable exemptions from duty make it quite clear that wealthy young Athenians were expected to undergo the ephebic training, after which all of them, theoretically, then became eligible for cavalry service.

The ephebic program obviously did not include training specific to cavalry service—that, as we are now well aware, was the exclusive province of the well-to-do. As the fourth-century sources make abundantly clear, particularly Xenophon (e.g., *Hipparch.* 1.12 and Athenaios 9.402 [Pheidon]), cavalry training came after admission to the corps and after the acquisition of a cavalry mount. It is quite a separate and distinct arm of the military. But it would have been very derelict on the part of the Athenians not to have also taken a close look at the organization of their cavalry in an effort to improve its military efficiency and the quality of its personnel. The dramatic success of Alexander and his cavalry at Chaironeia and the apparent inability of the Athenian and allied cavalry to stop him may have stimulated the Athenians to take this very action.

For one thing, the Athenians may have raised the amount of the *katastasis* to 1,200 drachmas after Chaironeia in order to attract a higher caliber of recruit and better horses. Lykourgos' fiscal success would have allowed this additional expenditure, particularly since the loan was expected to be paid back anyway. This is quite speculative, but if the amount of the *katastasis* did not amount to 1,200 drachmas from the start, i.e., mid-fifth century, then this period of reform might be even more appropriate than the problematic 350s and 340s (see above). Correspondingly, I believe it is at this time (if not in the 340s), that Athens chose to reform its recruiting procedure by the introduction of the *katalogeis*, attested for the first time at *AP* 49.2.[41]

It might be helpful to repeat the basic information contained in the two pertinent passages, Xenophon *Hipparchikos* 1.9–12 and Aristotle *AP* 49.2.

Writing in the 360s, Xenophon advises his hipparch to recruit from among those most qualified by their wealth and physical fitness. The hipparch has recourse to two means to accomplish this—legal action or persuasion. He is urged to bring those men to court who are most likely to be suspected of having bribed him to pass over them. If the most powerful men escape service, Xenophon remarks, then the less wealthy will have cause to avoid service, too. These surely are not hypothetical cases Xenophon is referring to, but recruiting malpractices at that time. He then advises the hipparch to persuade the young men to join by emphasizing the illustrious aspects of horsemanship and thereby fire them up with an enthusiasm to serve. Persuasion is also to be applied to their parents. The hipparch should inform the parents (or guardians) that sooner or later they will be required to maintain a horse because of their wealth, if not by him, then by a hipparch in some other year. If their boys join during his hipparchy, he will stop their extravagant purchasing of expensive horses and quickly teach them to become horsemen.

In *AP* 49.2, however, the Athenian people elect ten *katalogeis*, each year presumably, to enroll the cavalrymen. After they have composed the list (*katalogos*) of new recruits, these "enrollers" give this list to the hipparchs and phylarchs, who in turn present the list to the Council of Five Hundred. The hipparchs and phylarchs open a tablet (*pinax*) on which are inscribed the names of the cavalrymen of the previous year and delete the names of those who petition to be exempted because of bodily incapacity. Those who have been enrolled (*tous kateilegmenous*, i.e., by the "enrollers") are then summoned (to step forward), and any of them who swears that he possesses neither the financial means nor bodily capability is granted exemption from service. The councillors decide by vote whether any of the new recruits who does not request an exemption is not fit to serve. Those qualified are inscribed on the *pinax* and the others are dismissed.

Nowhere in his account of the process does Xenophon mention the *katalogeis*. It is tempting simply to assume that the two accounts are

[41] For the argument in concise form, see my "Introduction of the *katalogeis* of the Athenian Cavalry," *TAPA* 112 (1982): 23–32.

the same,[42] that Xenophon's account reflects the stage in the process after the *katalogeis* have given the *katalogos* to the hipparchs and phylarchs. I do not think this is the case—it is a glaring omission on Xenophon's part, the more so when one considers his recommendations at 9.3–6 for keeping the force up to 1,000 with mercenaries and metics. Why would he make such comments if there existed *katalogeis* whose only apparent job was to "enroll" potential recruits? The procedures are not the same.[43] There is no indication in *AP* 49.2 that the cavalry commanders had such independent authority nor were so vulnerable to bribery nor that such flexibility existed in the decision to join or not to join. For Xenophon, persuasion appears to be the primary means of recruiting, the courts only a last resort, particularly among families of the hipparch's same social class and outlook. Genuine concern did exist about bribery.

The *katalogeis*, if they had existed in the 360s, would certainly have been in attendance at the meeting of the Council when the potential recruits were introduced; if certain names of those "enrolled" had curiously been erased (and the young men were not present at the summons), the *katalogeis* would have known it and brought the cavalry commanders to task. The turnover each year could not have been *so* large. Would they have risked that?

Along these lines of thought, I have reconstructed the procedure described at *AP* 49.2 in the following way: the *katalogeis* probably initially examined the financial and physical status of each of the potential recruits and then selected those liable for service. The hipparchs and phylarchs were probably still expected to prosecute recalcitrant recruits in the courts, but now they acted in conjunction with the *katalogeis* and the Council of Five Hundred, not as independent recruiters of first instance. The role of the Council was to confirm the selection made by the *katalogeis*, by test of oath and by autopsy.

What I am suggesting is that sometime between the composition of the *Hipparchikos* in the 360s and the *Athenaion Politeia* in the late 330s or 320s, the Athenians created the board of ten cavalry enrollers. There is compelling reason to associate this cavalry reform with the overall military shakeup, i.e., defenses, navy, army (*ephebeia*), following the disaster at Chaironeia.[44] It had one primary purpose: to insure

[42] As does F. Borowski, "Dokimasia: A Study in Athenian Constitutional Law" (unpubl. diss., University of Cincinnati, 1975), 42–43.

[43] So also A. Martin, s.v. "Hipparchos," *Dar.-Sag.* (1899) 3.189, n. 8.

[44] As a referee of my article, "Introduction of the *katalogeis*," p. 32, n. 23, suggested, an idea which now seems more compelling to me than to date the reform in the 340s by

that the Athenian cavalry maintained itself as close as possible to the canonical 1,000-man level. Specifically, the *katalogeis* were intended to eliminate bribery, or the hint thereof, and to achieve greater success in compelling the wealthier citizens to resume their traditional responsibility of cavalry service.

Though much less certain, perhaps the clause citing "bodily incapacity" as the only legitimate excuse for retirement from cavalry service was introduced at the same time as the *katalogeis.* As I have suggested earlier, it seems to have been possible to retire on other grounds in the past, but now the state may have eliminated those options, the most obvious one being a plea of financial hardship, which, considering the ability of some to hide their assets in the fourth century, would have been difficult to contest.

As I have argued in Chapter Two, I think that in the Classical period most *hippeis* did not serve beyond fifteen years. The restrictive clause in *AP* 49.2, then, would have had the effect of increasing the average length of cavalry service, in turn facilitating the maintenance of the canonical size of the cavalry. But even if this clause has not been recently introduced, I continue to hold that the state assumed that most cavalrymen would be retiring after ten to fifteen years and that the prohibition against all exemptions except physical incapacity was primarily directed at cases of "early" retirement, that is, at young men in their twenties and early thirties who had not served a reasonable length of time. It might be germane to note that Pseudo-Demosthenes does not cite a specific Athenian law that prohibits Phainippos from retiring from cavalry service, only that it is an outrage that this rich, young, virile man has done so.

The newly enrolled young man, of course, was still allowed to claim exemption on the grounds of insufficient wealth, but now he had to face his tribal *katalogeus* and openly swear on oath that the state's evaluation of his family wealth was inaccurate. No doubt some members of the upper classes supported cavalry reform, as they certainly supported the general Lykourgan program;[45] but now there existed more precise institutional guarantees for its success—applicable to *all* members of the upper classes.

Perhaps one will argue that surely Aristotle would have indicated that the *katalogeis* were a fairly recent innovation, if that were indeed the case. But he describes the *ephebeia* as if it had always operated this

associating it with the concern over Philip's continued encroachments and the irregularity of Meidias' hipparchy, with its new-fangled cavalry tax and other debilitating laws.

[45] Humphreys, "Lycurgus," p. 204.

way, yet it seems likely that a good chunk of it postdated 338. More to the point, at *AP* 49.1, Aristotle talks about *prodromoi* and *hamippoi*, as if they had always been a part of the Athenian cavalry organization, yet it seems certain that the *prodromoi* had replaced or supplanted the mounted archers (*hippotoxotai*), who do not appear after the early fourth century.[46] And the *hamippoi*, apparently a Boiotian innovation (Thuc. 5.57; Xen. *Hell.* 7.5.23), cannot be associated with the Athenian cavalry until Xenophon *Hipparchikos* 9.7, which talks about foot soldiers (*pezoi*) attached to cavalry, but does not use the term *hamippoi*. Thus, Aristotle describes what is current practice, but he cannot be counted on to inform us as to how current.

The Athenian Cavalry in Action after Chaironeia

The success of the cavalry reform can perhaps be seen in two military campaigns waged by the Athenians in the closing decades of the fourth century B.C.

In 323 B.C., word arrived that the great Alexander was dead. The time seemed ripe to cast off the Macedonian hegemony in Greece. So began the Lamian War (323–322). Despite some opposition from certain "men of property" (Diod. 18.10.1), Athens chose to commit itself to the cause. Its contribution to the allied Greek army was significant: 5,000 citizen infantrymen, 500 cavalrymen, and 10,000 mercenaries (Diod. 18.9.1, 18.11.3). This was one of the largest detachments of horsemen that Athens ever committed to a foreign campaign, no doubt a reflection of the respect held for the Macedonian cavalry and its Thessalian allies, numbering 2,000 (18.5.2), but also perhaps a reflection of the renewed confidence in its own cavalry. Under the command of Leosthenes, the citizen-mercenary army first defeated the Boiotians, then advanced north of Thermopylai to meet Antipater, Alexander's "general of Europe." When the Thessalian horsemen defected to the allies, Antipater was easily defeated in the battle and withdrew to safety at Lamia.

Athens had wisely elected not to commit more of its cavalry. While the campaign was going on, Mikion landed at Rhamnous on the northeast coast of Attika with a force of Macedonians and mercenaries and proceeded to pillage the surrounding countryside. Phokion came out from Athens with an army, including cavalry, killed Mikion, and

[46] See Appendix D.

routed his force.[47] Athens always held back a portion of its cavalry to defend Attika against attack.

Diodoros' narrative of the Lamian campaign makes it very clear how much the allies looked to the Thessalian cavalry for leadership (18.15.2, 17.2). The Thessalian and allied Greek cavalry, 3,500 in all, quickly demonstrated their superiority over their Macedonian counterparts, first against the combined forces of Antipater and Leonnatos (18.15.1–4), and then again at the battle of Krannon in Thessaly in 322 against the forces of Antipater and Krateros who now had 5,000 horse at their disposal (18.16.4–17.1–5). According to Diodoros, it was only when the allied infantry wilted before the charge of the Macedonian phalanx led by Antipater that the Greek cavalry, "having gained the advantage (*proterêma*)," likewise withdrew from the fray to rejoin their infantry forces. Although special honor is accorded throughout to the skill and courage of the Thessalian horsemen, one cannot help but conclude that the other Greek cavalries, including the Athenian, acquitted themselves well.

The inconclusiveness of the battle and the fact that the Athenian fleet had already been defeated in the Aegean convinced the Greeks not to continue the war. When Antipater advanced upon Attika, even Athens was forced to come to terms. This meant the abolition of the democracy, a war indemnity, and the imposition of a Macedonian garrison at Munychia. The government was to be entrusted to those whose wealth exceeded 2,000 drachmas—Antipater blamed the common folk of Athens for having voted for the war against the better counsel of the upper classes (see Diod. 18.10.1–2). Those who were disenfranchised were allowed to expatriate to new homes in Thrace. According to Diodoros, those excluded by this census numbered 12,000, leaving 9,000 citizens.[48] In 317 a new constitutional arrangement expanded the franchise to include those whose property exceeded 1,000 drachmas, but at the same time placed the supervision of Athens into the hands of one Athenian, Demetrios of Phaleron (Diod. 18.74.3), who was to rule for ten years. Needless to say, the members of the Athenian cavalry, by the

[47] Plut. *Phok.* 25. For Phokion's actual route to Rhamnous, see the suggestion by J. Ober, "Edward Clarke's Ancient Road to Marathon," *Hesperia* 51 (1982): 453–58. Phokion may have belonged to the deme Potamos: L. A. Tritle, "Phokion Phokou Potamios?" *AJAH* 6 (1981): 118–32.

[48] 18.18.4–5; Plut. *Phok* 28.4. Many of the disenfranchised Athenians succeeded in returning to Athens in 318 (*Phok.* 32–33.1; cf. Diod. 18.56; 65.6–67.); for recent discussion of this period, see J. M. Williams, "A Note on Athenian Chronology, 319/8–318/7," *Hermes* 112 (1984): 300–305.

wealth implied by that fact, were unaffected by these changes in the constitution. Timocracies never exclude horsemen.

Moreover, an inscription, dated ca. 315/4, records a decree by the troops stationed at Eleusis, Panakton, and Phyle honoring Demetrios of Phaleron as *strategos*; in addition two crowns attest that the Council of Five Hundred and the *hippeis* each honored him as a former hipparch, while three other crowns bear witness to his four-horse chariot victories at the games of the Panathenaia, Delia, and the Hermaia. This notable equestrian tradition continued in his homonymous descendant (grandson?) who held the offices of phylarch (possibly), hipparch, and *strategos* in the 250s.[49]

Plutarch (*Phok.* 37.1) records a last poignant note. Caught in the middle of the intrigues and power struggles between Kassander and Polyperchon, the successors of Antipater, Phokion, the old Athenian general, was condemned by the now restored and bitter democratic assembly in the spring of 318. It happened that the Athenian *hippeis*, participating in a procession in honor of Zeus, had to pass by the prison where Phokion was awaiting execution. As they passed, some of the horsemen removed the garlands from their heads, while others openly wept.

We learn of a second campaign involving the Athenian cavalry from a brief reference by Pausanias (1.15.1) to a trophy commemorating an Athenian cavalry victory over Pleistarchos, commander of the cavalry and mercenaries for his brother Kassander. This trophy was erected near the Stoa Poikile and the Hermes Agoraios (bronze statue of Hermes of the Marketplace). Recently, S. Burstein[50] has linked the captured horsemen whose release was secured by Oxythemis of Larissa (*IG* II² 558, lines 33–34) with the battle, presumably in the Peloponnesos, that led to the defeat of Pleistarchos at the hands of Demetrios Poliorketes in 303 B.C. and in turn to the trophy mentioned by Pausanias.[51] In addition, T. L. Shear, Jr.,[52] has suggested that the Hellenistic

[49] *IG* II² 2971; Ditt. *SIG*³ 519; *SEG* XXV.206; XXVIII.101; *APF* 3455, pp. 108–9, dates the hippic victories ca. 320 or soon after; the date of his hipparchy, before or after 317 B.C., is more problematic; for his grandson(?), see *IG* II² 1285, line 6; *SEG* III.123; Hegesander in Athenaios 4.167 E–F (see R. E. Wycherley, *The Athenian Agora*, vol. III [Princeton, 1957], no. 302, p. 105); *APF* 3455, pp. 109–10; for his phylarchy (queried by Davies), see the restoration of lines 4–7 = phylarch under Antimachos by Habicht, *Studien*, pp. 49–51.

[50] "The Date of the Athenian Victory over Pleistarchus: A Note on Pausanias 1.15.1," *CW* 71 (1977): 128–29 (= *SEG* XXVI.89).

[51] For the names of Pleistarchos, Kassander, and Demetrios of Phaleron on curse tablets dated 313 to 307 B.C., see D. Jordan, "Two Inscribed Lead Tablets from a Well in the Athenian Kerameikos," *AM* 95 (1980): 229–36 (= *SEG* XXX.325). See, however, Ch.

gate recently discovered in the excavations next to the Stoa Poikile may have supported this hippic trophy. If this is true, it signals a splendid architectural statement of the city's pride in its cavalry.

Summary Remarks

Xenophon bears witness to the fact that the state was experiencing difficulty in recruiting a sufficient number of qualified recruits for the cavalry and that the reputation of the cavalry had suffered after the Peloponnesian War. While it is true that "in the fourth century B.C. the cavalry was obviously a much less brilliant affair than in the fifth," it cannot be claimed to be true for the precise reasons that Anderson goes on to suggest, to wit, that many

> of the old families had been destroyed or impoverished by the wars of 431–404 B.C., during which the farmland of Attica and the country villages had been thoroughly laid waste. So in the fourth century B.C. the proportion of rich men (as distinct from the moderately well off) in the cavalry may have been comparatively small.[53]

First of all, Hanson (see Chapter Three) has argued that the effects of damage in Attika have been exaggerated. Secondly, there is no hint in 404/3 that the composition of the cavalry was significantly different from what it had been in the preceding decades. We should not forget that the 300 *hippeis* lost to the plague in the early 420s were quickly replaced, as Aristophanes' *Acharnians* and *Knights* would lead us to believe, and likewise, presumably, however many of the 250(280?) *hippeis* who failed to return to Athens from Sicily in 413 B.C. No doubt, some old Athenian families did disappear and others did not recover from the effects of the long war, but not in sufficient numbers to explain the apparent diminution of the quality and quantity of the cavalry personnel in the fourth century. This phenomenon has more to do with specific historical events and the resulting negative attitudes towards cavalry service, topics which I have already treated in Chapter Four.

By the mid-fourth century, however, it would be perilous to argue that the taint of service in the Athenian cavalry under the Thirty still

---

Habicht, *Pausanias' Guide to Ancient Greece* (Berkeley, 1985), 77–82, who argues that the trophy probably commemorated a cavalry victory near the Dipylon Gate in Athens and that the battle took place in 304 B.C., not 303.

[52] T. L. Shear, Jr., "The Athenian Agora: Excavations of 1980–1982," *Hesperia* 53 (1984): 19–24.

[53] *AGH*, p. 134.

carried the direct and immediate force it had had in the early decades of the century—we are now nearly two generations away from that critical year. But the damage had been done and the cavalry commanders had not won back sufficient numbers of the old equestrian families. Now other forces take precedence, particularly the phenomenon of the mercenaries and the appearance of a relentless foe to the north.

The 350s to the 330s were dominated by the personality and ambitions of Philip of Macedon. His expansionist policies could not be resolved by one decisive battle in the tradition of rival city-states. The ever-present mercenary soldier seemed ideally suited to the task. If it is true that the Athenian citizen body at large was displaying a disinterest in military service, particularly in seemingly endless conflicts abroad, why should we expect a different attitude among those likely recruits for cavalry service? In a way, cavalry service resembled mercenary service: both were paid by the state for year-round service. If the state chose to send its cavalry abroad for long terms of service, who could cry foul? But Demosthenes' recommendations for a joint citizen-mercenary force to harass Philip make it plain that it was asking too much of the horsemen to serve long periods under war conditions away from home (not a problem with mercenaries), thus explaining his proposal for rotation of personnel. To be cynical, I suggest that Demosthenes knew his city's horsemen all too well. Despite the unique nature of their military corps in Athens, they thought like citizen-hoplites—not too much duty and not too far away.[54]

Xenophon himself called for more sweeping reforms in the cavalry organization than even Demosthenes would have accepted, particularly with regard to the use of mercenaries. He proposed the establishment of a permanent mercenary force of 200 horsemen, a pretty good hint that he thought that 800 was about the upper limit of citizens that Athens could hope to keep in service from year to year (and even less demand from citizens if the metics had been allowed to join). And we should not forget that one of the reasons for the mercenary cavalry force was to serve as a model for the Athenian *hippeis*, to make them better horsemen and the cavalry more effective.

Some of the upper-class families must have welcomed Xenophon's modest proposal: if implemented, it would mean that many of them could now avoid the burdens of cavalry service; in essence they maintained their status as citizen-hoplites—still subject to occasional lev-

[54] For evidence of the Athenian *hippeis* serving a tour of duty on the island of Salamis, ca. 320 B.C., see *IG* II[2] 1955. For further discussion and the possibility that they also served on the island of Lemnos, see Appendix B.

ies—but freed from extended military duty. It is more than likely that Xenophon intended the 200 mercenary *hippeis* to be deployed primarily for emergency service abroad, freeing the citizen cavalry for major engagements, e.g., Mantineia, and the usual business of home defense. The Athenian *demos* must have concluded—if this proposal ever reached public debate—that this was just another case of the rich trying to avoid their "liturgical" responsibilites. Of course, they would be paying for the exemption. Opposition to these proposals may also have come from old aristocratic families who felt that their cavalry pedigree would be correspondingly diluted by the admission of foreigners and resident aliens. Xenophon, for one, was more concerned about military efficiency than aristocratic sensitivities.

The troubled hipparchy of Meidias cannot have served the cavalry well. Meidias seems to typify the hipparch whom Demosthenes describes as more concerned about festival activities than those of war. In view of the events of the year 349/8, it would have been better to have had a mercenary hipparch command the cavalry in place of someone like Meidias. We cannot know all the details nor the whole truth of it, but there is sufficient evidence to suggest that the reputation of the cavalry was diminished and that certain new laws may have allowed rich citizens, like Meidias, to escape cavalry service by means of a financial payoff. An echo of Xenophon?

There is a natural tendency to attach too much importance to the battle of Chaironeia, but it did bring home to the Athenians the need to reexamine their military preparedness. The cavalry could not have escaped this self-scrutiny. Xenophon's concerns may have received renewed attention, but not, apparently, his specific recommendations. The desired goal was to reinvigorate the cavalry as an effective citizen force, as the ephebic training was intended for the hoplites, and to compel the wealthy classes to resume their traditional participation in it. After Chaironeia, no discerning Athenian could fail to appreciate how important cavalry was, especially in light of the fact that Thessaly had added its redoubtable cavalry to that of the Macedonians.

This brings us to the *katalogeis*. As these elected officials were to insure the quality and quantity of the Athenian cavalry personnel, perhaps even the *demos* now took more care in its choice of cavalry officers—ones for war, not for the marketplace. Philokles, son of Phormio, was elected hipparch three or four times (!) and *strategos* more than ten times, a military career presumably spanning the 330s and 320s.[55]

[55] Dein. 3.12; for prosopography, see *APF* 14521, pp. 539–41. Cf. the incredible career of Phokion, elected *strategos* forty-five times (Plut. *Phok.* 8.2)

And no matter how hard Deinarchos tries to denigrate Philokles' achievement, e.g., he was elected over "good and noble men" (*andrôn kalôn k'agathôn*), it is clear that his military record was impeccable. We recall that there were only two regular hipparchs elected each year—this reiteration of hipparchies can be interpreted as a sign of confidence on the part of the Athenian people, a deliberate effort to select an experienced officer. Of course, it might also imply a declining number of interested applicants.

Prosopographical Considerations

Of course, it would be wrong and unfair to indict every cavalryman or officer in the fourth century. At all times the cavalry consisted of both enthusiastic members and not-so-enthusiastic members, from all strata of the upper classes, but I am convinced that proportionally there were more of the latter in the fourth century than had been the case in the halcyon days of the fifth.

On the one hand, we have the two Xenophons, the historian and his sons (recall Polystratos and his sons); and Xenophon, son of Euripides, of the deme Melite, who served as one of the three hipparchs in the 450s or 440s and whose paternal grandson is attested as a *hippeus* in the mid-fourth century;[56] the three winning phylarchs in the *anthippasia*, Demainetos and his two sons, Demeas and Demosthenes, recorded on the Bryaxis base; Philoktemon Euktemonos Kephisieus, *hippeus* and trierarch many times, who died off Chios in the 370s;[57] and Lykophron in the 330s. In Hypereides' defense (1.16–18), we learn that Lykophron had been an avid horse-breeder (*hippotrophos*), even though this occupation had greatly taxed his strength and estate. The defendant is not talking about racehorses here; he goes on to say that he was crowned for bravery by his fellow cavalrymen and his fellow officers. He was elected phylarch and then Hipparch to Lemnos. In the last office he held command for an unprecedented two consecutive years and then an additional one beyond that. Considering the importance that Demosthenes attaches to the hipparch to Lemnos as the only cavalry officer who sees any sort of active military duty (4.27), perhaps here with Lykophron, as with Philokles, the Athenian *demos* is exercising better judgment in its elected military officials. We will have occasion to speak of Lykophron further (Appendix B), but it might be

[56] *IG* I² 400; Kroll, "Archive," no. 24, pp. 111–12; *APF* 5951, pp. 199–200.

[57] Is. 6.5; for prosopography, see R. F. Wevers, *Isaeus, Chronology, Prosopography, and Social History* (The Hague, 1969), 61 and *APF* 15164, pp. 562–66 with stemma. See also, W. K. Lacey, *The Family in Classical Greece* (Ithaca, N.Y., 1968), 127–28.

useful to add that Lykophron obviously did a good job on Lemnos, being crowned three times by the Athenian cleruchs of Hephasteia and Myrina.

On the other hand, we have wealthy Meidias who is likely never to have served in the cavalry as a simple *hippeus* and when elected hipparch neglects his command, antagonizes the cavalry, and tries to avoid cavalry service; and young Phainippos, strong, rich, a *hippotrophos*, who exchanges his war-horse for racehorses and chariots; and Dikaiogenes III who belonged to "one of the wealthiest and most distinguished families of democratic Athens" (*APF* 3773, p. 145), but who did not maintain racing teams nor spend more than 300 drachmas on a cavalry mount,[58] yet whose grandfather (albeit by adoption), Menexenos I, lost his life serving as a phylarch in the Spartolid campaign of 429 (Is. 5.42–43, 46). And then there are the friends and supporters of Leokrates in 330, who breed horses (*hippotrophêken*), provide choruses, and the like, but who contribute nothing to making the city militarily secure, like serving as trierarchs or financing the building/repair of the city walls (Lyk. *Against Leokrates* 139). Obviously, the orator could not make these accusations if these *hippotrophoi* had actually served as *hippeis*, hipparchs, or phylarchs. Lykourgos may not be completely fair, however; the friends of Leokrates could have been too old to be enrolled by the *katalogeis*. Phainippos is another matter entirely (see Chapter Two).

I suspect that it is impossible to present for scrutiny a typical Athenian cavalryman of the fourth century B.C. Dexileos catches our artistic and historical eye, but we know too little about him and his family to say much beyond that he was young and brave and his parents were apparently well-off financially. To be honest, the prosopography of cavalry personnel for this period is simply not full enough to make authoritative statements. For example, the three longest lists of Athenian *hippeis*, (1) *IG* II² 5222, recording the cavalry casualties in the campaigns of 394 (including Dexileos); (2) the mid-fourth century deposit of lead tablets from the Agora; and (3) *IG* II² 1955, an incomplete list of *hippeis* serving on Salamis, ca. 320 B.C. along with their hipparch, Theogenes of Eleusis, son of Theomedes, provide us with names of twelve, twenty-six, and twenty-one *hippeis*, respectively. Yet, of these fifty-nine cavalrymen we know surprisingly few beyond a nodding acquaintance.

[58] Perhaps Dikaiogenes received his *katastasis* loan from the state and then proceeded to invest most of it in other enterprises, alloting only a modest but adequate sum for the actual purchase of his war-horse. An opportunistic businessman? How many others did the same, to the detriment of the cavalry?

With the casualty list of 394 this is not unexpected inasmuch as most probably died too young to have yet achieved public recognition. Furthermore, only the *nomina* are provided, so linkages with known families are mere guesses. As for the cavalry tablets, we encounter the same problem. Although handicapped by being forced to work only with the *nomina* of the *hippeis*, Kroll has been able to make some tentative prosopographical connections here and there, but they tend not to be very informative. Only Komaios (no. 13), perhaps to be identified with Komaios Komonos Semachides, trierarch between 356 and 346/5 and councillor in 334/3; and Xenophon (no. 24), see above, allow us to see beyond a simple name.

With the *katalogos* of ca. 320, we possess all the identifying elements for each *hippeus*, i.e., nomen, patronymic, and demotic, but still the results are disappointing. For the hipparch Theogenes, we can identify as his son the Theomedes who served as a phylarch, ca. 286–261 B.C.,[59] confirming strong cavalry traditions in the family. And we do know that two families had more than one member in the cavalry, i.e., (a) Nikias Euktaiou Xypetaion (line 5) and Euktaios Nikiou Xypetaion (line 20) were probably cousins; Nikias was in his early thirties, since he was an ephebe in 334/3;[60] and (b) Charamantides Sokleou Euonymeus (line 9) and Sokles Aleximachou Euonymeus (line 22) were also probably cousins.[61] And Phileas Diokleous Elaiousios (line 17) is the son or brother of Diokles, who was trierarch in the 330s.[62] There are other prosopographical scraps from these three lists, but nothing sufficient to flesh out these *hippeis*, nothing to hint how strong were the cavalry traditions in their families nor just how wealthy they were nor what they thought about cavalry service.

In addition, very few of the litigants in Lysias, Isaios, or Demosthenes claim or admit to have served in the cavalry. In some cases this must have been a deliberate omission so as not to jeopardize a client's case—*hippotrophia* was an obvious sign of conspicuous consumption, even though it might only mean owning one horse, as with the phylarch Theophon;[63] at the very least, knowledge of the fact was sure to

[59] Threpsiades-Vanderpool, no. 2, line 25, p. 111.

[60] See *APF* 10814, pp. 407–8 with stemma. Nikias as ephebe: *IG* II² 1156, line 21; see Reinmuth, *Ephebic Inscriptions*, no. 2, p. 8.

[61] *PA* 13070.

[62] *IG* II² 1624, lines 63–64; *APF* 4021, p. 157.

[63] Is. 11.41–42, 45. It was, however, a very fine horse (*hippos lampros*, 11.41). For the prosopography, see Wevers, *Isaeus*, p. 66 and *APF* 2921, p. 84: died ca. 369, phylarchy in the 370s. For economic considerations related to cavalry service in the fourth century, see Anderson, *AGH*, pp. 134–35.

tag you with more wealth than you might want to admit to the court. Even in cases where we would expect the litigant to accuse his opponent of *hippotrophia*, it appears infrequently. Why? Two reasons come to mind: (1) they simply were not *hippotrophoi* and everybody knew it; or (2) for the desired effect, it has to be demonstrated that *hippotrophia* is a bad thing, because it can be defended as a service to the state, e.g., in the case of Lykophron. Furthermore, to be identified even as a simple *hippeus*, depending on the date of the litigation, could be used in association with the events of 404/3.

These attitudes, all too human, confound attempts to construct a data base of *hippeis* in the Classical period—you simply cannot assume, even if a man is unmistakably wealthy or even clearly identified as a *hippotrophos*, that he was a *hippeus*. For example, Timokrates, son of Antiphon, like Demades (see above), belonged to a well-known family of the fourth century, served as a trierarch, and won the two-horse chariot race at Olympia, yet there is no evidence that he actually served in the cavalry.[64] Or the case of Philomelos II Philippidou I Paianieus, who was related by marriage to the family of Xenophon, son of Euripides, had a long and distinguished liturgical record (particularly trierarchic), and was apparently a *hippotrophos*—it was this Philomelos who purportedly loaned a horse to Meidias for use in processions.[65] But we have no indication that he had ever been a *hippeus*. Like Timokrates, maybe yes, maybe no. We cannot know. In lieu of adequate prosopographical data, what I have attempted to do is suggest historical trends and fluctuating attitudes, both society general and cavalry specific, to explain problems affecting the enrollment and performance of the Athenian cavalry in the post-imperial period of the fourth century.

In his *Oikonomikos*, written as late as 360, Xenophon cites examples of public-spirited men who served in the cavalry. Xenophon has Socrates discourse with the wealthy Kritoboulos, a young man in the 420s, whose career included liturgies, trierarchies, war-taxes, and cavalry service (2.6). And likewise with the splendidly wealthy Ischoma-

[64] *IG* II² 3127, for the victory, see L. Moretti, *Olympionikai. I vincitori negli antichi agoni olympici* (Rome, 1957), 124, no. 440; for prosopography, see *APF* 13772, pp. 513–14 and Kyle, *AAA*, Appendix B, A65. On Demades, if it could be demonstrated that he was closely related to the family of Demainetos I Demeou II Paianieus (*APF* 3276), famous for providing three phylarchs in the fourth century (see Figure 6), his interest in *hippotrophia* would be easily explained and would make his participation in cavalry service at least arguable. But Davies is skeptical of a close association between these two families in the fourth century (*APF* 3276, p. 103). See also, Kyle, *AAA*, p. 121.

[65] Dem. 21.174; for prosopography, see *APF* 14670, pp. 548–50 and 5951, p. 203 (stemma).

chos, both civic-minded and a *hippeus*, who must have performed his cavalry service in the 430s and possibly the 420s (11.14–18, 20). Was Xenophon just invoking the Socratic age and its contemporaries out of respect for his master and the force of his authority, as Plato did, or could he have also been gazing back to the period before and during the Peloponnesian War to draw deliberate contrast with his present age? Considering his comments in other works, notably his *Hellenika* and *Hipparchikos*, surely the old historian is saying, where are such men now? He means men like the father (incidentally, also the husband of the daughter of our Xenophon, son of Euripides) of the speaker of Lysias' *On the Property of Aristophanes*, who was also linked to the Socratic circle (died 389 or 388 at the age of seventy), and who not only spent over nine talents on public liturgies, including seven trierarchies, numerous *choregiai*, and frequent and large *eisphora* contributions, but also served as a cavalryman (no doubt in his youth) and raised fine racehorses which won at the Nemean and Isthmian Games.[66]

[66] Lys. 19.57, 59, 62–63; see *APF* 5951, pp. 200–203.

CHAPTER SIX

The Horsemen of Hellenistic Athens

"They [hipparchs and phylarchs] 'established' one hundred additional horsemen, thereby increasing the total to three hundred."
(Threpsiades-Vanderpool, p. 104, lines 7–10)

A "Voluntary" Athenian Cavalry?

The third century B.C. marks a very difficult and confused period in Athenian history. In foreign policy, the Athenians had to deal with the ambitions of a succession of Macedonian dynasts and the presence of Macedonian garrisons on Attic soil. In domestic politics, never unrelated to foreign policy, democracy, oligarchy, tyranny, and even a disguised monarchy each took a turn at ruling. It serves no good purpose to record all of these machinations; there are many excellent works, some quite recent, which do.[1] In this chapter we rather concentrate our efforts on the fortunes of the Athenian cavalry, to illuminate how it was different from its Classical counterpart. It will not spoil the ending to reveal from the outset that the Hellenistic *hippeis* probably had more in common with the aristocratic horsemen of Archaic Athens than they did with the majority of those of the Classical period, with one important exception: the state still continued to support them with the system of *katastasis* and the *sitos* allowances.

The Hellenistic period is epigraphically rich for information on the

[1] On the history of Hellenistic Athens, see the classic study by Ferguson, *HA* and the brief survey by Cl. Mossé, *Athens in Decline 404–86* B.C., trans. by J. Stewart (London and Boston, 1973), Chapters 5–7. The foundations for a new general study can be found in Ch. Habicht, *Untersuchungen* and his *Studien*, along with Heinen, *Untersuchungen*; and T. L. Shear, Jr., *Kallias of Sphettos and the Revolt of Athens in 286* B.C. (Princeton, 1978; *Hesperia Supplement*, 17). For economic considerations, see J. Day, *An Economic History of Athens under Roman Domination* (New York, 1942; Arno reprint, New York, 1973), and generally, the monumental M. Rostovtzeff, *Social and Economic History of the Hellenistic World*, 2d ed., 3 vols. (Oxford, 1953). The best political history of the period is E. Will, *Histoire politique du monde hellénistique (323–30 av. J.-C.)*, 2d ed., 2 vols. (Nancy, 1979–81); see now also *CAH*[2] VII, pt. 1, 1984 (to the end of the third century B.C.). For relations with Rome, see two new works: A. N. Sherwin-White, *Roman Foreign Policy in the East 168* B.C. *to* A.D. *1* (Norman, Okla., 1983), and E. S. Gruen, *The Hellenistic World and the Coming of Rome*, 2 vols. (Berkeley, 1984).

basic operation of the Athenian cavalry. The lead tablets recording the "evaluations" of the cavalry mounts published by Braun and Kroll and several detailed inscriptions, all from the third century, bear witness to the fact that Athens continued to view its cavalry as important. The most significant change will come in the number of Athenian *hippeis* serving, or required to serve. Sometime between the end of the fourth century and 282/81, the size of the cavalry dropped to 200 men. Our much discussed cavalry inscription from that year records that the *hippeis* praised their hipparchs and phylarchs for having "established" (*proskatestêsan*) 100 *hippeis* to add to the existing 200.[2]

The reasons for this reduction could be associated with the abandonment of military conscription by the Athenians at the end of the fourth century. On this change, W. S. Ferguson has written (though before much of the relevant evidence was available):

> The abandonment of an aggressive foreign policy made preparations for attack senseless. This was the view taken by the moderates in 301 B.C. Hence they made a profound change in the whole life of the people in that they abandoned conscription as a system of national defense, and instituted in its stead volunteer service. Henceforth the fighting force of Athens consisted ordinarily of twelve select bodies of both infantry (*epilektoi*) and cavalry, and of detachments of paid mercenaries. It was no longer necessary for all, or, as prior to 338 B.C., the propertied young men of Attika to gather in the Piraeus and the frontier forts for training and service in the art of war; and, the compulsion being removed, the ephebe registration fell from about eight hundred to about thirty. The corps of ephebes was, thereafter, composed of young men of good families; and while it remained like the cavalry, a source of civic pride and an ornament to religious processions, it became almost a negligible military factor.[3]

There are many good points in this classic statement, particularly on the linkage with the ephebic program. We might note, however, that Reinmuth has argued that the evidence for the reduction of ephebic service from two years to one can be found in an inscription dated to 305/4.[4] And there is no doubt that a conscious decision had been made to allow such a significant reduction in the size of the cavalry and I am

[2] Threpsiades-Vanderpool, p. 104, lines 7–10.

[3] *HA*, pp. 127–28.

[4] O. Reinmuth, *The Ephebic Inscriptions of the Fourth Century* B.C. (Leiden, 1971; *Mnemosyne Suppl.* 14), no. 17 (*IG* II² 478), pp. 86–117. Cf. Ch. Pélékidis, *Histoire de l'éphébie attique des origines à 31 av. J.-C.* (Paris, 1962), 159–72.

convinced that it is not unrelated to a redefining of Athens' international position.

But did cavalry service become fully voluntary? I think not. The recent epigraphic evidence from the third and early second centuries affirm an administrative system unchanged from that of the fourth century, that is to say, council, *dokimasia, katastasis, timêsis, sitos.* Furthermore, a cavalry inscription from 188/7 offers proof that Athens still selected *katalogeis*, although now numbering twelve to keep pace with the addition of two new tribes, Attalis and Ptolemais.[5] Now unless we assume that the board of "enrollers" was abolished in the third century and then reintroduced in the second century—an unlikely development—we must conclude that the *katalogeis* continued to execute their duties throughout this period.

At any rate, if all the key ingredients of cavalry recruitment were retained, how can one insist that cavalry service became "voluntary"? The answer is simple; it did not. The Athenians chose to maintain a quota of 200 men and to do that required some administrative regulation. It is probably true that many of the 200 *hippeis* were to a degree "voluntary" in that they approximated those propertied families of Athens who had legitimate equestrian traditions, but it is also true that this fact did not guarantee that all of those eligible always felt enthusiastic about committing themselves to a cavalry career. The *katalogeis* (along with the cavalry officers) ensured that those reluctant well-to-do young men still did their part, as they had been compelled to do in the late Classical period. With the smaller numbers and a more willing pool of recruits, however, maintaining the cavalry quota must have been made considerably easier. To these young men of equestrian families, cavalry service might not have seemed such a burden.

Size of the Third-Century Athenian Cavalry

The size of the cavalry in this period requires further discussion. In the Threpsiades-Vanderpool inscription, unlike that of 188/7, the cavalry officers are praised for their management of the cavalry *katalogos* (or

[5] Ch. Habicht, "Neue Inschriften," *AM* 76 (1961): 129, lines 26–27 (= *SEG* XXI.435), though Habicht has curiously restored *deka* (ten) instead of *dôdeka* (twelve). Two tribes, Antigonis and Demetrias, which had been added to the original Kleisthenic ten in 307/6 in honor of king Antigonos and his son Demetrios Poliorketes, were abolished in 201/0. In the third century B.C., then, the *katalogeis* would have numbered twelve, until a thirteenth tribe, Ptolemais, in honor of Ptolemy III Euergetes of Egypt, was added (presumably along with another *katalogeus*) in 224/3. Attalis, in honor of Attalos I of Pergamon, was introduced in 200 B.C.; this restored the number to twelve, which remained unchanged for the rest of the Hellenistic period.

katalogê, line 6) and for the "establishment" of the additional 100 horsemen (lines 6–10), but the *katalogeis* are mentioned nowhere in the inscription. At first glance, it would appear that the hipparchs and phylarchs had resumed full responsibility over the enrollment-recruitment process in much the same manner as Xenophon describes in his *Hipparchikos*. But the omission of the *katalogeis* in the inscription can be explained in another way. This inscription attests a critical moment in Athenian history when the usual enrolling-recruiting methods were not sufficient. The key phrase is *eis to dynaton epi tou parontos*, "as much as possible under the present circumstances" (lines 8–9). This signals unsettled times and refers to the ongoing revolt of Athens from Macedonian rule in the 280s[6] and the continued presence of a Macedonian garrison in Attika (Threpsiades-Vanderpool, p. 106). In a state of emergency, I suspect that the Athenians decided to increase their cavalry to 300 and that while the *katalogeis* had dutifully composed the *katalogê* to maintain the quota of 200 men, the cavalry officers had been given the task of augmenting the force by another 100 whom they personally and independently succeeded in recruiting. Because of this successful supplementary "draft," they received pointed praise from the *hippeis*; the *katalogeis* had performed a routine task and did not warrant mentioning.

It might also be worth noting that the cavalry apparently received some special legislative attention in 282/1. In lines 11–14, the inscription records that in order for there to be phylarchs for each of the tribes, a law (*nomos*) which prevented a phylarch from being elected at large from all the *hippeis* was abolished. The editors (p. 106) are probably right to interpret this as a sign that some tribes had not been able to find sufficient candidates to fill the phylarchy—proof that cavalry command was not always sought after by those eligible among the propertied families; henceforth, one tribe could supply more than one phylarch.

I would also suggest that this law was precipitated by the failure to find a phylarch for one of the tribes in 282/1: in the list of cavalry officers, lines 25–33, the name of the phylarch for what would have been the tribe of Aigeis (the phylarchs are listed in the official tribal order) is missing. A stonemason's error? I think not. Aigeis had not been able to supply a phylarch for 282/1—this law guaranteed that this

[6] On the dispute over the exact date of Athens' revolt from Demetrios of Macedon, see Shear, *Kallias of Sphettos*, pp. 14–15, n. 23; M. J. Osborne, "Kallias, Phaidros, and the Revolt of Athens in 287 B.C.," *ZPE* 35 (1979): 181–94; and Ph. Gauthier, "La réunification d'Athènes en 281 et les deux archontes Nicias," *REG* 92 (1979): 348–99. For our purposes, the issue is not really critical.

would not happen again. This reform is hardly a sign that the cavalry was viewed as militarily negligible. Moreover, it has been suggested that another inscription, *IG* II² 3210, which records the council honoring the ephebes and *kosmêtês* with a crown, belongs not to 296/5, but to 282/1 and can perhaps parallel the cavalry reform of that year.[7] Later, at the outbreak of the Chremonidean War in 267/6, ephebes were pressed into military service and used as garrison troops at the Mouseion.[8]

In 279 B.C. a large body of Gauls invaded Greece and approached the pass at Thermopylai. A number of Greek states sent military forces, including Athens. Pausanias (10.20.5; cf. 1.4.2) reports that the Athenians sent a fleet, 1,000 hoplites, and 500 horsemen under the command of Kallippos who was also appointed the supreme commander of the allied army. Part of this information is confirmed by an inscription, *IG* II² 680 (now dated to 247/6 or 246/5), concerning the games instituted to celebrate the salvation of Greece from the Gauls; lines 11–13 provide the detail that the Athenian hoplites sent were the *epilektoi* and that *hippeis* accompanied them. Unfortunately, no exact number is given for either group.

Pausanias' account has been called into question, particularly the curious role of the Athenian fleet in the narrative and the fact that Kallippos was appointed as supreme commander.[9] Even the number of Athenian *hippeis* has been disputed on the grounds that only with some difficulty had the Athenians increased their cavalry to 300 men just three years before.[10] Others, however, have found no reason to reject the reported size of the Athenian cavalry at Thermopylai.[11] Unless it can be shown conclusively that Athens could not have increased its cavalry once again in the intervening three years or that Pausanias is simply wrong in his troop count, we can accept that the Athenian cavalry numbered at least 500 in 279 B.C.

If we could confidently identify the *hippeis*, mentioned in an inscription honoring Komeas, Hipparch to Lemnos in 280/79, as Athenian horsemen and not Lemnian cleruchic horsemen,[12] we might suggest

[7] See Shear, *Kallias of Sphettos*, p. 29, n. 59.

[8] *IG* II² 665, lines 11–13; this inscription is now downdated from the *Corpus* (282/1) to 266/65 (B. D. Meritt, "Athenian Archons 347/6–48/7 B.C.," *Historia* 26 [1977]: 174).

[9] See G. Nachtergael, *Les Galates en Grèce et les Sôtéria de Delphes: Recherches d'histoire et d'épigraphie hellénistique* (Brussels, 1977), 142–45, with earlier literature.

[10] Habicht, *Untersuchungen*, p. 91 and n. 21.

[11] Ferguson, *HA*, p. 158; W. W. Tarn, *Antigonos Gonatas* (Oxford, 1913), 151; and Shear, *Kallias of Sphettos*, p. 29.

[12] *IG* II² 672, lines 21–25. For discussion, see Ferguson, *HA*, pp. 155–56 and Ch. Habicht, *Gottmenschentum und griechische Städte* (Munich, 1956; *Zetemata* 14), 89–

that the restoration of Athenian control over the island after the defeat of Lysimachos in 281 and the renewed dispatch of a hipparch necessitated an additional increase of Athenian *hippeis*, some of whom would have been expected to serve with this commander during his tenure on Lemnos. At any rate, no source mentions the exact (or ideal) number of Athenian *hippeis* serving in a given year for the rest of the Hellenistic period. But we do possess some related evidence.

In her analysis of the Kerameikos cavalry archives, Braun dated the tablets to the period after the cavalry reform in 282/1 and before Athens' defeat in the Chremonidean War (261). She detected a correspondence between the total number of different horsemen inscribed on the tablets, i.e., 256 from the Kerameikos and 24 from the Agora, and the expanded cavalry of 300.[13] Kroll, on the other hand, rightly criticized her methodology, as the tablets do not represent a single year, but a span of at least a decade. In addition, he dates the tablets to the mid-third century, in other words, after the Chremonidean War. According to his analysis, each of the twelve tribes in a given year can claim only a modest enrollment of less than 20 *hippeis*; therefore a total quota near the old 200, pre-282/1 mark. I quote him in full:

> One concludes that the tablets belong to a time when the strength of the cavalry had fallen well below the 300 men attested in 282/1 B.C. With only 14 or 15 men in each of twelve tribal contingents, the full strength would be in the vicinity of 170–180 men, which is close enough to 200 to suggest that the latter (involving a quota of 16 men per tribe) may have been the effective theoretical total intended. Having increased the size of the cavalry to 300 in 282/1 B.C. and apparently having anticipated a still greater increase in the future, [here Kroll seems to have missed the testimony of Pausanias on the cavalry at Thermopylai] the Athenians would hardly have allowed the corps to shrink to less than 200 during the ensuing period of national revival that culminated in the Chremonidean War. Accordingly, the reduction of the cavalry's size may be plausibly attributed to Antigonos Gonatas' stern treatment of the city upon the conclusion of the war in 261 B.C. (pp. 96–97)

Because of the unique nature of these cavalry archives, many questions arise. For one thing, I am not quite as confident as Kroll in that we possess enough tablets to give an accurate count on a tribal series; if

90. In a letter of May 14, 1984, Professor Habicht expressed the view that Athenian horsemen probably ordinarily accompanied the Hipparch to Lemnos, as may be the case here. See Appendix B.

[13] Braun, "Funde," p. 269, n. 44.

for some reason the Athenians dumped these tablets in two different locations, why not a third? Or how many were retained in the archives of the Hipparcheion and not discarded? But if he is correct, then his conclusions are compelling.

At this juncture, it might be useful to recap the evidence concerning the size of the Athenian cavalry in the third century. At the end of the fourth century, the Athenians recognized that imperial ambitions were out of step in an age of competing Hellenistic monarchs who could finance huge mercenary armies. A reasonable goal was to retain a modicum of independence and the ability to defend the homeland. The 1,000-man cavalry had been a grandiloquent statement of imperial pride, that had allowed Athens to dispatch respectable cavalry forces of 300 to 600 men to distant battlefields and yet retain sufficient horsemen at home to defend Attika. And it was expensive, forty talents a year for the grain allowance. Now it was adequate to possess a modest cavalry force of 200 men to be available to protect and patrol the frontier and the coasts, to keep a watchful eye on the slaves in the silver mines at Laureion, to be prepared to act quickly against displaced, marauding mercenaries and pirates,[14] and to lend a hand in case of actual invasion. The forts at Eleusis, Panakton, Phyle, Sounion, Rhamnous, and of course, the Peiraeus, assume great strategic importance in the third century B.C.,[15] whether they be garrisoned by Athenians, mercenaries, or Macedonians. The Athenian cavalry was a complement to this defensive network; by the nature of its mobility, it was expected to secure the surrounding countryside.

But again, military considerations played a key role. In the revolt of the 280s Athens saw the need to maintain a larger cavalry force, not of imperial proportions, but one to contain the annoying Macedonian garrison at the Peiraeus and probably also the mercenaries stationed in the border forts. The Athenians therefore increased the cavalry to 300 men by 282/1, and to (at least?) 500 by 279 B.C., though this last may have been purely an emergency expeditionary force, not meant to be maintained after the crisis. We cannot know. In any event, with the disaster of the Chremonidean War (267–261), if we follow Kroll, it reverted to its pre-282/1 level, i.e., 200 men.

The Athenian cavalry probably never dipped much below that level

[14] Used by Antigonos Gonatas to harass Attika during the Chremonidean War, 267–61 B.C.: *SEG* XXIV.154; *CAH*² VII. 1, p. 238.

[15] The epigraphical evidence from these forts is relatively abundant in this period; for discussion, see J. R. MacCredie, *Fortified Military Camps in Attica* (Princeton, 1966; *Hesperia Supplement* 11); J. Pouilloux, *La forteresse de Rhamnonte* (Paris, 1954); Habicht, *Studien*, pp. 43–63; and Heinen, *Untersuchungen*, pp. 159–67.

for the rest of the Hellenistic period. We do know that in the Pythais of 106/5, there were about ninety *hippeis* in attendance commanded by one hipparch, one tarantinarch (see below), and two phylarchs (*FD* III.2, no. 28). If we knew the method of selection, we might be able to estimate the total number of Athenian *hippeis* in that year on the basis of the relative size of the officer staff, representing as it does, only a small percentage of the cavalry officers, i.e., two hipparchs, at least two tarantinarchs (see *IG* II² 958.I.57–59), and twelve phylarchs. Certainly it is safe to say that a good number of Athenian *hippeis* must have remained at home. On general principles, if one claims that the Athenian cavalry numbered much below 200 in this period, there would seem to be an overabundance of cavalry officers—too many chiefs and too few Indians, so to speak. If we look at the tally for the Pythais of 128/7, it becomes even more obvious how careful we must be with this data: for approximately sixty *hippeis* in attendance, the officer staff included one hipparch, two tarantinarchs, and four phylarchs (*FD* III.2, nos. 24, 27, 46).

The Athenian Cavalry in Defense of Attika

The primary reason for the existence of the Athenian cavalry in the Hellenistic period remained the same as in earlier periods: to act as a mobile home guard to protect Attika. It is this military *raison d'être* that I think Ferguson slights unfairly. In the struggle of the great powers, Athens found itself frequently sandwiched in between.

In the 240s, it became necessary for the Athenians to solicit emergency contributions to survive an economic and military crisis. According to a contemporary inscription, the military treasurer, Eurykleides of Kephisia, collected approximately four to six talents for the emergency fund.[16] Some of this money probably found its way into the hands of the Athenian *hippeis*, primarily for grain allowances—we note with special interest the stated reason for the contributions: "for the salvation of the city and the protection of the countryside" (*[eis tên sôteria]n tês poleôs kai tên phylakên tês chôras*: lines 16–17, 30–31). It requires no temerity to suggest this. Habicht has calculated that no fewer than twenty-four of the sixty-five citizens named in the inscription are attested in the mid-third century cavalry tablets.[17] Therefore,

[16] *IG* II² 791 and B. D. Meritt, "Greek Inscriptions," *Hesperia* 11 (1942): 287–92, no. 56, now dated to 245/4 (Meritt, "Mid-Third-Century Athenian Archons," *Hesperia* 50 [1981]: 78–79), or to 244/43 (Nachtergael, *Les Galates*, p. 77); for the tally, see Habicht, *Studien*, pp. 26–33; *SEG* XXXII.118.

[17] *Studien*, pp. 32–33; *SEG* XXXII.118.

almost half of the contributors were either retired or actively serving *hippeis*.

In the 230s, as an ally of Demetrios II, successor to Antigonos Gonatas on the Macedonian throne, Athens suffered its Thriasian and Eleusinian plains to be ravaged by Aratos and the forces of the Achaian League. And from their position in neighboring Megara, the Achaians raided and destroyed year after year, reminiscent of the Archidamian phase of the Peloponnesian War or of the occupation of Dekeleia. To make matters worse, the Aitolian League, ally of Aratos, harassed the coasts of Attika. In 232 B.C. Boukris of Naupaktos raided the Attic coast and carried off a number of Athenians to be sold on the slave market in Crete.[18]

Even though the Athenian cavalry does not figure prominently in the events of the 240s and 230s (at least in our meager sources), we simply cannot assume that they did not "render good service in the countryside" (*tas chreias parechôntai têi chôrai*) for which the hipparchs and phylarchs earned praise in the cavalry inscription of 282/1 (lines 10–11).

After the Athenians succeeded in removing the last of the foreign garrisons in 229 B.C., under the leadership of Eurykleides and his brother Mikion, they apparently began a concerted effort to strengthen their military preparedness. Detachments of Athenian troops served continuously in garrison duty at the border forts, and the city walls of Athens and the Peiraeus were repaired. Just as in the post-Chaironeia period in Athens, we may be sure that the cavalry did not go unnoticed, but there is no evidence of any attempt to increase its size again. In fact, we possess an oblique piece of evidence that may just hint at the continuing modest size of the cavalry.

Around 217/6 B.C., an inscription records a personal gesture of munificence towards the cavalry.[19] Theophrastos, one of the hipparchs of 220/19, was crowned by the council, the *demos*, and the *hippeis*. Apparently, Theophrastos, while hipparch, took special care of the cavalry by providing maintenance support for all the horsemen (*[hip]peas pantas hippotrophôn*, lines 14–15):[20] and presented a gift

[18] *IG* II² 844, lines 4–7; for discussion of the events of the 230s, see Ferguson, *HA*, pp. 200–204; Mossé, *Athens in Decline*, pp. 132–35; and *CAH*² VII.1, pp. 446–53.

[19] See *IG* II² 1303; W. S. Ferguson and S. Dow, "The Decree of the Athenian Garrisons Honoring Theophrastus, *IG* II² 1303," *Hesperia* 2 (1933): 447–49; A. Wilhelm, *Theophrastos, Hipparchos der Athener* (Vienna, 1946; *Sonderabdruck aus dem Anzeiger der phil.-hist. Klasse der Akademie der Wissenschaften*, no. 11), 115–27; L. Moretti, *Iscrizioni storiche ellenistiche*, vol. 1 (Florence, 1967), no. 31, pp. 69–71; *SEG* XXV.157.

[20] Moretti, *Iscrizioni*, pp. 70–71, suggests that this is probably just a reference to the

to the *lochagoi* (?) of ten minas (= 1,000 drachmas) so that the horsemen would be equipped with arms in the finest way (*tois hoplois ôs arista kateskeuasmenoi*, lines 15–16).[21] By way of comparison, for the fourth century, Xenophon remarks in his *Hipparchikos* (1.22–23) that whereas the *hippeis* were expected to arm themselves from their pay (*misthos*),[22] it was advised that the phylarchs arm their *prodromoi* in the best possible way (*kosmêsais . . . hoplois ôs kallista*, 1.25) with their own money. And Xenophon recommends that the hipparch have available extra leather straps to offer to his horsemen should their bridles or saddlecloth straps break in the field (*Hipparch.* 8.4). To return to the point, it is hard to imagine Theophrastos making such a generous gift if the Athenian *hippeis* numbered many more than a couple of hundred. Of course without additional information about this munificence, we can only speculate.

In 200 B.C., after a raid by the troops of Philip V, king of Macedon, into Attika and the seizure of some of their ships, the Athenians prepared to break the neutrality which they maintained with effort since 229 B.C. An embassy was dispatched to Rome. The Romans sent a legation to Athens, and Attalos, king of Pergamon, enemy of Philip, hurried to the Peiraeus to join them. Polybios records a marvelous scene in which the magistrates of Athens, the Athenian *hippeis*, and the Roman legation escorted the king in great pomp from the Peiraeus to Athens (16.25.5). Won over by the entreaties of Attalos and the ambassadors from Rhodes, the Athenians declared war on Philip in 200 B.C.

While the Roman envoys tarried in Athens, Philip ordered Nikanor, his general, to invade Attika. He had penetrated as far as the Akademy when the Romans sought an audience with him and delivered an ultimatum to convey to his king. Nikanor withdrew his army, but Philip

hipparchs' role in supplying the horsemen with their *sitos* allowances (cf. Threpsiades-Vanderpool, p. 104, lines 19–20), but it seems to be more than that. Of course, if he had really supplied the *sitos* allowances to "all the horsemen," it would have entailed a huge sum, even for a cavalry of 200 men (about twelve talents). But it need not mean for the whole year nor for the total allowance sum—understood may be a supplementary grain gift or one for only part of the year. At any rate, his generous grant for cavalry equipment in the next line (15–16), cannot have been customary practice.

[21] J. H. Kroll, "Some Athenian Armor Tokens," *Hesperia* 46 (1977): 141–46, pl. 40, suggests that the nine lead tokens depicting various military arms found with the third-century cavalry tablets may have been provided to some of the *hippeis* by the state to be used to requisition appropriate arms for cavalry service. Also, he suggests that *lochagoi* may be a slip for *phylarchoi* (p. 146, n. 15).

[22] Kroll, "Athenian Armor Tokens," p. 145, believes that here *misthos* is a reference to the *katastasis*.

rejected the terms. The Second Macedonian War (200–197) had begun.

Livy (31.16.1; 24.26) records the military operations of Philip and his general Philokles in Attika. Basically, the Macedonians expended their efforts in unsuccessful attempts to take Athens and the forts at Eleusis and the Peiraeus. The Athenians elected to remain behind the walls, with good reason. Early in the campaign, a force of Athenians, garrison troops of Attalos, and their own mercenary force under the command of Dioxippos, marched out of the Dipylon Gate to engage Philip. Philip, mounted, led his army to victory and drove the Athenians back inside the gate. Philip then proceeded to pillage, destroy, and burn the gymnasia areas at Kynosarges and the Lykeion. Unable to secure further success, Philip went off to the Peloponnesos in search of allies (31.25). Upon his return, he and Philokles continued to plunder the Attic countryside and assault the fort at Eleusis. The only bright moment for the defenders occurred when Philip decided to test the defenses of the Peiraeus. Between the old and now delapidated Long Walls which had once linked Athens to the port, Philip was defeated by a sudden infantry and cavalry sally from the city (31.26.8), a temporary setback for the king. Having failed to take any strong place in Attika, Philip contented himself by systematically destroying and burning temples, shrines, and statuary in the country demes, an act for which he was soundly censured by Livy. After this, Philip departed.

Philip's war in Attika is reminiscent of the military operations in the Peloponnesian War. The Athenians bided their time behind the city walls, not chancing a (second) major engagement with what was viewed as a superior force, 5,000 light infantry and 300 horse[23] commanded by Philip (31.24.1–2) and 2,000 infantry and 200 horse under Philokles (31.16.1; 26.1). A sudden sally, however, could win a small victory, just as in the earlier war the Athenian cavalry had harassed the Peloponnesian army and made it dangerous to approach too near the walls. But the final events of 200 B.C. made one thing clear: the enemy controlled the countryside. The *hippeis* of Hellenistic Athens were too few in number to seriously contest the movements of a major army, an army which included a numerically superior cavalry. Hit and run, harass, single out the enemy foragers—these modest but important military actions we may suspect the Athenian cavalry carried out against Philip V.

The wars with the Macedonian kings would not end with Philip's defeat at the hands of the Romans at Kynokephalai in Thessaly in 197

[23] Ferguson, *HA*, p. 274, records 3,000 cavalry, but this must be a typographical error.

B.C., but Athens' enduring friendship with Rome would insure that no major Hellenistic power would again invade Attika or garrison its forts. Favored with peace in the second century B.C., it is not surprising to find the activities of the Athenian cavalry attested primarily—at first glance—in nonmilitary ways, i.e., in the many festivals and processions.

Athens under Roman Protection

The epigraphical evidence for the participation of the Athenian cavalry in the agonistic festivals of this period is quite full, particularly after 166 B.C. In the Panathenaia of ca. 166/5, we observe competitions strictly for the cavalry, i.e., for the phylarchs: the double-distance war-horse race in armor; the same race without armor; and the single-distance horse race. And for the *hippeis*: the war-horse race and the double-distance and single-distance horse races.[24] In the Theseia of 161/60 and every four years thereafter into the 140s, the events for the cavalry were numerous, including "the most manly" (*euandria*) and "the best equipped" (*euhoplia*) of the tribal squadrons,[25] along with double- and single-distance war-horse races and simple horse races for both phylarchs and *hippeis*, and a torch race for the latter.[26] They even opened up the equestrian competition to noncavalry members in the single- and double-distance horse races, the hurling of the javelin while mounted, and perhaps the event for "best-looking horse" (*lampros hippos*), to be interpreted as "most showy performance," e.g., in the procession[27]—good evidence that in any age, there were always those who actively engaged in *hippotrophia*, but who did not actually belong to the cavalry.

In the Pythaides of 138/7, 128/7, 106/5, and 98/7, ephebes and members of the cavalry (at least for the last three years) escorted the

[24] *IG* II² 2316, lines 28–41; see Martin, *Cavaliers*, pp. 227–58.

[25] *IG* II² 956.I.56–60; 957.I.32–41; 958.I.53–59; 960.I.19–24; 961.I.19–24. On these events, see Martin, *Cavaliers*, pp. 211–25 and Pélékidis, *Ephébie*, pp. 229–33. I am in the process of preparing a new study of these Theseia inscriptions.

[26] *IG* II² 956.II.79–88; 957.II.64–79; 958.I.67–68, II.85–89; 960.I.33–34; 961.I.34–35.

[27] Cf. Xen. *Peri Hippikes* 11.1; for the second century B.C., see *IG* II² 956.II.88–91; 957.II.80–83; 958.II.91–96; for *lampros hippos*, see *IG* II² 956.II.87–88; 957.II.64; 958.II.79–80. For two cases of Athenians who won equestrian events in the "open competition" (*ek pantôn*) in the Pythaids, yet do not appear in the lists of *hippeis*, see Theopeithes Theophilou (*FD* III.2, no. 41; 128/7); and Straton Boulonos (*FD* III.2, no. 42; 128/7).

procession to Delphi and then competed in various equestrian events.[28] These competitions included the double-distance war-horse with and without armor for both the phylarchs and horsemen, single- and double-distance horse races, and javelin throwing. Again, some of the equestrian events were open to noncavalry participants.

Of course, all of this agonistic activity could be interpreted as having no legitimate military purpose. After all, the period between 168 and 86 B.C. was one of peace and prosperity for Athens. Athens had reaped obvious and tangible benefits from its loyalty to Rome during the Third Macedonian War (171–168): the reacquisition of its lost cleruchies, Lemnos, Skyros, and Imbros, as well as the outright possession of the sacred island of Delos.

Yet, while no Athenian misunderstood the international situation nor forgot that Rome had made all of this possible, one cannot help but feel that the return of the old "imperial provinces" stirred the Athenians with memories of their glorious past. And Rome's military presence had not yet come permanently to Greece (Macedonia became a Roman province in 148 B.C.) and Athens assumed the status of Rome's special friend in the east. That is why, I believe, we now have a proliferation of quasi-military events in the festivals, particularly in the Theseia, not only for the cavalry, but for the *epilektoi* and the mercenaries and the ephebes, and even mock hoplite-fighting for youths. E. N. Gardiner has commented on this heightened interest in military displays:

> There is a certain pathos in the existence of these elaborate military reviews and competitions at an age when Athens had no more any freedom to defend, and when her military service was of no practical value. It may be that with the loss of the reality she clung the more closely to the empty form and semblance of an army. But it seems to me more probable that these competitions were not the futile invention of her decadence, but were the survival of the great outburst of patriotism and militarism in the fifth century.[29]

[28] *FD* III.2, nos. 23–26 (ephebes), 26–28, 46 (*hippeis*), 34–35 (victory crowns). For discussion see *FD*, pp. 1–46; A. Boethius, *Die Pythais* (Uppsala, 1918); G. Daux, *Delphes au IIe et Ier siècle* (Paris, 1936); S. V. Tracy, "Notes on the Pythais Inscriptions," *BCH* 99 (1975): 185–218 (with convenient tables of participants for each Pythais, pp. 215–18); and Tracy, *First Fruits*.

[29] *Greek Athletic Sports and Festivals* (London, 1910), 247. For the argument that increased military influence resulted in the addition of new "military" events to the agonistic program of the major Athenian festivals, e.g., Panathenaia, Olympieia, and possibly Theseia, in the late fifth and early fourth centuries, see Kyle, *AAA*, pp. 50, 186–90, *passim*.

To be sure, Athens no longer enjoyed the imperial stature of that bygone age, but it did now control extraterritorial possessions which required administering and protecting, not to speak of Attika itself. And O. W. Reinmuth has argued that the *ephebeia*, after the 180s and at least to the sack of Athens by Sulla in 86, emphasized the military branches of instruction: "The instruction which was given by the hoplomachos, akontistes, toxotes, and katapaltaphates was still primarily for practical and military use and not, as later, for primarily educational purposes."[30]

With respect to the cavalry itself, we can detect signs to confirm this trend. First of all, the Hipparch to Lemnos reappears after 168 B.C., though now apparently subordinate in command to a *Strategos* to Lemnos.[31] Secondly, the Theseia and Pythais inscriptions attest a reorganization of the Athenian cavalry with the introduction, probably in the second century, of additional cavalry officers called tarantinarchs, and the *hippeis* under their command, *tarantinoi*. In 153/2 or 149/8, we find that the winning tribe of the "best-equipped" category was that of Aiantis, commanded by the two tarantinarchs, the brothers Euthoinos and Polynikos, sons of Moschion of Marathon.[32] In two later Theseia inscriptions from the 140s, the *tarantinoi* compete in a torch race, Andron Zoilou of the tribe Pandionis being victorious in the second.[33] In the Pythais of 128/7, the two tarantinarchs, Diogenes Aropou and Lyson Demokratous, accompanied the procession to Delphi;[34] in 106/5, one tarantinarch, Lysander Glaukou, was in attendance.[35] The *tarantinoi* are represented by one Kallistratos Kallistratou, who won the war-horse race *ek tôn tarantinôn* in the Pythais of 128/7.[36] The Athenians apparently adopted a popular military component of Hellenistic armies, that of light-armed cavalry which perhaps origi-

[30] "Ephebic Texts from Athens," *Hesperia* 30 (1961): 15; for full discussion of the epigraphic evidence in the late second century, see Pélékidis, *Ephébie*, pp. 183–209.

[31] *IG* II² 1224, frag. a, lines 10–11, 17; *SEG* XXVIII.104; for two inscriptions dated after 167/6 by reference to a hipparch (restored) to Myrina, one of the two cleruchies on Lemnos, see M. Crosby, "Greek Inscriptions," *Hesperia* 6 (1937): 457–61, no. 7, line 17; no. 8, line 6. See Appendix B.

[32] *IG* II² 958.I.56–59. The corpus dates the archonship of Phaidrias to 155/4, Pélékidis, *Ephébie*, pp. 295–300 to 153/2, but B. D. Meritt, *Hesperia* 33 (1964): 205–206, and reaffirmed in his list of archons (*Historia* 26 [1977]: 184), dates his archonship to 149/8. I think there are some prosopographical problems with Meritt's redating, but that issue, not especially critical for my present purpose, awaits the completion of my study of the Theseia.

[33] *IG* II² 960.I.31–32; 961.I.34–35.

[34] *FD* III.2, no. 24, line 16; no. 35, lines 3–4; no. 46, lines 1–2, 6.

[35] *FD* III.2, no. 28, lines 7–8.

[36] *FD* III.2, no. 37.3.

nated at Tarentum in southern Italy, to complement the increasingly heavy-armored regular horsemen.[37] Now why would the Athenians have troubled themselves with such a reform of the cavalry organization if they found the cavalry to be so militarily "negligible"? For variety in festivals and processions? Surely not.

In fact, the Athenian cavalry may have provided military assistance at the end of the second century B.C. when the slaves at the silver mines at Laureion revolted, slaughtered the guards, seized the acropolis at Sounion, and ravaged Attika "for some time" (*epi polun chronon*).[38] This slave revolt has been dated to the years just prior to 100 B.C.[39] It must have interrupted the silver production, still important for the so-called New Style Silver coinage of Athens, which confirms its prosperity in this period.[40] In addition, the eastern Mediterranean was being victimized by pirates who had been operating more or less at will out of their bases in Cilicia in southern Asia Minor during the second half of the second century.[41] In 102 B.C. the Roman Senate was forced to send M. Antonius to try to suppress them. But Tracy argues that the former event and the operations of the pirates may not be directly connected; if this is true, then the slave revolt in Attika need not be dated closely to 102 B.C., but can have occurred a few years later. Noting that only five *hippeis* accompanied the Pythais of 98/7, yet almost 100 of their number had been present at the previous Pythais of 106/5,[42]

[37] See Martin, *Cavaliers*, pp. 418–23; G. T. Griffith, *The Mercenaries of the Hellenistic World* (Cambridge, 1935), 246–51; and Snodgrass, *AAG*, pp. 128–29.

[38] Poseidonios in Athenaios 6.272 E–F (= *FGrH* 87 F 35).

[39] For discussion, see S. V. Tracy, "Athens in 100 B.C.," *HSCP* 83 (1979): 232–35.

[40] The basic study is M. Thompson, *The New Style Silver Coinage of Athens*, 2 vols. (New York, 1961). I follow the lower chronological dating, i.e., starting ca. 165 B.C., of the coinage as suggested by D. M. Lewis, "The Chronology of the Athenian New Style Coinage," *Num. Chron.*, 7th ser. 2 (1962): 275–300; and defended by H. B. Mattingly, "Some Third Magistrates in the Athenian New Style Silver Coinage," *JHS* 91 (1971): 85–93 and his "Some Problems in Second Century Attic Prosopography," *Historia* 20 (1971): 26–46.

[41] On the piracy laws of 101 or 100 B.C. found in inscriptions from Delphi and Knidos, see *FD* III.4, no. 37 and M. Hassall, M. Crawford, & J. Reynolds, "Rome and the Eastern Provinces at the End of the Second Century B.C.," *JRS* 64 (1974): 201–7; for translation, bibliography, and brief commentary on these important documents, see R. K. Sherk, *Rome and the Greek East to the Death of Augustus* (*Translated Documents of Greece and Rome*, no. 4), no. 55, pp. 59–66.

[42] *FD* III.2, no. 26.III.14–18; no. 45 (98/7) in contrast to *FD* III.2, nos. 28, 43, 44 (106/5). Though Tracy ("Athens in 100 B.C.," p. 233) has assumed that those sixteen soldiers under the heading of *systratiôtai* (no. 28.III.41–44, IV.27–39) were also members of the cavalry, the heading "Hippeis" with additional names immediately following in column IV indicates that they are to be differentiated. Professor Tracy has acknowl-

Tracy compellingly suggests that in 98/7 most of the *hippeis* "remained behind in Attica to help ensure the safety of the countryside."[43] Faced by an endemic problem with pirates who could suddenly sweep down upon coastal communities and now a mutinous slave population, Athens naturally turned to its standing mobile force, the cavalry.

Athens and Sulla

In 88 B.C. Athens committed a disastrous error in judgment—it cast its lot with Mithridates VI, king of Pontos, in his efforts to remove Roman influence from the eastern Mediterranean. In a confusing series of events,[44] first Athenion and then Aristion assumed power in Athens, the former with the support of the *demos* and the latter with the aid of 2,000 Pontic mercenaries provided by Mithridates. Archelaos, his general, proceeded to garrison the Peiraeus. Sulla, the Roman commander, then lay both Athens and the Peiraeus under siege through the winter of 87/6. In the following spring, he finally penetrated the defenses of Athens and allowed his troops to massacre large numbers of the now destitute and famine-stricken Athenians. Archelaos withdrew from the Peiraeus, which Sulla subsequently destroyed. The two generals soon met in Boiotia at Chaironeia and Orchomenos. Sulla's victories proved the end of Mithridates' grand campaign in Greece.

To the remaining Athenians Sulla granted freedom, but abolished their electoral prerogatives (App. *Mith.* 38). The tyrannies of Athenion and Aristion had created a number of pro-Roman exiles (*Mith.* 28), like Medeios, an escort *hippeus* in the Pythais of 106/5,[45] and Kalliphon, who would intercede with Sulla to spare the city total depopulation (Plut. *Sulla* 14.5). We can only guess how many of the exiles belonged to the cavalry. Poseidonios' lengthy account of Athenion's reign of terror (Athenaios 5.214 A–D) definitely leaves the impression that the oligarchic party, which had governed Athens for a number of years prior to the Mithridatic War and whose leaders belonged to the

edged this probability *per litteras*. The *systratiôtai* were probably infantry attached to the cavalry for this occasion.

[43] "Athens in 100 B.C.," p. 233.

[44] For the basic primary accounts for Athens in the Mithridatic Wars, see App. *Mith.* 28–41, Plut. *Sulla* 12.1–3; 13–14; and Poseidonios in Athenaios 5.211 E-215 B. For discussion, see Ferguson, *HA*, pp. 440–59; E. Badian, "Rome, Athens, and Mithridates," *AJAH* 1 (1976): 105–28; Ch. Habicht, "Zur Geschichte Athens in der Zeit Mithridates' VI," *Chiron* 6 (1976): 127–42; and now Sherwin-White, *Roman Foreign Policy*, pp. 134–38.

[45] *FD* III.2, no. 28.IV.51 (see above); for full prosopography, see Tracy, *First Fruits*, p. 210.

propertied classes with pro-Roman leanings, were the principal targets of the purge. When many of them tried to escape Athens, Athenion dispatched his hippic bodyguard in pursuit, horsemen whom Poseidonios specifically identifies as "fully-armored cavalrymen," the so-called cataphracts (5.214 A). I suspect that these horsemen were not Athenians; cataphracts are unattested among the Athenian *hippeis*, but are well-known in the armies of the eastern kings.[46]

The only other possible references to Athenian cavalry occur when Sulla ambushed a body of infantry and horsemen from the Peiraeus who had ventured out with the plan of catching Sulla by surprise (*Mith.* 31) and when Sulla ordered a ditch dug around his camp at Eleusis to protect his position from the enemy horsemen (*Mith.* 33). But there is no reason to believe that these horsemen were not simply members of Archelaos' army stationed at the Peiraeus, particularly as the cavalry arm was viewed as one of the chief strengths of the Mithridatic army (Plut. *Sulla* 15.2). Among them may have been Athenian *hippeis*, but surely only a token force. The political crisis created by Athenion and Aristion saw to that.

We may be certain, however, that Sulla's murderous troops did not pause to inquire as to their victims' political or military commitment to Mithridates' cause. This was a disaster for the Athenians collectively, an irreparable blow to their "imperial" exuberance of the past 100 years and a fitting end to our study of the Athenian cavalry. We cannot know what changes were made to the Athenian cavalry in the post-Sullan period, but as we might expect with the returning oligarchic exiles, it did continue to exist. At the end of the first century B.C., the *hippeis* honored Echedemos Mnesitheou Kydathenaieus for his excellence and *philotimia* towards them and the Athenian *demos*.[47] Even more interesting is a certain Sosis Sosibiou Oethen who was only an ephebe in 38/37, yet is also described as having served as gymnasiarch, phylarch, and choregos for three days.[48] If this is correct, then we have finally entered an era when the Athenian cavalry was militarily negligible.

[46] On the origins of the cataphracts, see W. W. Tarn, *Hellenistic Military and Naval Developments* (Cambridge, 1930; Ares repr., Chicago, 1984), 73–83. Antiochos III of the Seleucids had 3,000 cataphracts in his army that met the Romans at Magnesia in 190 or 189 (Livy 37.40.5, 11; cf. 35.48.3). His eldest son Antiochos commanded the cataphracts at the battle of Panium in 200 B.C. (Polyb.16.18.6) and his younger son Antiochos IV Epiphanes included cataphracts in a splendid display at the games at Daphne in 166 B.C. (Polyb. 30.25.9).

[47] *IG* II² 3000.

[48] *IG* II² 1043, lines 1, 65. A. W. Gomme and F. H. Sandbach, *Menander. A Commentary* (Oxford, 1973), 546–47, accept this evidence.

Prosopographical Considerations

Up to this point we have concentrated our efforts on administrative and military considerations. But inasmuch as people are eminently more interesting than institutions, it behooves us to talk a little about the *hippeis* themselves, particularly as the third and second centuries B.C. offer us decent prosopographical documentation. But we should not forget the limitations, already noted in the previous chapter. The data are irritatingly insufficient to allow broad, sweeping conclusions about the patterns of change in the personnel of the Athenian cavalry from the Classical into the Hellenistic period. But certain profiles can be drawn on the basis of the available evidence and we can speak intimately of at least some of the *hippeis*.

Many of the faces are new to us, but the profile of an Athenian *hippeus* remains pretty much the same. There can be no doubt that eligibility or liability, considering the fact that the mechanisms of selection remained intact, was determined by wealth and physical fitness. But because the required quota of personnel was now only one-fifth to one-third that of the Classical period, the task of recruitment was made easier. Also, these *hippeis* probably served in the cavalry, on the average, longer than their Classical counterparts because of interest and tradition. With the following data, I offer a paradigmatic selection of Athenian *hippeis* from the Hellenistic period; a more complete register with select bibliography for these men and for the rest of the identifiable *hippeis* from the Archaic to the Hellenistic period can be found in the *Catalogus Hippeum* at the end of this book.

Of those who served in the cavalry as *hippeis* or cavalry officers as part of a full military and civic career—characteristic of certain men of property and influence in any period—we might mention Aristophanes, phylarch ca. 238/7, but also gymnasiarch and *strategos*; Glaukon Eteokleous Aithalides, phylarch (whose tribe Antigonis won the *anthippasia* in the Olympieia and the Panathenaia), hoplite general, and agonothete, ca. 280/79, and brother of Chremonides who gave his name to the Chremonidean War; Epichares, hipparch in the 270s and *strategos* in the Chremonidean War, and known to us by a lengthy inscription recently unearthed at Rhamnous; Theophrastos, already mentioned as a munificent hipparch for 220/19 and also a gymnasiarch and *strategos*; Thrasykles of Dekeleia, phylarch for 282/81 and agonothete in 271/70; Kallisthenes Kleoboulou Prospaltios, phylarch, hipparch, and *strategos* in the 230s. From the later period, we mention three men with powerful Delphian and Delian connections: Medeios II Medeiou I Peiraeus, escort *hippeus* for the Pythais of 106/5; Sarapion II Sarapionos I Meliteus, his fellow escort *hippeus* in 106/5 and twice hoplite general; and Menophilos Lykophronos of Sounion, possibly *hippeus* in two Pythaids, 128/7 and 106/5, confirming a lengthy cavalry career.

This sampling is about what we would expect from the propertied families of Athens, particularly for those elected to high office. But the reason why I continue to caution against expecting too much prosopographically can best be illustrated by the cavalry inscription of 282/1 B.C. There we have recorded a complete catalog of the twelve phylarchs (except for one, see above), yet for over half of them we know next to nothing, e.g., Antimachos of Acharnai, Kalliades of Steiria, Kallidemos of Euonymon, Lysippos of Kettos, Python of Alopeke, and Charias of Kephale. And the other cavalry inscription, dated 286–261 B.C.,[49] provides a list of the two hipparchs and the twelve phylarchs—unfortunately, the left-hand margin of column 1 containing the names of six of the phylarchs is broken away, making identification of these men impossible.[50] But of the remaining officers, we know nothing about the hipparchs, Diokles of Teithras and Philion of Phrearrhioi, and nothing significant about the three phylarchs, Leon of Pallene, Sosigenes of Myrrhinous, and Sosistratos of Sounion. Yet all of these *hippeis* must be presumed to have been wealthy and well-known figures in their own time. To us now they are just names.

We do have some informative evidence in other areas, for instance, that of continuity. Cavalry service, as we know, assumes a certain level of wealth and station; therefore, attestation in the cavalry in the Hellenistic period of "liturgical" families known from the Classical period, would prove that the families had not only not died out, but had retained their social and economic position. For example, the following: (1) Hagnias of Oion, phylarch ca. 286–261 B.C., belongs to the famous family of Bouselos from the fourth century; (2) an Alkibiades, recorded as a *hippeus* on a mid-third century cavalry tablet, may be linked with the famous Alkibiades; (3) Konon of Alopeke, *hippeus* in the mid-third century, may be a descendant of either the Alkmeonidai or of the famous Konon of the fifth and fourth centuries and also perhaps the son or the father of Hippokrates, attested as a *hippeus* on another tablet of the same period—a father-son combination, if true, would place the age of the father in his forties at least; (4) Kleochares,

[49] Threpsiades-Vanderpool, no. 2, pp. 109–11.

[50] My examination of the stone in the apotheke of the Library of Hadrian in 1977 convinces me that the *nomen* of the fifth phylarch listed in column 1 of Threpsiades-Vanderpool, no. 2, p. 110, line 19, has been incorrectly published as [---]os Oethen; it should read [---]ôn Oethen.

hippeus in the mid-third century, was the grandson of Kleophanes Glauketou of Kephisia, trierarch in 341/40, 334/3, and 325/4; (5) Olympiodoros of Anaphlystos, *hippeus* in the mid-third century, was a member of a well-attested fourth- and third-century family; (6) Prokleides of Aphidna, *hippeus* in the mid-third century, was the grandson of a trierarch ca. 323/2 B.C.; and (7) Sostratos of Lamptrai, *hippeus* in the mid-third century, was probably the grandson of Sostratos Deinippou Lamptreus, trierarch several times in the 320s.

For specific family traditions of cavalry service, we can name (1) Antikrates, *hippeus* in the mid-third century, whose ancestor was Antikrates Sokratous Hermeios, a hipparch in the fourth century; (2) Theomedes of Eleusis (mentioned in Chapter Five), a phylarch, ca. 286–261, whose father was Theogenes Theomedous Eleusinios, hipparch on Salamis, ca. 320 B.C.; and (3) Nikomachos of Lamptrai, *hippeus* in the mid-third century, whose father was perhaps Kineas Nikomachou Lamptreus, hipparch, *strategos*, and trierarch, sometime between 350 and 270 B.C.

Then there are the families with strong cavalry and agonistic traditions during the Hellenistic period, reminiscent of the Alkmeonidai and Philaidai. (1) The family of Habron of Bate was particularly illustrious: Ophelas, son of Habron III, was hipparch at the Theseia of 157/6, and his son Drakon won the *hippos lampros* competition in the same festival. Habron's great-grandson, Kallias VI, son of Habron IV, was a *hippeus* in the Pythais of 106/5; (2) Thoukritos, *hippeus* in the mid-third century, was the son of Alkimachos Kleoboulou Myrrhinousios, hipparch in 282/1; (3) Achaios Kekropidos was a *hippeus* in the Pythais of 106/5 and his father won four chariot events at the Theseia of 157/6; (4) Glaukos II Lysandrou I Peiraeus was a polemarch in the Pythais of 128/7; his son Lysander Glaukou was one of the tarantinarchs of 106/5. The Glaukos Hippothontidos attested as a *hippeus* in 128/7 belongs to the family, but to avoid multiple military offices, he probably should be identified as a homonymous son of the polemarch; (5) Eubios Telesidemou, phylarch in the Theseia of 157/6, may have been the son of Telesidemos Aminiou Hekalethen, twice Hipparch to Lemnos, ca. 166 B.C. (but there are problems with this last date, see Appendix B); (6) of special interest is Eukles III Herodou I Marathonios, phylarch in the Pythais of 106/5, whose descendant in the fifth generation was the famous second-century A.D. sophist and philanthropist, Herodes Attikos; and (7) the very famous hippotrophic family of the brothers Eurykleides and Mikion, dominant political figures in Athens in the last three decades of the third century B.C., whose descendants included Eurykleides II and his son Mikion IV, both chariot winners at the Panathenaia of 182/1 and 178/7, though no members of the family are actually attested as cavalrymen; (8) in the same Panathenaia, one of the other great Athenian families of the third and second century placed two of its members, Arketos I and Mnesitheos III, sons of Echedemos III, in chariot victories. Arketos' son, Echedemos IV, was a multiple winner in open equestrian events in the Theseia of 153/2 or 149/8, as well as being a cavalryman. A descendant, Echedemos V Mnesitheou, was praised by the *hippeis* in the Augustan era (see above). (9) Theodoros Charikleous of Phaleron was a winning phylarch in the Theseia of 157/6 and his son or cousin, Charikles, served as a phylarch in the Pythais of 128/7; (10) the well-documented family of Leon and Kichesias of Aixone supplies Kichesias III Leontos as the orator of the decree of the *hippeis* in honor of their hipparch in 188/7. His grandson, Kichesias IV, was a *hippeus* of the Pythais of 128/7, while Leon, *grammateus* of the horsemen in the same Pythais, may perhaps be his son, Leon IV; (11) Arkesas, *hippeus* in the mid-third century, was the son of Nikogenes Arkesantos Euonymeus, hipparch in 282/1; (12) Nikogenes Nikonos of Philaidai, agonothete of the Theseia in 161/60, returned as hipparch in 157/6. His son, Lyander, was a winner of *lampros hippos* in 161/60; Lyander's son, Nikogenes, was a *hippeus* in the Pythais of 106/5; and finally (13) Polykleitos II Philokratou, *hippeus* in the Pythais of 128/7, whose father Philokrates had been enormously successful in the Panathenaia of ca. 165 B.C., winning nine (!) chariot events. Philokrates' brother, Hagnias, was also a chariot winner in the same festival. This was clearly one of the most illustrious hippotrophic families in the second century.

The following families have such a strong cavalry commitment that more than one of their members was serving in the cavalry at the same time: (1) the two sons of Athenobios, Dionysios and Niketes, both served as *hippeis* in the Pythais of 106/5 and he himself had been a *hippeus* in the Pythais of 128/7; (2) the two cousins, Buttakos Buttakou and Pyrrhos III Pyrrhou II of Lamptrai, served as *hippeis* in the Pythais of 128/7, and Buttakos IV Pyrrhou II, brother of Pyrrhos III, was a *hippeus* in the Pythais of 106/5; (3) the cousins, Alexander II Nikagorou I and Polykleitos I Alexandrou I of Phlya, served as *hippeis* in the Pythais of 106/5; (4) Alkidamos and Eumareides, sons of Euphanes of Euonymon, were *hippeis* in the Pythais of 128/7; (5) the brothers Hagias and Straton, sons of Boulon of Euonymon, were involved with the cavalry of the Pythais of 128/7, Hagias as a winning phylarch and Straton as a winner of an open hippic competition, though he is not actually listed in the catalogue of participating *hippeis*; (6) Apollonides II Kratonos is attested as pythaist for the *hippeis* in the Pythais of 106/5 and won the open competition hippic event; his father Kraton Apollonidou I was the hipparch in the same procession; (7) the cousins, Iason Seleukou and Seleukos Iasonos, were *hippeis* in the Pythais of 106/5; and (8) Euthoinos and Polynikos, sons of Moschion, were tarantinarchs of the tribe Aiantis at the Theseia of 153/2 (or 149/8). The most prolific cavalry family, however, is that of Diodoros of Halai, who had three sons, Diopeithes, Theophilos, and Philanthes, all serving as *hippeis* in the Pythais of 106/5.

Summary Remarks

In the Hellenistic period the Athenian cavalry was recreated in the image of its Archaic predecessor, whose primary function had been to patrol and protect Attika. Whatever may have been its effectiveness in the earlier period, the cavalry of Hellenistic Athens was not militarily negligible. It could even be dispatched on a foreign expedition, as to Thermopylai to stop the Gauls. To maintain this cavalry, the state continued the administrative apparatus introduced in the Classical period, to wit, the grants of *katastasis* and *sitos*, though now to a much reduced corps.

The decision not to maintain the Athenian cavalry at the canonical 1,000-man level can be explained in several interrelated ways. First of all, Hellenistic Athens was no longer a power of international stature; an "imperial" cavalry was out of step with the times. Certainly, after the reduction at the end of the fourth or early third century, to have restored the cavalry to its canonical size could not help but have been interpreted by the Macedonians as a provocative act. Secondly, the economic factor. To be sure, a 1,000-man cavalry is expensive to maintain. One could argue that Athens was no longer economically capable of supporting a cavalry at that level or even at times a mere 200 men, if the extraordinary measures taken by Eurykleides in the 240s and Theophrastos in 220/19 are spotlighted. But Athens still managed to pay for its mercenaries and finance a building program even in that period. And we should not forget that the Athenians were willing to maintain 1,000 *hippeis* in the fourth century, at a time when they no longer received imperial tribute. In my mind, monetary distress at specific times or even a general overall economic decline are certainly factors, but not primary in explaining the reduction of the cavalry or the decision to maintain it at several hundred men.

The primary reason is societal. In an age of mercenaries and voluntary military service for most of the citizens, it was not an easy task to convince many of the well-to-do Athenian families *without* cavalry or hippotrophic traditions that cavalry service was worth the commitment of money, time, and effort. This would help explain why the recruiting of 100 additional *hippeis* by the cavalry officers in 282/1 was viewed as such a notable achievement—the 200 horsemen probably represent about the normal maximum attainable quota from the pool of available and interested equestrian families. Although incomplete and problematic, the prosopographical evidence (see Table 1) leads me to suggest that there were now proportionally more older *hippeis*, i.e., in their thirties and forties, in the cavalry than had been the case in the Classical period—an aging of the corps, so to speak, as more men served longer. This was the direct result of interest and tradition. For most of these *hippeis*, the question of continuing in the cavalry beyond the serviceable lifetime of their *katastasis*-furnished mounts was not raised; whether illustrious *hippotrophoi* or possessed of strong cavalry traditions or both, they thought nothing of purchasing another horse or horses for an extended cavalry career after the exhaustion of their original *katastasis* loans. I am also inclined to think that if any of the administrative procedures fell into disuse in the Hellenistic period, it was probably that associated with the restrictions on retirement contained in *AP* 49.2. In this sense, cavalry service did take on a "voluntary" character. A *hippeus* could probably retire from the cavalry after a reasonable length of service—there has to be some continuity—without interrogation or fanfare; the fact that a sufficient number stayed on made the issue inconsequential.

The cavalry of Hellenistic Athens stands as a direct heir to those several hundred families with hippotrophic and cavalry interests who formed the nucleus of the canonical 1,000-man Classical cavalry. It would not be too far afield to suggest that this reality explains the number of the 300-man cavalry in the 450s which the Athenians ultimately realized was inadequate for an imperial power. In a sense, the Athenian cavalry of the Hellenistic period closely reflected both its Archaic and Classical roots, aristocratic in outlook and personnel, democratic in institution.

APPENDIX A

Ages of the Horsemen of the Pythais

1.	Achaios (Bou[---Me]liteus) Kekropidos, 106/5	29–31
2.	Athenobios (Dionysiou Eupyrides) Leontidos, 128/7	under 38
3.	Basileides Apollophanou Ptolemaidos, 106/5	at least 40
4.	Boulon (Leostratou Palleneus) Antiochidos, 106/5	29–39
5.	Buttakos (Pyrrhou Lamptreus) Erechtheidos, 106/5	at least 30
6.	Demokleides [Timok]leous Erechtheidos, 98/7	early 20s
7.	Demophilos (Polykleous Alopekethen) Antiochidos, 128/7	at least 25
8.	Dio[dotos] Diodotou Kekropidos, 128/7, 106/5	at least 40
9.	Dionysios (Athenobiou) Leontidos, 106/5	29–39
10.	Eirenaios (Eirenaiou Skambonides) Leontidos, 106/5	at least 30
11.	Eudemos (Sarapionos Meliteus) Kekropidos, 128/7	at least 30
12.	Eudemos (Eudemou Eiteaios) Antiochidos, 106/5	at least late 30s
13.	Eudoxos (Eudoxou) Hippothontidos, 128/7, 106/5	at least early 40s
14.	E[umareid]es (Euphanou Euonymeus) Erechtheidos, 128/7	ca. 29
15.	Glaukos (Lysandrou) Hippothontidos, 128/7	at least 57?
16.	Herakleides (Dionysiou Phlyeus?) Ptolemaidos, 128/7	at least 30
17.	Hipparchos (Timoleous Peiraieos) Hippothontidos, 128/7	over 40?
18.	Iason (Seleukou Paianieus) Pandionidos, 106/5	ca. 45
19.	Kallias Oineidos, 128/7, 106/5	over 40?
20.	[Kallias] Habronos Aigeidos, 106/5	over 30
21.	Kallitheos (Lysiadou Berenikides) Ptolemaidos, 128/7	over 30
22.	Kephisodoros (Demetriou Aixoneus) Kekropidos, 106/5	ca. 40
23.	Kratermos (Athenadou Rhamnousios) Aiantidos, 106/5	ca. 30
24.	Krates (Marathonios) Aiantidos, 106/5	over 30
25.	Lysias Artemonos (Paianieus) Pandionidos, 98/7	27–29
26.	Medeios (Medeiou Peiraieos) Hippothontidos, 106/5	29–39
27.	Menekrates Oineidos, 128/7, 106/5	at least 40
28.	Menophilos (Lykophronos Sounieus) Attalidos, 128/7, 106/5	ca. 42–49
29.	Nikanor (Nikanoros Leukonoeus) Leontidos, 106/5	ca. 30
30.	Niketes (Athenobiou Eupyrides) Leontidos, 106/5	ca. 30

31.	Phanoleos (Charisandrou) Aiantidos, 106/5	over 30
32.	Philemon (Philemonos Meliteus) Kekropidos, 106/5	ca. 25–35
33.	Polykleitos (Alexandrou Phlyeus) Ptolemaidos, 106/5	29–39
34.	Polykleitos (Philokratou) Ptolemaidos, 128/7	20–27
35.	Pyrikles (Asopokleous Agrylethen) Erechtheidos, 106/5	ca. 40–50
36.	Rhadamanthos (Attinou) Antiochidos, 128/7	20–27
37.	Sarapion (Sarapionos Meliteus) Kekropidos, 106/5	over 40
38.	Seleukos (Iasonos Paianieus) Pandionidos, 106/5	early–mid 40s
39.	Sosibios (Themakeus) Ptolemaidos, 128/7	over 30
40.	Sosigenes (Sosikratous) Leontidos, 128/7	39–42
41.	Sotion (Nikanoros Phrearrhios) Leontidos, 128/7	over 25
42.	Sthenelos Sthe[nelou] (Meliteus) Kekropidos, 106/5, 98/7	8 years' service
43.	Theopeithes Theophilou Antiochidos, 128/7	20–27
44.	Theophrastos (Herakleitou Acharneus) Oineidos, 128/7	39–43
45.	Timokrates (Theodorou Cholargeus) Akamantidos, 106/5	40–42
46.	Zenion (Marathonios) Aiantidos, 106/5	at least 40s
47.	Zopyros (Onesandrou Oethen) Oineidos, 128/7	28–30

APPENDIX B

The Hipparch to Lemnos

THE BEST discussion of the cleruchy of Lemnos (along with the commonly associated islands of Skyros and Imbros) can be found in A. J. Graham, *Colony and Mother City in Ancient Greece*, 2d ed. (Chicago, 1983), 175–90. We possess the names of only four hipparchs to Lemnos in all of Athenian history: (1) Pheidon of Thria, known as a phylarch in Mnesimachos' play *Hippotrophos* (Athenaios 9.402) and by twenty-six clay tokens stamped with the letters *Hipparchon* (or *Hipparchos*) *eis Lemnon Pheidona Thri* (Figure 8) from shortly after the mid-fourth century B.C. (J. H. Kroll and F. W. Mitchel, "Clay Tokens Stamped with the Names of Athenian Military Commanders," *Hesperia* 49 [1980]: 86–96, esp. 89–91); (2) Lykophron in the 330s, first phylarch, then Hipparch to Lemnos, the first to have held the last office for two consecutive years, and then again a third (Hypereides *In Defense of Lykophron* 16–18); (3) Komeas Chaireous of Lamptrai, perhaps an Athenian of a Lemnian cleruch family (Ferguson, *HA*, pp. 155–56), who was sent as Hipparch to Lemnos to settle affairs after the island was returned to Athens and democracy restored shortly after 281 B.C. (*IG* II[2] 672, now with additional fragments [E. Schweigert, "Inscriptions from the North Slope of the Acropolis," *Hesperia* 7 {1938}: 307–8, frags. a, b, no. 32] joined to lines 9–16 (providing patronymic) and 17–21 by Dr. D. Delmouzou, Director of the Epigraphic Museum in Athens); for a fragmentary copy of this decree, see E. Schweigert, "A Decree in Honor of Komeas, 280/79 B.C.," *Hesperia* 10 (1941): 338–39; for the date of *IG* II[2] 672 in 280/79, see Schweigert, "Decree in Honor of Komeas," pp. 338–39 following B. D. Meritt, "Greek Inscriptions," *Hesperia* 4 (1935): 578, but now 277/6(?): B. D. Meritt, "Athenian Archons 347/6–48/7 B.C." *Historia* 26 (1977): 173; for a date of 279/8 (archonship of Anaxikrates, as restored and dated in the *Corpus*), see Kroll and Mitchel, "Clay Tokens," p. 90; and (4) Telesidemos, son of Aminias of Hekale, twice Hipparch to Lemnos, ca. 166 B.C. in decree of cleruchs of Myrina (*IG* II[2] 1224, frags. a, b, lines 10–11; but now downdated to ca. 148/7–135/4 on basis of Cutter of Agora I 6006: S. Tracy, "Five Letter Cutters of Hellenistic Athens," *Hesperia* 47 (1978): 261–66 (= *SEG* XXVIII.104).

Beyond this basic information two questions are of interest: (a) Why did Athens commit a hipparch to Lemnos and to no other cleruchy (excluding for the moment the problematic hipparch on Salamis: *IG* II^2 1955); and (b) were the *hippeis* that the Hipparch to Lemnos commanded cleruchs or members of the Athenian cavalry?

(a) Lemnos is a large volcanic island, about 476 square kilometers, capable of supporting two cleruchies, Myrina on the southwest coast and the larger Hephaisteia on the north-central coast (for detailed bibliography, see L. Bernabò-Brea, s.v. Lemno, *Enciclopedia dell'arte antica, classica, e orientale* [Rome, 1961], 542–45 and more briefly, R. Stillwell, ed., *The Princeton Encyclopedia of Classical Sites* [Princeton, 1976], 496–97). Like Imbros and Skyros, it lay on a natural grain trade route from the Black Sea and consequently assumed strategic importance (E. F. Bloedow, "Corn Supply and Athenian Imperialism," *AC* 44 [1975]: 20–29 and R. J. Hopper, *Trade and Industry in Classical Greece* [London, 1979], 71–92), but Lemnos also produced about three times the amount of barley and wheat as the other two islands combined, 305,275 to 108,600 *medimnoi* and even more striking is the fact that Attika produced only 371,161 *medimnoi*, less than 20 percent more than Lemnos (these figures from the first fruits [*aparchai*] given to Eleusis in 329/8 [*IG* II^2 1672], as calculated by A. Jardé, *Les céréales dans l'antiquité grecque* [Paris, 1925], 41; see also R. Seager, "Lysias Against the Corndealers," *Historia* 15 (1966): 172 and S. Isager and M. H. Hansen, *Aspects of Athenian Society in the Fourth Century* B.C., trans. by J. H. Rosenmeier [Copenhagen, 1975], 17, 19). A. Philippson and E. Kirsten, *Griechische Landschaften: Das Aegaeische Meer und seine Inseln*, vol. 4 (Frankfurt am Main, 1959), 227, quotes C. Fredrich, editor of *IG* XII.8 (inscriptions of Lemnos, Imbros, Skyros) that Lemnos is the flattest island in the northern Aegean and also the most fertile; particularly productive are the regions around the Bay of Purnia on the north and the Bay of Mudros on the south and also on the east coast: "hier kann man eine Stunde und länger durch Getreidefelder reiten." This I can personally attest from my travels on Lemnos in the summer of 1981. Apollonios Rhodios refers to the rich soil for ploughing on Lemnos (*Argonautica* 1.868). Not only was Lemnos a significant grain producer, some of which must have found its way to Athens, but its harbors were highly regarded, an island perfect as a base of operations, as the allies realized in their ill-fated Gallipoli campaign against the Turks in World War I. Demosthenes tells his fellow Athenians that Lemnos, Thasos, and Skiathos and the neighboring islands, can make good winter bases for a standing army to check Philip II because they possess harbors, grain (*sitos*), and whatever an army needs (4.32). Can there be any doubt that Lemnos heads the list because it most admirably satisfied all of these criteria?

Lemnos, then, is a rather large fertile island, strategically important for economic and military reasons. Its size and prosperity accommodated two distinct communities, about 30 km distant from each other on opposite sides of the island. Certainly, its broad stretches of level countryside and its extensive coast line required protection. During the Social War (357–355), the Chians, Rhodians, and Byzantines and their allies sacked Lemnos and Imbros and then proceeded to do the same to Samos, at that time (365–321) also an Athenian cleruchy (Diod. 16.21.2). Just a few years later (before 353), Philip II began operations in the northern Aegean, invading Lemnos and Imbros, and took a number of the cleruchs captive (*aichmalôtous politas*; Dem. 4.34; cf. Aischin, 2.72). In the years shortly before the Euboian campaign of 349/8, the three islands were again threatened by Philip ([Ps. = Dem.] 59.3; see A. J. Patteson, "Commentary on [Demosthenes] LIX: Against Neaera" [unpubl. diss., University of Pennsylvania, 1978], 28–29). When in 351 Demosthenes singles out the Hipparch to Lemnos as the only Athenian cavalry officer who is really engaged in military duties (Dem. 4.26f), this is no casual observation on his part; Lemnos was now a theater of war.

Though it is always dangerous to employ the *argumentum e silentio*, I am inclined to think that it was not until the military crisis of the 350s that Athens introduced the Hipparch to Lemnos. The only earlier reference to this hipparch is the very tentative restoration by the *Corpus* editor of an inscription of 387/6, *IG* II^2 30, line 5: [----*arch*] *ontos* (vel *hipparchontos* vel *strategontos*) *en Lêmn[ô] i egenet* [o----], which has now been rejected in favor of an eponymous archon of Athens or of Lemnos (R. S. Stroud, "Inscriptions from the North Slope of the Acropolis, I," *Hesperia* 40 (1971): 170–71; he provides a new edition of *IG* II^2 30 with additional fragments, pp. 162–73, no. 23; see *SEG* XXV.63; for another inscription detailing confiscated property on Lemnos, dated to 370/69, see B. D. Meritt, "Greek Inscriptions," *Hesperia* 29 (1960): 25–28, no. 33). All of our evidence, then, including the clay tokens of Pheidon, postdate the mid-fourth century B.C. But again, we cannot flatly dismiss the possibility that the Hipparch to Lemnos was introduced sometime after the creation of the imperial cavalry in the mid-fifth century (though probably not the third hipparch listed in *IG* I^2 400; see Chapter Two) or perhaps after the apparent reacquisition of the island from Spartan control shortly after the naval battle of Knidos in 394 (cf. And. 3.11–12, 14; Xen. *Hell.* 4.8.15) and in the official recognition of Athenian sovereignty in the King's Peace of 386 (Xen.

Hell. 5.1.31; Stroud, "North Slope Inscriptions," pp. 170–71). But whatever date one chooses for the introduction of this cavalry officer, it is clear that Lemnos, more so than any of the other cleruchies of Athens, was eminently suited for cavalry (*hippasimon*). And where there is cavalry, there is a need for a cavalry commander.

(b) Now we have a hipparch, popularly elected like the other two hipparchs, and sent from Athens each year to command a cavalry (*AP* 61.6), and he, like the other magistrates sent to Samos, Skyros, and Imbros, received money for food (*eis sitêsin argyrion*; *AP* 62.2). But did he command a cleruchic cavalry or one sent from Athens? *AP* 61.6 seems to imply cleruchic cavalry: the Athenians elect a Hipparch to (*eis*) Lemnos, who takes charge "of the horsemen *on* Lemnos" (*tôn hippeôn tôn en Lemnôi*), and this is how T. L. Shear, Jr., "The Athenian Agora: Excavations of 1971," *Hesperia* 42 (1973): 178 and Kroll and Mitchel, "Clay Tokens," p. 90, interpret this passage. But even this passage has been interpreted as cavalry "stationed" on Lemnos (J. E. Sandys, *Aristotle's Constitution of Athens* [London, 1912; Arno repr., New York, 1973], 244). In addition, by the *comparandum* of *IG* II^2 1955 (see Chapter Five), which records Athenian horsemen serving on Salamis ca. 320 B.C. along with their hipparch Theogenes Theomedous Eleusinios, U. von Wilamowitz-Moellendorff, *Aristotles und Athen*, 2 vols. (Berlin, 1893) 1: 195–96, n. 18, C. Horner, *Quaestiones Salaminiae* (Basel, 1901), 32, and Ch. Habicht, *per litteras*, May 14, 1984, favor the position that if the Athenians could station Athenian *hippeis* on Salamis, they probably did likewise for Lemnos. We should perhaps not press this analogy too far—the case of Athenian *hippeis* on Salamis may be exceptional: (1) there is no attested *Hipparchos eis Salamina*, and we would expect Aristotle to have mentioned this office with the other cavalry personnel if it had existed (unless one assumes that it was a very recent innovation, postdating the composition of the *Athenaion Politeia*); Theogenes, more than likely, is one of the two regular hipparchs for that year and the stationing of the horsemen on Salamis may have been a temporary war measure, perhaps connected with the Lamian War, 323–322 B.C. (see Chapter Five). An analogy may be found in the decree recorded in Demosthenes *On the Crown* (116, 114). Although these passages are bracketed by editors as spurious to the text, the historical information related need not be. In the 330s(?), Charidemos was sent to Salamis (*eis Salamina*) in command of the infantry (*ho epi tôn hoplitôn*) and Diotimos in command of the cavalry (*ho epi tôn hippeôn*) and a battle subsequently took place involving these forces. Diotimos was probably one of the two hipparchs for that year. (2) Salamis lies just off the coast of Attika, Lemnos many miles and days sailing away—it would have been relatively easy to transport horsemen across the straits, but much less so to Lemnos. Perhaps relevant, an interesting deme pattern emerges from *IG* II² 1955: of the twenty-one horsemen listed, only one of them comes from an inland deme, that is to say, twenty out of twenty-one *hippeis* come from south Attic coastal or city demes. Were these men selected on the basis of proximity of residence to Salamis and were they stationed for periodic, temporary duty on the island? Such an arrangement, convenient for returning home, would have been much less practical for Lemnos because of the distance involved. Only six Athenian tribes are represented in this *katalogos*: Hippothontis (6 + 1 hipparch), Akamantis (2), Kekropis (4), Aigeis (3), Erechtheis (5), and Aiantis (1). Even if H. G. Lolling (original editor, "Inschrift aus Salamis," *AM* 7 [1882]: 40) was correct when he estimated that the complete stone could have accommodated about eighty names (we cannot judge this, as now even *IG* II² 1955 is lost), there is no reason to assume that the stone was completely inscribed (as Wilamowitz noted). In fact, eighty seems high—on the basis of a rough parity of tribal representation, say approx. five *hippeis* per tribe, the total would yield not many more than fifty horsemen. And the lower the total number of names, the less coincidental the deme distribution becomes in proximity to Salamis.

I have sought to be fair about the usefulness of *IG* II² 1955 as a *comparandum* for Lemnos; at the same time, I do believe that a case can be made, even without this inscription, that Athens did send cavalry to Lemnos. First of all, *IG* II² 30 provides administrative guidelines for the cleruchic allotments on Lemnos in 386 B.C., most significantly, line 12 records the word *pentakosiomedimnôn*. Luria, by analogy with the Brea colony decree of the mid-fifth century (*IG* I³ 46, lines 39–42; ML 49, pp. 128–33), which stipulates that the colonists were to be selected from the two lowest census classes, the thetes and *zeugitai*, restored line 12 to read *[plên hippeôn ka] i pentakosiomedimnôn*, "except for the hippeis and the pentakosiomedimnoi" (*SEG* III.73). Stroud, "North Slope Inscriptions," pp. 171–72, repeats this restoration without objection. We know that the cavalrymen of Athens did belong, generally speaking, to these two upper classes. If the restoration is correct, this inscription (admittedly very fragmentary) would seem to bar cavalrymen from cleruchic allotments on Lemnos, which makes perfect sense because the cavalrymen can hardly be considered ones deserving of allotments of land and a new start elsewhere. If they were included, it would defeat the apparent primary purpose of a cleruchy—to provide land for the needy and at the same time perhaps increase the numbers of hoplites for the Athenian army. (Recently, however, J. Cargill has argued on prosopographical grounds that many of the residents on Athenian cleruchies, particularly Samos but also Lemnos, appear to have come from the wealthier classes: "Fourth-Century Athenian Citizen Colonies in the Aegean: An Aspect of Naval/Military Policy," in *Naval History*. The Sixth Symposium of the U.S. Naval Academy, ed. D. M. Masterson [Wilmington, Del., 1987], 32–38. But the case for Lemnos, at least, is quantitatively slight and, until Professor Cargill publishes his awaited major study on the Athenian cleruchies, it would be premature to discard the *communis opinio*.) That the cleruchs were expected to fulfill military obligations to Athens can be demonstrated from the well-known late sixth- or early fifth-century inscription concerning the cleruchy on Salamis (*IG* I³ 1, lines 8–10; ML 14, pp. 25–27; Graham, *Colony and Mother City*, pp. 189–90; A.H.M. Jones, *Athenian Democracy* [Oxford, 1957], 174–77). And there is the reference in *IG* II² 30, line 34, to regulations for the cleruchs of Salamis: *[kathap]er tois es Salam[ina]*, "just as for the cleruchs(?) on Salamis." Stroud, "North Slope Inscriptions," pp. 172–73, properly cautions against concluding too much from this, but it may well encompass conditions of military liability, as seen in the earlier Salamis decree with its clear implication of *infantry* service: each cleruch is to provide *ta hopla* to the value of thirty drachmas. Now, if this is true, what sort of cleruchic cavalry force could have been mustered in the 380s, if one argues that the Hipparch to Lemnos was first dispatched at that time, or even a generation later, when we definitely do have a hipparch to the island and assuredly cavalry as well? It does not seem likely that a sufficient number of cleruchs in a single generation would have acquired the necessary wealth and property holdings (at the expense of their fellow cleruchs) to man a decent cavalry force. Admittedly, the cleruchies of Lemnos in the fifth century may have been a mixture of Athenian colonial families from the early sixth century and Athenian cleruchs dispatched around the mid-fifth century (see Graham, *Colony and Mother City*, pp. 175–85), a situation in which one must assume a differentiation of wealth and status and the possibility of the existence of older hippotrophic families. But if *IG* II² 30 bears witness to a fresh evaluation of property holdings on Lemnos, then it seems inescapable that any decent sized cavalry, at least 50 to 100 men to justify a hipparch, would have to originate from Athens.

The case of Lykophron, Hipparch to Lemnos in the 330s, may provide a piece of information which supports this line of argument. Lykophron boasts that he had held command on the island an unprecedented two consecutive years and even stayed on a third. Why? "As I did not wish, in exacting the pay for the horsemen rashly, to burden citizens in financial straits," *ou boulomenos politas andras epi kephalên eisprattein ton misthon tois hippeusin aporôs diakeimenos* (Hyp. 1.17; Loeb). I interpret (as does Habicht, *per litteras*) these lines to mean that the *politas andras* were the cleruchs and that they were expected to pay (*misthos*) for the services of the cavalry on their island (unlike the 2,000 Athenian hoplites sent to help the cleruchs of Imbros and expecting to be paid by Athens: Dem. 18.115). Now, why would Lykophron have been so concerned about the financial welfare of the cleruchs so as to prolong his hipparchy if his cavalry personnel, presumably from well-to-do families, resided on the island and could return to their estates? This strikes me as very odd unless the cleruchic cavalry was organized on a model radically different from that of Athens. What this suggests to me is that cavalrymen from Athens, chosen by lot and/or a call for volunteers, accompanied the Hipparch to Lemnos and at the end of the normal year stint, expected to receive their pay from the cleruchs upon their departure with him from Lemnos. This *misthos* constituted a subsidy, above and beyond the *sitos* allotments, furnished by the two cleruchies, Myrina and Hephaisteia. This "hardship pay," as in modern submarine service, would have been only reasonable in view of the extended time involved for cavalry service on Lemnos and the separation from Athens (for comments, see F. Blass, *Die attische Beredsamkeit*, 3 vols. 3d ed. [Leipzig, 1887–88; Hildesheim repr., 1962], 3.2, p. 69). The case of Lykophron is, of course, unusual in that he was elected a second term, and presumably a third. Perhaps his troopers returned to Athens after the first year, or some or all elected (or were required to) remain a second year with their commander, and when the cleruchs appeared to be experiencing some financial crisis, Lykophron allowed his name to be nominated for a third term, knowing that by this reiteration the payment of the requisite *misthos* due his men upon completion of his second hipparchy would be postponed. This magnanimity would help explain why no one brought any legal actions, public or private, against him and why he was crowned three times by each cleruchy (Hyp. 1.18). I admit that this reconstruction is quite speculative, but on general principles, if a military officer performed extended service abroad, why should his men not be required to do the same? On the other hand, we cannot dismiss the possibility that normally there was a rotation system for the horsemen of periods of service of less than a year (cf. Dem. 4.21) and that the case of Lykophron was exceptional by the required length of service beyond the norm. Perhaps, then, the clay tokens of Pheidon were part of the administrative appartus of cavalry rotation to Lemnos. Kroll and Mitchel concluded that the tokens functioned as *sym-*

bola, passports of official recognition, "to be carried by persons sent out from Athens to the commanders, such as official couriers on state business and private individuals (travelers, merchants, ships' captains and the like) who had obtained a guarantee of safe passage through or favored treatment within an officer's command" ("Clay Tokens," pp. 95–96). This is probably correct; only the fact that Kroll and Mitchel assumed that the cavalry under the command of the Hipparch to Lemnos was cleruchic prevented them from suggesting that these mass-produced clay tokens might also have been provided to and carried by the replacement *hippeis* from Athens for their journey to Lemnos.

In Chapter Six, I briefly touched upon *IG* II² 672, dealing with Komeas, Hipparch to Lemnos, ca. 280/79 B.C. The inscription contains three decrees, the first by the Athenian *demos* (lines 1–17), the second and third by the Athenian cleruchs on the island of Lemnos (lines 17–46). If we could restore lines 21–25, referring specifically to *hippeis* (lines 21, 23, 25), the issue of the proper identification of the cavalry on Lemnos might be settled once and for all. According to the existing text, Komeas was involved in "examinations" (*tas exetaseis*), presumably of horsemen, and the horsemen dutifully obeyed the commands given them *[tous hipp] eis eis to euktaktôs hypakouein to parangel [lomena . . .]* (line 25). But again, are these *hippeis* cleruchs or Athenians? At first reading, one is almost tempted to reply cleruchs because the relevant lines belong to the decree of the Lemnian cleruchs and it would make reasonable sense to assume that Komeas is being praised for his exemplary command of their *own* horsemen, rather than of those who might have accompanied him to Lemnos. Perhaps this is the case, particularly in light of presumed dislocations of wealth and property that must have followed the struggles of Alexander's successors over the island and also in view of the modest size of the Athenian cavalry at this date. In other words, what may have been normal operating procedure for the 1,000-man cavalry of the Classical period may not have been reintroduced on Lemnos when Athens recovered the island in 281/80. Certainly no cavalry and no hipparch could have been sent from Athens during those periods when Athens no longer had complete control of the island: when Kassander held the island, 318 to 307 (Diod. 20.46) nor when Lysimachos did, ca. 287/6–281 (Phylarchos in Athenaios 6.254 = *FGrH* 81 F 29), and possibly not even when Demetrios Poliorketes was dominant in the intervening years–Diodoros 20.46.4 reports that Demetrios withdrew his garrison from Imbros and gave it back to Athens. There is no mention of Lemnos here. It would have been foolish on his part to have relinquished military control over the prosperous and strategic island of Lemnos.

On the other hand, if Schweigert, "Decree in Honor of Komeas," p. 339, following Meritt ("Greek Inscriptions," p. 578), is correct that lines 17–39, particularly line 28, of the second decree "is concerned with the confusion on Lemnos and the dissension between the cities Hephasteia and Myrrhina after the restoration of democracy following the death of Lysimachos in the summer of 281 B.C., and the immediate appointment of an Athenian hipparch to settle the quarrels," then it seems unlikely that Komeas would be commanding some sort of joint cleruchic cavalry to settle the affairs of the two cleruchies. Rather, he arrives with an Athenian cavalry force (now expanded back home, see Chapter Six) to restore order, for whose judicious use to that end, the cleruchs praise Komeas. One could argue, though I think it is less credible, that the decree records the reorganization of the cleruchic cavalry and is praising Komeas for that accomplishment. Komeas would still have had to arrive with some military force, and he is a hipparch after all.

This incident hints at another reason why Athens would have sent a hipparch to Lemnos. The island is not one cleruchy, but two, acting like two little *poleis*, whose interests did not always coincide, perhaps concerning divisions of land in the rich countryside north of the Bay of Mudros, where the boundary between them is probably to be located, and perhaps on general political issues. It is inevitable that disagreements would arise between the two cleruchies, as appears the case in *IG* II[2] 672, particularly lines, 8–9, 28, requiring an objective, outside authority, to resolve the matter. To have commanded what must be assumed to have been a joint cleruchic cavalry would have placed the hipparch, particularly one only on a year's appointment from Athens, in a precarious position and would have led to doubts as to the loyalty of his horsemen. All of this could have been, and probably was, avoided by dispatching a cavalry from Athens, not one recruited from among the cleruchs. And the logistical problems of recruitment for a hipparch not from Lemnos (with the exception, perhaps, of Komeas) beg to be explained. Finally, what better way to preserve or restore *eunomia* than to have had a military arm whose mobility insured its speedy arrival over the 30 km distance between the two cleruchies?

The Hipparch to Lemnos was much more than a military official; he was the chief administrator of the island. In the fourth century, two civil administrators (*epimelitai*) are attested, one for each of the two cleruchies (*IG* XII.8.4, lines 8–9 [Myrina]; 5, line 6 [Hephaisteia]). These men no doubt were cleruchs who governed their respective cities under the authority of the hipparch. After the middle of the second century B.C., some sort of administrative/military reorganization took place in which a *strategos epi Lemnon* appears, apparently as a superior, to the hipparch, in this case Telesidemos tou Aminiou Hekalethen (*IG* II² 1224, frag. a, b, lines 10–11). At about the same time, we have two documents which have been restored to read *[hipparchos eis Myr] inan* and *[hipparcho]s eis Myrinan* (M. Crosby, "Greek Inscriptions," *Hesperia* 6 [1937]: 457–61, no. 7, line 17; no. 8, line 6). I have examined the last inscription closely and cannot offer an alternative restoration for *hipparchos*, except perhaps for a snug *[stratego]s*, on the *comparanda* of the *strategoi e Skyrou, eg Myrinês, ex Hephaistias*, recorded in *IG* II² 1672, lines 275–77, for 329/8 B.C. But Kroll and Mitchel are probably correct in assuming that these were "members of the board of ten *strategoi* assigned specific island commands" ("Clay Tokens," p. 91). We can only guess at what happened in the second century B.C. But in all probability, the *Strategos epi Lemnon* took residence as supreme commander of the island at Hephaisteia, undeniably the more important and prosperous of the two cleruchies, to judge by the difference in the tribute quotas in the fifth century (Hephaisteia, 3 talents; Myrina, 1 1/2; see R. Meiggs, *The Athenian Empire* [Oxford, 1972], pp. 424–25) and the grain production of each in 329/8 (28,650 to 28,100 in wheat; 151,325 to 97,200[!] in barley; combined grain total: 179,975 to 125,300 *medimnoi*, see A. Jardé, *Les Céréales dans l'antiquité grecque* [Paris, 1925], 41 from *IG* II² 1672), while the *Hipparchos eis Lemnon* resided at Myrina, where the cavalry was perhaps stationed. In any event, this arrangement did not survive the catastrophe of the Sullan sack in 86 B.C. In an inscription of the first century A.D. from Hephaisteia, a hoplite general, the strategos for Lemnos, and a general for the city are all mentioned, but no hipparch to Lemnos (*IG* XII.8.26; see D. Geagan, *The Athenian Constitution after Sulla* [Princeton, 1967; *Hesperia Suppl.* 12], 27; for a general of Hephaisteia, in the second century A.D. perhaps only a local magistrate, and an epimelete of Lemnos, see S. Follet, "Inscription inédite de Myrina," *ASAtene* 52–53 (1974–75): 309–12 = *SEG* XXVIII.718; for comment see J. and L. Robert, "Bulletin épigraphique," *REG* 92 (1979): no. 343). Unless we assume that a hipparch to Myrina still existed, and there is no attestation to that fact after the second century B.C., it appears that the era of a hipparch to Lemnos has now come to an end. This should not surprise us; the era of the Athenian cavalry was over.

Some summary remarks directed to my original questions. I personally favor the position that the Hipparch to Lemnos was not introduced until the middle of the fourth century B.C. and that the horsemen under his command were not cleruchs, but members of the Athenian cavalry. And why to Lemnos and not elsewhere? It, unlike the other cleruchies, was truly a *hippasimon* island.

APPENDIX C

The Hipparcheion

IN AN inscription of 187/6 honoring an unidentified hipparch of the preceding year, there is mentioned the *Hipparcheion*, headquarters of the hipparchs (*IG* II² 895, line 6; *SEG* XXI.436). Although this is our only reference to this building and it has not yet been located, there is now little doubt that it was situated close to the Stoa of the Herms in the northwest corner of the Athenian Agora. This linkage was first argued by Habicht in his publication of another inscription honoring one of the two hipparchs for 188/7 wherein the secretary is ordered to have the decree inscribed on a stone stele and erected *pros tois Hermais*, "at the Herms" (p. 128, line 11; "Neue Inschriften aus dem Kerameikos," *AM* 76 [1961]: 127–41; *SEG* XXI.435) and advocated most recently by Kroll, "Archive," pp. 83–84 and R. L. Pounder, "A Hellenistic Arsenal in Athens," *Hesperia* 52 (1983): 246. And now that the Stoa Poikile, linked geographically by Harpokration with the Herms and the Royal Stoa (see T. L. Shear, Jr., "The Athenian Agora: Excavations of 1980–1982," *Hesperia* 53 [1984]: 40–43) and made famous by the "Stoic" philosopher Zeno, has been located by the American excavators in that area, further archaeological work immediately to the west of the exposed foundation of this stoa should reveal the Stoa of the Herms and in turn, hopefully, the *Hipparcheion*.

Xenophon recommended that the Athenian cavalrymen perform a processional ride around the Agora starting and ending at the Herms; then they should gallop by tribes from the Herms to the Eleusinion on the far side of the Agora (*Hipparch.* 3.2–4). From his comments at 3.5, it is clear that the cavalry did not ordinarily engage in such displays in the marketplace, but that is unimportant. What stands as fact is the overwhelming evidence, of all shapes and sizes, for cavalry activity near the Herms in the northwest corner of the Agora: (1) the Bryaxis base with its three phylarchs from the same family (*IG* II² 3130) found *in situ* there; (2) at the southeast corner of the Stoa of Zeus, a tribal victory of Oineis (probably in the *anthippasia*) with reference to its phylarch (B. D. Meritt, "Greek Inscriptions," *Hesperia* 15 [1946]: 176–77, no. 24); (3) found reused in front of the north end of the Stoa of Attalos, but perhaps originally standing nearer the Herms, a statue base dedicated by the hipparchs and phylarchs with the hipparchs named, Antikrates Sokratous Hermeios and Patroklees Hierokleous

Philaides (H. A. Thompson, "Excavations in the Athenian Agora, 1952," *Hesperia* 22 [1953]: 49–51); (4) the cavalry relief of the Leontis victory in the *anthippasia* (T. L. Shear, Jr., "The Athenian Agora: Excavations of 1970," *Hesperia* 40 [1971]: 271–72; (5) the enormous cavalry archives of lead tablets, lead armor tokens (Kroll, "Archive," pp. 83–140, 141–46) and clay tokens of Pheidon, Hipparch to Lemnos (J. H. Kroll and F. W. Mitchel, "Clay Tokens Stamped with the Athenian Military Commanders," *Hesperia* 49 [1980]: 89–91); (6) the fragment of Mnesimachos in Athenaios (9.402 E–F) which speaks of the area of the Herms which the phylarchs frequent, where Pheidon teaches his youthful (*hôraioi*) charges how to mount and dismount; (7) the two cavalry inscriptions published by Threpsiades and Vanderpool from the third century; the first (282/1) designates that a stele of the decree be set up "in the Stoa of the Herms"; the second (286–261 B.C.), "next to the Herms," *pros tois Hermais* (p. 105, lines 43–44 and p. 110, lines 8–9). Moreover, in the first inscription, another stele is to be erected in the Agora at the *Poseidonion*, which was the sanctuary of Poseidon, patron god of the cavalry, at *Kolonos Hippios* (see R. E. Wycherley, *The Athenian Agora III. Literary and Epigrahical Testimonia* (Princeton, 1957), 99–100; (8) a note by Hegesander in Athenaios (4.167 E–F) that Demetrios, grandson of Demetrios of Phaleron, as hipparch at the Panathenaia had raised above the Herms a special viewing platform for his Korinthian mistress, Aristagora (for Demetrios, see *IG* II^2 1285 and discussion, Ch. Habicht, "Neue Inschriften," *AM* 76 [1961]: 137); (9) Pausanias (1.15.1) reports seeing a trophy commemorating an Athenian cavalry victory over the Macedonian Pleistarchos, perhaps in 303 B.C., which may have stood atop a Hellenistic gate constructed adjacent to the Stoa Poikile (for discussion, see Chapter Five); (10) [Plut.] *Moralia (Lives of the Ten Orators)* 844 B mentions the orator Kallistratos, son of Empedos of Aphidna, commander of the cavalry, who dedicated an altar to Hermes Agoraios, but [Plutarch] has apparently confused the fourth-century orator Kallistratos with the man who was general in 418/7 and died while serving as a hipparch on Sicily in 413 (see Chapter Three and *APF* 8157, p. 278, n. 1); (11) a Lesbian spear butt may have been stored in the *Hipparcheion* (J. Mck. Camp II, "A Spear Butt from the Lesbians," *Hesperia* 47 [1978]: 192–95); and finally, (12) three unpublished inscriptions from the recent excavations (1981–82) in the area of the Stoa Poikile: a cavalry decree (Agora I 7551), a very fragmentary horse relief (Agora I 7515), and a horse relief with a fragmentary list of phylarchs (Agora I 7518); my deep appreciation to T. L. Shear, Jr., and J. Mck. Camp II for permission to examine these three inscriptions.

APPENDIX D

Hippotoxotai and Prodromoi

FOR A CONCISE survey of the literary evidence, see R. Cagnat, s.v. *sagitarii Dar.-Sag.* 4.2 (1911), 1003–4 and A. F. Pauli, s.v. *prodromoi, RE* 23.1 (1957): 102–3. Despite the ambiguity found in And. 3.7; Aischin, 2.174; and *AP* 24.3 to the effect that Athens established a force of 1,200 *hippeis*, it is clear that from Thucydides 2.13.8 that 200 of that number were really mounted archers, *hippotoxotai*. Moreover, [Lys.] 15.5–6 reveals that it was acceptable to refer to a *hippotoxotês* as a *hippeus*, when one was not overly worried about precise military terminology. There can be little doubt that the *hippotoxotai* were introduced as part of the same reform that resulted in the increase of the Athenian cavalry from 300 to 1,000 men, sometime in the 440s or 430s. This was probably an acknowledgement that the Persian mounted archers had been an effective fighting arm in the Persian Wars, and with the dramatic expansion of the Athenian cavalry, *hippotoxotai* seemed an appropriate addition. We ought to disassociate the 300 Skythian archers mentioned by Andokides (3.5) from the *hippotoxotai* because it seems that the latter were Athenian citizens: Alkibiades the Younger became one, albeit under questionable circumstances, in the Haliartos campaign of 395 and the Lysianic account seems to confirm that they were citizens (15.5–8, 11). The fragmentary name beginning Alechs--, under the heading of *hippo [tochsotai]* in a casualty list dated before the end of the fifth century B.C., could be a barbarian mounted bowman inasmuch as the preceding heading reads "barbarian bowman" (perhaps Thracian, see Ar. *Birds* 1179–87 with Thuc. 2.96.1). But its heading does not repeat "barbarian" and there exists a *vacat* between the two lists, making it probable that this man is an Athenian (see B. D. Meritt, "Greek Inscriptions," *Hesperia* 21 (1952): 341–42; *SEG* XII.73). As we have noted in Chapter Three, literary references to these Athenian *hippotoxotai* in military action appear only twice in the Peloponnesian War: twenty accompanied the Athenian force against Melos in 416 (Thuc. 5.84.1–2) and thirty were dispatched to Sicily in 414 (along with 250 *hippeis*) as part of the reinforcements requested by Nikias (Thuc. 6.94.4). From a conversation (probably fictitious) between Socrates and an unidentified young hipparch, we learn that the *hippotoxotai* were accorded the privilege of charging the enemy first and that they preceded the hipparchs (Xen.

Mem. 3.3.1). This has led some (e.g., E. Ekman, *Zu Xenophons Hipparchikos* [Uppsala, 1933], 47) to allocate 100 *hippotoxotai* to each hipparch. Mathematically and ideally this may be correct, particularly for peacetime cavalry displays, but it is evident that Athens did not send its full complement of mounted archers on foreign expeditions, to say nothing of the fact that it is unlikely that both hipparchs were sent abroad for the same campaign.

Sometime between 395 and the 360s, the Athenians apparently decided to abolish the corps of mounted bowmen and replace them with more traditionally equipped *hippeis* called *prodromoi*, "front-runners" literally, who functioned as skirmishers and perhaps scouts. It has been suggested that this change is to be associated with the military innovations of the Athenian Iphikrates which brought the light-armed infantryman, the peltast, into prominence in the early fourth century B.C. (Ekman, *Xenophons Hipparchikos*, p. 47 following A. Brückner, "Zu athenischen Grabreliefs," *JDAI* 10 [1895]: 209). Whatever the exact reason, it is clear that by the time of the composition of the *Hipparchikos, hippotoxotai* no longer figure in the military organization of the Athenian cavalry. But Xenophon does talk about *prodromoi* (1.25). He urges the phylarchs to arm their *prodromoi* in the best way possible and to instruct them in the use of the javelin (*akontion*). There is no evidence as to their number; to argue (as Pauli, s.v. *prodromoi*, col. 103) that they, like the *hippotoxotai* who preceded them, numbered 200 is only a guess. Also, it is quite unlikely that the *prodromoi* are simply the *hippotoxotai*, as G. Gilbert, *Constitutional Antiquities of Sparta and Athens* (London and New York, 1895), 324, n. 3, suggests. One thing is certain—like the *hippotoxotai*, they apparently did not belong to the same social class as the regular *hippeis*. This is clearly implied in a gloss by Photios, s.v. *prodromoi. adoxoi*, "undistinguished" or worse "obscure." Aristophanes speaks of 1,000 good horsemen, not 1,200, and clearly intends to leave the impression that these 1,000 were aristocratic young men with flowing locks, even if that did not in fact characterize all of them (*Knights* 225, 580). The conclusion seems inescapable: the *hippotoxotai* were not meant to be included in the elite ranks of the 1,000. Alkibiades the Younger insinuated himself into what was viewed as a less distinguished military body (cf. Plato *Laws* 834 B–D). And now we are aware of an important piece of information overlooked by previous commentators—in the Lysianic fragment of the speech *Against Theozotides* dated to 403/2, we recall that the *sitos* allotments for the Athenian *hippeis* were reduced from one drachma to four obols, but that of the *hippotoxotai* were increased from two obols to eight obols! Much can be culled from this testimony. First of all, the *hippotoxotai* had not been implicated in the actions of the cavalry of 404/3 and therefore had not suffered reproach—in fact, I might suggest that they were simply disbanded by the Thirty Tyrants because they belonged to a lower class than the *hippeis* and by that fact had not been welcomed to the oligarchic cause. Secondly, the fact that their *misthophoria* was so dramatically increased lends credence to the claim that they came from the more needy of Athenian society. Thirdly, as I have suggested in Chapter Four with respect to Pamphilos literally taking away Alkibiades' horse, it seems very likely that the state ordinarily maintained the horses for the *hippotoxotai* (analogous perhaps to the Spartan *hippotrophoi* who provided horses for designated troopers), thus explaining why the *hippotoxotai* had only to be granted a *per diem* allotment of two obols for themselves. With this measure perhaps the Athenians thought to offer the mounted bowman a greater personal involvement in this maintenance, while still providing the horse, as implied in [Lys.] 15.5. As for the *prodromoi*, it is obvious from Xenophon's comments that whereas the regular *hippeis* were to equip themselves from their own *misthos*, the phylarchs were expected to equip their *prodromoi* with their own money (*Hipparch.* 1.23, 25), again accentuating their lower economic status. Finally, the account in *AP* 49.1 certainly differentiates the *hippeis* from the *prodromoi* (and the *hamippoi*) in the *dokimasia* by the Council of Five Hundred.

With the discovery of the cavalry archives of the third century B.C., we can now say more about the *prodromoi*. On two tablets, Kroll, "Archive," no. 62, pp. 125–27 and Braun, "Funde," no. 565, there appear the names in the genitive case, Antimachou *prodromou* and Chariou *prodro(mou)*, respectively. Although these could be explained as tablets for the two *prodromoi*, Antimachos and Charias, Kroll has convincingly argued that these refer to horses provided for the *prodromoi of* Antimachos and Charias, whom he identifies as cavalry officers. Kroll suggests that by the third century the *prodromos* had become "a kind of specialized aide attached to any cavalry officer" (p. 127). Thus, a cavalry officer was granted the right of two evaluations, one for his own horse and the other for that of his *prodromos*. This may also be the case for the tablets listing *hypêretai*, "attendants," of Hieron Anagyrasios (Braun, "Funde," no. 236), Epikrates (Braun, "Funde," no. 140), and possibly (though the term does not appear), Epigenes Euonymeus (Kroll, "Archive" no. 40), all of whom Kroll suggests may be synonymous with *prodromoi* at this date (p. 127; see n. 65 for *hypêretai* in the Classical period). In the *Hipparchikos* (4.4–5), Xenophon recommends that some of the *hypêretai* be sent out in advance of each tribe to find safe routes through hostile territory. These are probably the same as the "grooms," *hippokomoi* (*Hipparch.* 5.6; *Hell.* 2.4.6). Xenophon goes on to suggest that another advanced guard (*proodoi*) be sent further out to reconnoiter the position of the enemy.

Some troubling questions remain, however. It is still rather curious that we have so few references to *prodromoi* in the tablets, even if they can be equated with the *hypêretai.* If this option was available to all the cavalry officers, why do we not see more evidence of it? As I have noted in Chapter Six, one wonders whether we really do have an accurate tally of the yearly quota of the *hippeis* in this period. But an answer may lie in an extension of Kroll's theory. If we can agree that the horse for the *hippotoxotai* and *prodromoi* must have ordinarily been provided directly by the state and not indirectly through the system of *katastasis*, then the state will have been required to replace the horses for these men in case of loss or injury as a matter of course. It may be that certain cavalry commanders, very few to judge by the evidence of the archives, were bona fide *hippotrophoi* who chose to provide their *prodromoi* (or *hypêretai*) with one of their own (more expensive?) horses which the state agreed to evaluate and replace at value in case of loss or injury. A final note. Considering those special differences in the cavalry of the Hellenistic period from that of the much larger Classical one, we should not be disinclined to suggest that this option was a Hellenistic innovation.

CATALOGUS HIPPEUM

Abbreviations Found Only in Catalogus

AAA Kyle, *Athletics in Ancient Athens*, Appendix B
B Braun, "Funde"
DAA Raubitschek, *Dedications from Athenian Acropolis*
GHI Tod, *Greek Historical Inscriptions*, vol. 2
ISE Moretti, *Iscrizioni storiche ellenistiche*
K Kroll, "Archive"
TV Threpsiades-Vanderpool
W Davies, *Wealth and Power of Wealth*

Notes: all references to Appendixes and Chapters are to this book. The asterisk for entry II.A.16 signifies that the horseman's name is unknown. The Theseia of *IG* II² 958 is dated 149/8, but may be 153/2.

I. Cavalry Officers

A. Classical Period

1. Anacharsis Me[---] of Kydathenaion, phylarch, mid-4th century (*IG* II² 3135, 3136; *W*, p. 153)
2. Antidoros of Thria, hipparch, 4th century (Kroll and Mitchel, *Hesperia* 49 [1980]: 92–93; *W*, p. 155)
3. Antikrates Sokratous of Hermos, hipparch, 4th century (H. Thompson, *Hesperia* 22 [1953]: 49–51; *W*, p. 155; homonymous descendant was *hippeus* in mid-3d century (B 24–25)
4. [An]timachos [---]o of Pelekes, phylarch or taxiarch, 373/2 (Schweigert, *Hesperia* 8 [1939]: 4, no. 2; *W*, p. 153)
5. Antiphanes (Akamantis?), phylarch in 394 (*IG* II² 5222; *PA* 522; *GHI* 104)
6. Char[ide]m[os] [E]uni[---] (Hippothontis), phylarch or taxiarch, 373/2 (Schweigert, *Hesperia* 8 [1939]: 4, no. 2)
7. Demainetos Demeo of Paiania, phylarch, mid-4th century (*IG* II² 3130; *APF* 3276, pp. 103–6; Bryaxis base)
8. Demeas Demaineto of Paiania, phylarch, mid-4th century (*IG* II² 3130; *APF* 3276, pp. 103–6; Bryaxis base)
9. Demetrios Phanostratou of Phaleron, hipparch, *strategos*, ca. 315/4 (*IG* II² 2971; *APF* 3455)
10. Demetrios [--4--]phanous of Alopeke, phylarch, ca. 330 B.C. (*SEG* III.115; *W*, p. 154)

11. [Democh]ares (Demonos) of Paiania, [*hippa*]*rchos*, 375–369 B.C. (*IG* II^2 102, line 19; *APF* 3737, p. 144)
12. Demosthenes Demaineto of Paiania, phylarch, mid-4th century (*IG* II^2 3130; *APF* 3276; Bryaxis base)
13. Dieitrephes (Nikostratou) of Paiania, phylarch and hipparch, before 415/4, *strategos* (Ar. *Birds* 799 w/ scholia; 1440–43; Paus. 1.23.3–4; Thuc. 8.64.2; 7.29.1; *IG* I^3 110, line 6; ML 90; Connor, *New Politicians*, pp. 156–58)
14. Diotimos III Diopeithous I of Euonymon, "*ho epi tôn hippeôn*" on Salamis, 330s (Dem. 18.114, 116; *APF* 4386, pp. 163–65)
15. Epilykos Nikostrat[ou of Gargettos], hipparch, before 325 B.C. (Crosby, *Hesperia* 6 [1937]: 462–63, no. 10; *APF* 8429, p. 312)
16. [Eue]tion Pythangelou of Kephisia, hipparch, before 325 B.C. (Crosby, *Hesperia* 6 [1937]: 462–63, no. 10; *APF* 5463, p. 190)
17. Euthykrates Perg[asethen], phylarch, ca. 380 B.C. (Meritt, *Hesperia* 9 [1940]: 57, no. 6; *W*, p. 153)
18. [Ha]gnodemos of Paiania, phylarch, ca. 380 B.C. (Meritt, *Hesperia* 9 [1940]: 57–8, no. 6; *W*, p. 154, though it lists his *nomen* incorrectly as [Ha]gnodoros)
19. Kallisthenes Ch[---], cavalry officer(?) along with *hippeis*, post mid-4th century (Meritt, *Hesperia* 30 [1961]: 247, no. 44; *SEG* XXI.701)
20. Kallistratos Empedou of Oe, hipparch killed on Sicily in 413 B.C.; *strategos* in 418/17 (Paus. 7.16.4–5; *APF*, p. 278, n. 1; *W*, p. 154; ML 77, line 20; *IG* I^3 370)
21. Kallistratos, hipparch in cavalry victory over Pleistarchos, ca. 303 B.C. (Paus. 1.15.1; Appendix C)
22. Kephisodoros of Marathon, hipparch who died at Mantineia in 362 B.C. (Paus. 8.9.10; 8.11.5; D.L. 2.54; *FGrH* 70 F 85 [Ephoros]; Harp. s.v. Kephisodoros; *PA* 8376; *W*, p. 154)
23. [Kineas Nikom]achou of Lamptrai, hipparch, *strategos*, *anthippasia*, 350–270 B.C. (Vanderpool, *Hesperia* 43 [1974]: 311–13, no. 1; *APF* 12883, pp. 491–93; K 48a, p. 119 [for Nikomachos as *hippeus*])
24. Kleonymos Kleoxeno of Marathon, phylarch or taxiarch, 373/2 (Schweigert, *Hesperia* 8 [1939]: 4, no. 2; *W*, p. 153)
25. Kleophanes, possibly phylarch in Euboian campaign at Tamynai in 349/8 (Plut. *Phok*. 13.6; *W*, p. 153)
26. Kratinos, hipparch in 349/8 (Dem. 21.132 with Scholion 557.14; *PA* 8753; *W*, p. 155)
27. Lakedaimonios Kimonos II of Lakiadai, hipparch (ca. 450s), *strategos* (433/2) (*IG* I^2 400; *DAA* 146, no. 135a, b; *APF* 8429, XIII; *W*, p. 154)
28. Lykophron, Hipparch to Lemnos, three consecutive years(?), phylarch, 330s (Hyp. 1.17–18 [Loeb, OCT]; *PA* 9486; *W*, pp. 154, 155)
29. Lysimachos, hipparch implicated in Eleusis treachery and Aixone brutality, 404/3 (Xen. *Hell*. 2.4.8, 26; *PA* 9486; *W*, p. 154)

30. Meidias I Kephisodorou I of Anagyrous, hipparch, 349/8 (Dem. 21.164,166, 171–72; *APF* 9719, pp. 385–87; *W*, p. 155)
31. Menexenos I Dikaiogenes I of Kydathenaion, phylarch killed at Spartolos in 429 B.C. (Is. 5.42; *APF* 3773, pp. 145–49: born before 470 B.C.)
32. Moiragenes Ika[rieus], phylarch, ca. 380 B.C. (Meritt, *Hesperia* 9 [1940]: 57–58, no. 6; *W*, p. 154)
33. Moschion, very young phylarch in Menander's play, *The Samian Woman* (*Sam.* 15; Gomme & Sandbach, *Menander. A Commentary*, pp. 546–47)
34. Nike[--8--] hipparch for oath-swearing of treaty between Leukas and Athens, 368 B.C. (*IG* II² 104; *GHI* 134; *W*, p. 154)
35. Orthoboulos (of Kerameis?), phylarch or hipparch in 395 B.C. (Lys. 16.13; *IG* II² 41, line 17; *GHI* 121; *APF*, pp. 364–65; Wevers, *Isaeus*, p. 79 [phylarch]; *W*, p. 153 [phylarch])
36. Pamphilos of Keiriadai, hipparch in 395 B.C. ([Lys.] 15.5; *APF* 9667, p. 365; *W*, p. 154)
37. Panaitios, hipparch, 425/4 (Ar. *Knights* 243 with scholia; *W*, p. 154)
38. P[ant]aleon, phylarch acting as *strategôn* in casualty list of Erechtheis, 459/8? (*IG* I² 929, lines 6–7; restored *ph[ylar]chos*, but now suggested as a name Ph[ryni]chos: Daux, *BCH* 99 [1975]: 150–54 and tacitly accepted by Lewis in Bradeen & Lewis, *ZPE* 34 [1979]: 240)
39. Patroklees Hierokleous of Philaidai, hipparch, 4th century (H. Thompson, *Hesperia* 22 [1953]: 49–51; *W*, p. 155)
40. Pheidon of Thria, Hipparch to Lemnos and phylarch, ca. 340s (Athenaios 9.402 F; Kroll and Mitchel, *Hesperia* 49 [1980]: 89–91; *W*, pp. 154, 155)
41. Phi[. . .]on Philist[---], possibly a hipparch (on basis of heading HIPP[--]), late 4th–early 3d century (Vanderpool, *Hesperia* 43 [1974]: 311–13; the rasura in line immediately above was probably the other hipparch, [Kineas], no. 21 or another?)
42. Philinos [---]nos [Anaphlys]tio[s], phylarch or taxiarch 373/2 (Schweigert, *Hesperia* 8 [1939]: 4, no. 2; *W*, p. 153)
43. Philokles Phormionos of Eroiadai, hipparch (3 or 4 times), *strategos* (more than 10 times), ca. 325 B.C. (Deinarchos 3.12; *APF* 14541, pp. 539–41; *W*, p. 155)
44. Pronapes Pronapidou of Prasiai, hipparch, ca. 450s and winner of chariot races (*IG* I² 400; *DAA* 146, no. 135; *APF* 12250; *W*, p. 154; *AAA*, A57)
45. Pythodoros Epizelou of Halai Araphenides, hipparch, 420s (*IG* I² 816; *APF* 12402; *W*, p. 154)
46. Pythodoros of Hippothontis, phylarch, casualty in naval battle, 412/1(?) (*IG* I² 950, line 180; *W*, p. 153)
47. Simon, hipparch, 425/4 (Ar. *Knights* 242 with scholia; possibly same as *PA* 12689: Simon Athenaios who wrote work on horsemanship and equestrian statue dedicated to him at Eleusinion [Xen. *Peri Hippikes* 1.1]; *W*, p. 154)
48. Theogenes Theomedous of Eleusis, hipparch on Salamis, ca. 320 B.C. (*IG*

II² 1955; cf. TV, p. 111, line 25: son, Theomedes, phylarch, 286–261 B.C.)

49. Theophilos Eua[ngelo] of Hermos, phylarch or taxiarch, 373/2 B.C. (Schweigert, *Hesperia* 8 [1939]: 4, no. 2; *W*, p. 153)
50. Theophon, phylarch, ca. 370s (Is. 11.41–42, 45; Wevers, *Isaeus*, p. 66; *APF* 2921, XI, p. 84; Anderson, *AGH*, pp. 134–35)
51. Thrasymedes Kal[l]istrato of Acharnai, phylarch or taxiarch, 373/2 (Schweigert, *Hesperia* 8 [1939]: 4, no. 2; *W*, p. 153)
52. Xenophon Euripidou of Melite, hipparch, ca. 450s, and *strategos* (441/40, 430/29) (*IG* I² 400; *DAA* 146, no. 135; *APF* 5951, pp. 199–200; *W*, p. 154; cf. K 24, pp. 111–12 for descendant as *hippeus*, mid-4th century; *AAA*, A78)
53. Lost name, hipparch, ca. 320s (*IG* II² 3209; *APF* D 11, p. 594; *W*, p. 155)
54. [---]kles of Erchia, hipparch, oath-swearer and Athens, 375–359 B.C. (*IG* II² 102, lines 18–19; *GHI* 129, pp. 90–92; *W*, p. 154)
55. [--8--]ikles, hipparch, oath-swearer of treaty between Leukas and Athens, 368 B.C. (*IG* II² 104, lines 1–2; *GHI* 134; *W*, p. 154)
56. [-ca. 9-]n of Oe, phylarch in *anthippasia*(?), ca. 325 B.C. (Meritt, *Hesperia* 15 [1946]: 176–77, no. 24; *W*, p. 154)
57. [---] of Erechtheis, probably phylarch in *anthippasia*, 321/20 or 318/7 (*IG* II² 379, line 3; *W*, p. 154 citing W. B. Dinsmoor, *Archons of Athens in the Hellenistic Age* [Cambridge, Mass., 1931], 27 for 322/1, but see Meritt, *Historia* 26 [1977]: 170 for archonship of Archippos)
58. [. . . .]okle[s. . .]tio of Kedoi, phylarch or taxiarch, 373/2 (Schweigert, *Hesperia* 8 [1939]: 4, no. 2; *W*, p. 153)
59. [. . . .]as [. . . .]imo of [Myrrhi]nous, phylarch or taxiarch, 373/2 (Schweigert, *Hesperia* 8 (1939) 4, no. 2; *W*, p. 153)
60. [*ho deina* ---]enou Pro[spaltios or balisios], hipparch in *anthippasia* at Panathenaia, late 4th or early 3d century (Vanderpool, *Hesperia* 43 [1974]: 313)
61. [*ho deina* ---]ous A[*demoticum*], hipparch in *anthippasia* at Panathenaia, late 4th or early 3d century (Vanderpool, *Hesperia* 43 [1974]: 313)

B. Hellenistic Period, Third Century B.C.

1. Agathaios of Prospalta, phylarch, ca. 286–261 B.C. (TV, p. 110, line 26)
2. A(l)kmachos Kleoboulou of Myrrhinous, hipparch in 282/1 (TV, p. 104, line 25; for son Thoukritos, hipparch, now dated to 252/1, see later listing)
3. Antimachos of Acharnai, phylarch in 282/1 (TV, p. 104, line 31)
4. Antimachos (of Alopeke(?), Antiochis), possibly phylarch or hipparch by reference to *prodromos*, mid-3d century (K 56, 62b, pp. 125–27)
5. Aristophanes, phylarch honored by soldiers stationed at Panakton and Phyle in 239/8 (*IG* II² 1299, lines 55–56; F. G. Maier, *Griechische Mauerbauinschriften*, vol. 2 [Heidelberg, 1961], 22; also multiple generalships, gymnasiarchy, etc: II² 1299)

6. Aristophon of Oinoe, phylarch, 282/1 (TV, p. 104, line 33)
7. Charias of Kephale, phylarch, 282/1 (TV, p. 104, line 31)
8. Charias of Rhamnous, possibly hipparch or phylarch on basis of *prodromos*, mid-3d century (B 556–66; K 62b, pp. 125–27)
9. Demetrios of Phaleron, hipparch, phylarch(?), *strategos* several times, 250s (*IG* II² 1285; Athenaios 4.167 F; *APF* 3455, pp. 109–10; but on date of archon Antimachos [II² 1285, line 5] to 233/2, see Meritt, *Historia* 26 [1977]: 175, 177; Habicht, *Studien*, pp. 49–55 [=*SEG* XXXII.154])
10. Diokles of Teithras, hipparch, ca. 286–261 B.C. (TV, p. 110, line 13)
11. Dionysios of Gargettos, phylarch 282/1 (TV, p. 104, line 28, and p. 108; Reinmuth, *Ephebic Inscriptions*, pp. 88, 96)
12. Epichares, hipparch, 272/1 (Petrakos, *AD* 22 [1967]: 38–52; *SEG* XXIV.154; Heinen, *Untersuchungen*, pp. 152–59; Petrakos, *AE* [1979{81}]: 15–16 [*strategos*]; *CAH*² VII.1, p. 238)
13. Epigenes of Euonymon, possibly phylarch or hipparch on basis of attendant, mid-3d century (K 40, 41, pp. 125–27; B 135–37)
14. Epikrates, possibly phylarch or hipparch on basis of attendant, mid-3d century (B 139–43; K, pp. 125–27)
15. [Gl]auko[n E]teokleous of Aithalidai, phylarch, agonothete, hoplite general, ca. 270s–260s (*IG* II² 3079, now dated to 266/5, winning phylarch in *anthippasia* of Olympieia and Panathenaia; J. Pouilloux, "Glaucon, fils d'Étéocles, d'Athènes," *Le monde grec. Hommages à Claire Préaux* [Brussels, 1975], 376–82; Étienne & Pierart, *BCH* 99 [1975]: 51–75; *CAH*² VII.1, p. 236; *SEG* XXV.186: hoplite general, 266/5; brother of Chremonides)
16. Hagnias of Oion, phylarch, ca. 286–261 B.C. (TV, p. 110, line 27 and p. 111; for 4th-century family, see *APF* 2921)
17. Hi(g)eron of Anagyrous, possibly a phylarch on basis of attendant, mid-3d century (B 243–36; K 30, 43, pp. 125–27)
18. Kalliades of Steiria, phylarch, 281/0 (TV, p. 104, line 30, p. 108)
19. Kallidemos of Euonymon, phylarch, 282/1 (TV, p. 104, line 29)
20. Kallistratos Kleoboulou of Prospalta, phylarch, hipparch, *strategos*, ca. 260/59–257/6, but now downdated to 234/3–231/0 (*IG* II² 2854; Pouilloux, *La forteresse de Rhamonte*, pp. 121–22, no. 9; *ISE* I.26, pp. 56–58; *SEG* XXV.205; Meritt, *Historia* 26 [1977] and *Hesperia* 50 [1981]: 92–93)
21. Komeas Chaireou of Lamptrai, Hipparch to Lemnos, 280/79 or 279/8 (*IG* II² 672; see Appendix B)
22. Leon of Pallene, phylarch, ca. 286–261 B.C. (TV, p. 110, line 28)
23. Lysippos of Kettos, phylarch, 282/1 (TV, p. 104, line 30; p. 108; *APF* 9560)
24. Menon of Hippotomadai, phylarch, 282/1 (TV, p. 104, lines 28–29)
25. Nikoge[n]es Arkesantos of Euonymon, hipparch, 282/1 (TV, p. 104, lines 26–27; p. 108; for son, Arkesas, as *hippeus* mid-3d century, see K 27 and p. 104[i])

26. Philion of Phrearrhioi, hipparch, ca. 286–261 B.C. (TV, p. 110, line 13; p. 111)
27. Proxenos of Phlya, phylarch, 282/1 (TV, p. 104, line 32; p. 108)
28. Python of Alopeke, phylarch, 282/1 (TV, p. 104, line 33)
29. Sosigenes of Myrrhinous, phylarch, 286–261 (TV, p. 110, line 29; p. 111)
30. Sosistratos of Sounion, phylarch, ca. 286–261 (TV, p. 110 , line 24; p. 111; Reinmuth, *Ephebic Inscriptions*, p. 61)
31. Theomedes of Eleusis, phylarch, ca. 286–261 B.C. (TV, p. 110, line 25; p. 111; for father, Theogenes, hipparch on Salamis, see *IG* II² 1955, earlier listing, and Appendix B)
32. Theophrastos, hipparch, ca. 220/19, and gymnasiarch, *strategos* (*IG* II² 1303; *ISE* I.31, pp. 69–71; *SEG* XXV.157; and Chapter Six)
33. Thoukri[tos] (Alkimachou of Myrrhinous), hipparch in archonship of Philinos, 269/8, now downdated to 252/1 (*IG* II² 1279a; *SEG* XXII.125; Meritt, *Hesperia* 50 [1981]: 85–87; for father, Alkimachos Kleouboulou Myrrhinousios, hipparch in 282/1: TV, p. 104, lines 25–26, earlier listing; perhaps same as *hippeus* in mid-3d century [B 217–18])
34. Thrasykles of Dekeleia, phylarch, 282/1 (TV, p. 104, line 32; p. 108: agonothete in 271/0)
35. [--ca. 6--]n of Aithalidai, phylarch, ca. 286–261 B.C. (TV, p. 110, line 17)
36. [--ca. 4--]ikydes of Aphidna, phylarch, ca. 286–261 B.C. (TV, p. 110, line 15)
37. [--ca. 6--]nes of Athmonon, phylarch, ca. 286–261 B.C. (TV, p. 110, line 20)
38. [--ca. 6--]os of Erchia, phylarch, ca. 286–261 B.C. (TV, p. 110, line 18)
39. [--ca. 5--]les of Kephisia, phylarch, ca. 286–261 B.C. (TV, p. 110, line 16)
40. [*ho deina*--]nymou of Iphistiadai, *phy[larchos]*, honored by garrison troops at Eleusis, Panakton, Phyle (restored), end 3d century B.C. (*IG* II² 1307)
41. [--ca. 6--]on of Oe, phylarch, ca. 286–261 B.C. (TV, p. 110, line 19; the nomen of this man has been incorrectly printed as [-----]os)

C. Hellenistic Period, Second to First Centuries B.C.

1. [Aga]thokles tou [Zo]ilou (of Marathon?), winning phylarch at Theseia, ca. 140 B.C. (*IG* II² 961.I.22–24; for brother, agonothete for Theseia in 149/8: II² 958, lines 1–2, 24–25, 36–38)
2. Alexandros tou Alexandrou of Erchia, winning phylarch in Theseia, 157/6 (*IG* II² 957.I.32–35, II.72–73; Martin, *Cavaliers*, p. 281; *PA* 504)
3. Aration tou Simou of Myrrhinoutta, winning phylarch in Theseia, 161/0 (*IG* II² 956.I.57–60, II.79–82; Martin, *Cavaliers*, pp. 280–81; *PA* 1576)
4. Boularchos Damoklea Akamantidos, winning phylarch in Panathenaia, 160s B.C. (*IG* II² 2316, lines 29–31; Martin, *Cavaliers*, p. 281; *PA* 2910; Traill, *Hesperia* 51 [1982]: 205.12, identifies prytanis as same man, but on p. 204, incorrectly associates with Leontis tribe?)
5. Charikles III Theodorou II of Phaleron, phylarch for Pythais of 128/7 (*FD*

III.2, nos. 7.II.4; 24, lines 6–7; 35, line 5; 39, col. 1; and 46, lines 2–3; for relative, Theodoros, phylarch at Theseia, 157/6 (*IG* II² 957.II.71–71), whom Mattingly, *Historia* 20 [1971]: 34, identifies as cousin, not father of Charikles; *PA* 15422)

6. Deimachos tou Deim[achou] of Leukonoion, winning phylarch at Theseia, 157/6 (*IG* II² 957.I.37–38, II.65–66; *PA* 3253; Habicht, *AM* 97 [1982]: 172, n. 11, with discussion, pp. 176, 180)
7. Diogenes II tou Aropou (of Peiraieus), tarantinarch at Pythais of 128/7 (*FD* III.2, no. 24, line 16; no. 35, lines 3–4; no. 46, line 1; for stemma, see *NPA*, p. 54)
8. Diokles Diokleous Athenaios, *strategos epi to hippikon* (= *hipparchos*) at Pythais of 128/7 (*FD* III.2, no. 24, line 5, no. 27.I.4–5; no. 35, line 3; no. 41, col. 2; no. 46, lines 1, 5–6; Colin originally differentiated these two titles, but later identified them as same [*FD* III.2, p. 289] on basis of Ferguson's arguments in *Klio* 9, pp. 315, 317; *PA* 4023; *NPA*, p. 57; possibly same as [---]es Diokleous Kekropidos, *hippeus* in Theseia of 149/8 [*IG* II² 958.II.85–86]; if so, cannot be linked with family of Dromeas Dromeou of Erchia because tribe is Aigeis: *PA* 4023)
9. Echedemos V Mnesitheou of Kydathenaion, cavalry officer(?) in Augustan era (*IG* II² 3000; *PA* 6169; for this very important family, see listing on Echedemos IV Aretou I and MacKendrick, *AthArist*, pp. 37, 44–46; Martin, *Cavaliers*, p. 277)
10. Eubios Telesidemou of Ptolemais, winning phylarch in Theseia, 157/6 (*IG* II² 957.II.69–70; *PA* 5289; related to Telesidemos tou Aminiou of Hekale, Hipparch to Lemnos, ca. post 166/5, now downdated to 148/7–135/4: Tracy, *Hesperia* 47 [1978]: 261–66; see Appendix B)
11. [Euk]les tou Aristo[kle]ous [Peri]thoidou, winning hipparch of tribe Oineis at Theseia, ca. 140 B.C. (*IG* II² 961.I.20–21; *PA* 5730; *NPA*, p. 78; for discussion of family, see Meritt, *Hesperia* 11 [1942]: 293–98, no. 58)
12. Eukles III Herodou I of Marathon, phylarch at Pythais of 106/5 (*FD* III.2, no. 28.I.13; for family, see *NPA*, p. 78; MacKendrick, *AthArist*, p. 64)
13. Euthoinos tou Moschionos of Marathon, winning tarantinarch of Aiantis in Theseia, 149/8 (*IG* II² 958.I.56–58; *PA* 5506; Martin, *Cavaliers*, p. 281; for brother, Polynikos, also tarantinarch at same Theseia [II² 958.I.58])
14. Hagias Boulonos (of Euonymon), phylarch at Pythais of 128/7 (*FD* III.2, no. 24, line 6; no. 27.I.8; no. 34, lines 5–10; no. 35, lines 6–8; no. 39, col. 3; no. 46, lines 2, 7, 10; brother, Straton, equestrian winner in same year [*FD* III.2, no. 42])
15. Harmoxenos Harmoxenou of Kekropis, winning phylarch at Theseia, 157/6 (*IG* II² 957.II.74; *PA* 2238; Martin, *Cavaliers*, p. 281; identified as third mint magistrate for 128/7; second mint magistrate for 127/6, 124/3, 123/2 [Thompson, *New Style Silver Coinage*, p. 555] but now downdated to 95/4, 92/1, 91/90 [Mattingly, *JHS* 91 {1971}: 90], thereby voiding the identification)

16. Hegesandros Hegesandrou, cavalry officer(?), honored by Athenian *hippeis* with crown at Delphi, ca. 151 B.C. (*FD* III.2, 135; *SIG*[3] 654 A–B; Habicht, *AM* 76 [1961]: 137–38)
17. Herakon Charikleous Ptolemaiidos, phylarch at Pythais of 106/5 (*FD* III.2, no. 28, line 11; *NPA*, p. 89)
18. Hermon Dionysiou, *gonôi Antiphontos* (Aigeis), phylarch at Pythais of 128/7 (*FD* III.2, no. 24, line 5; no. 27.I.10; no. 35, lines 4–5; no. 36; no. 46, line 2; *NPA*, p. 73; cf. *IG* II[2] 2332, line 323; Meritt, *Hesperia* 36 [1967]: 89, line 21; and II[2] 2334, lines 45–49 with family stemma, p. 684; II[2] 2335, lines 18–21)
19. Kraton Apollonidou I, hipparch at Pythais of 106/5 (*FD* III.2, no. 28, lines 3–4; for son as pythaist *hippeus*[?], see *FD* III.2, no. 28, lines 5–6 [Apollonides Kratonos]; *NPA*, p. 115)
20. Lysandros II Glaukou II (of Peiraieus), tarantinarch at Pythais of 106/5 (*FD* III.2, no. 28, lines 7–8; *SIG*[3] 711; *NPA*, p. 120; Thompson, *New Style Silver Coinage*, pp. 572–73, identifies with mint magistrate, along with Glaukos, for 159/8, but Mattingly [*JHS* 91 {1971}: 90] to 127/6; for his father, Glaukos, as polemarch and *hippeus*(?) in Pythais of 128/7, see *NPA*, p. 44—there are problems with this apparent duplication of military offices, see later listing)
21. Lyson tou Demokratou (or -eos), tarantinarch at Pythais of 128/7 (*FD* III.2, no. 24, line 16; no. 35, line 4 [restored]; no. 46, line 2 *NPA*, p. 123; identified with winner of "men's wrestling" event in Theseia of 149/8 (*IG* II[2] 958.II.44–45; *PA* 9642; he would be at least 42 years old in 128/7)
22. Nikogenes Nikonos of Philaidai, hipparch at Theseia, 157/6 and agonothete at Theseia, 161/0 (*IG* II[2] 957.I.39–40; II[2] 956, lines 2, 29, 40–42; Martin, *Cavaliers*, pp. 274–75; TV, pp. 106–7 [*contra* restoration by Woodhead, *Hesperia* 29 {1960}: 78 and Habicht, *AM* 76 {1961}: 141–43)
23. [Ophel]as tou Habro[nos Bat]ethen, hipparch at Theseia, 157/6 (*IG* II[2] 957.I.41; exegete in Pythais of 128/7 [*FD* III.2, no. 24, line 9]; son, Drakon, in Theseia of 157/6 in *hippos lampros* event [II[2] 957.II.64]; see also *IG* II[2] 2334, lines 71–73; for family, see Martin, *Cavaliers*, pp. 275–76; *PA* 9 with stemma; and MacKendrick, *AthArist*, p. 52)
24. Pheidyllos Hieronos of Aithalidai, hipparch at Theseia, 149/8 and winner of *hippos lampros* event (*IG* II[2] 958.I.53–55; II.80; Martin, *Cavaliers*, p. 280; *PA* 7529; *NPA*, p. 162)
25. Platon Alexionos (Azenieus) Hippothontidos, winning phylarch at Theseia, 157/6 (*IG* II[2] 957.II.67–68; Martin, *Cavaliers*, p. 281; *PA* 11848); (560)
26. Polynikos tou Moschionos of Marathon, tarantinarch with winning tribe Aiantis in Theseia, 149/8 (*IG* II[2] 958.I.59; Martin, *Cavaliers*, p. 281; *PA* 12055; for brother Euthoinos, also a tarantinarch in Theseia of 149/8: II[2] 958.I.58 and earlier listing)
27. Satyros Hierokleous Kekropidos, phylarch at Panathenaia, ca. 166/5,

winner in two events (*IG* II² 2316, lines 31–36; Martin, *Cavaliers*, p. 281; *PA* 12583)

28. Sosis Sosidos of Oe, ephebe and phylarch(?) in 38/7 (*IG* II² 1043, line 65; *PA* 13280 [cites only status as ephebe]; Gomme & Sandbach, *Menander. A Commentary*, pp. 546–47)
29. [T]elesidemos tou Aminiou of Hekale, Hipparch to Lemnos twice, ca. 166 (*IG* II² 1224, frags. a, b, lines 10–11; downdated to 148/7–135/4 by Tracy on basis of cutter of Agora I 6006 (*Hesperia* 47 [1978]: 261–66; see Appendix B)
30. Theodo[ros] Charikleous (of Phaleron), winning phylarch in Theseia, 157/6 (*IG* II² 957.II.71–72; Martin, *Cavaliers*, p. 281; *PA* 6809; for son(?) Charikles III Theodorou II, phylarch in Pythais of 128/7, see *FD* III.2, no. 24, lines 6–7; *NPA*, p. 171; cf. *IG* II² 2452, lines 5, 9 [ca. 125/4] and discussion of relationship, Mattingly, *Historia* 20 [1971]: 34, who identifies them as cousins)
31. [T]imiades of Gargettos, possibly one of the two hipparchs, ca. 160 B.C. (Woodhead, *Hesperia* 29 [1960]: 78–80, no. 155 = *SEG* XVIII.31 as emended by TV, pp. 106–7 with possibility that nomen is Simiades)
32. Xenokles Demetriou Athenaios, phylarch in Pythais of 128/7 (*FD* III.2, no. 24, line 7; no. 35, lines 5–6; no. 39, col. 2; no. 46, line 3; *NPA*, p. 137)

II. Cavalrymen

A. Classical Period

1. Adeistos, ca. mid-4th century (K 16)
2. [A]gasias, ca. mid-4th century (K 15)
3. Agathinos, ca. mid-4th century (K 14)
4. Ais[---], ca. mid-4th century (K 5)
5. Alechs[---], hippotoxotês? (*SEG* XII.73; see Appendix D)
6. Alexiades (of Anaphlystos?), ca. mid-4th century (K 12)
7. Alkibiades III Kleiniou II of Skambonidai, *hippeus* at Delion, 424 B.C. (Pl. *Smp.* 221 A; Plut. *Alk* 7.2–3; Diog. Laert. "Socrates" 2.22–23; *APF* 600, VII, pp. 17–21; *AAA*, A4)
8. Alkibiades IV Alkibiades III of Skambonidai, illegal *hippotoxotês* in Haliartos campaign, 395 B.C. (Lys. 14; [Lys.] 15; Isok. 16; Plut. *Alk.*; *APF* 600, VIII, X, pp. 19–22)
9. Alkimos, ca. mid-4th century (K 1)
10. Amphis[th]enes, ca. mid-4th century (K 17)
11. Anthemion Diphilou, changed status from *thes* to *hippeus*, post 480 B.C.(?) (*AP* 7.4; Pollux 8.131; Rhodes, *CAP*, pp. 143–45; *APF* 1324, p. 40; Chapter One)
12. Antisthenes (=Antisthenes I Antiphatous of Kytheros?), "undistinguished" *hippeus*, elected to *strategia* instead of Nichomachides, early decades of 4th century (Xen. *Mem.* 3.4.1; *APF* 1194, pp. 38–39, for identification and stemma)

13. [Aris]to[.] e[--ca. 12--] of Anagyrous, stationed on Salamis, ca. 320 B.C. (*IG* II² 1955, line 13; Appendix B)
14. Aristokles, ca. mid-4th century (K 3)
15. Aristolochos, ca. mid-4th century (K 18)
16. *Aristophanes Nikophemou, father-in-law of Aristophanes and father of speaker of Lysias 19, name unknown, a *hippeus* who bought very fine horses (*lamproi hippoi*), won horse races at Isthmia and Nemea, and was a liturgist; moreover, this man married the daughter of Xenophon Euripidou of Melite, hipparch in the 450s; probably served in the cavalry in period of the Peloponnesian War (Lys. 19.14, 63; *APF* 5951, pp. 199–202, with stemma, p. 203; *AAA*, A78)
17. Aristophon, ca. mid-4th century (K 7)
18. Chairephon, reinscribed lead tablet, s.v. Stra[---], ca. mid-4th century (K 21 b)
19. Charamantides Sokleou of Euonymon, stationed on Salamis, ca. 320 B.C. (*IG* II² 1955, line 9; for relative, Sokles Aleximachou, also *hippeus* on Salamis, see II² 1955, line 22; later listing, and Appendix B)
20. Chares, ca. mid-4th century (K 26)
21. Cheimon, ca. mid-4th century (K 22)
22. Deinokles, ca. mid-4th century (K 19)
23. Demoklees (Akamantis?), killed at Korinth in 394 B.C. (*IG* II² 5222; *PA* 3492; *GHI* 104 and s.v. no. 23)
24. Dexileos Lysaniou of Thorikos, young *hippeus*, 19–20 years old, who died in the campaign at Korinth, 394 B.C. (*IG* II² 6217; II² 5222; *GHI* 104, 105, pp. 18–21; J. Labarbe, *Thorikos: Fouilles de Thorikos*, vol. 1: *les testimonia* [Ghent, 1977]; and Chapter Four; Clairmont, *Patrios Nomos* I, pp. 209–221)
25. Dikaiogenes III Dikaiogenous II of Kydathenaion, "unenthusiastic" *hippeus* in 390s (Is. 5.42–43, 46; *APF* 3773, pp. 145–47: born in late 420s; adopted son of Dikaiogenes II, son of Menexenos, phylarch killed at Olynthos in 429 B.C.; "one of the wealthiest and most distinguished families of democratic Athens" [*APF*, p. 145])
26. Diphilos II Philotadou of Lamptrai, stationed on Salamis, ca. 320 B.C. (*IG* II² 1955, line 4; see also *IG* II² 6710 for father; *PA* 4483 and *SEG* XXIX.211)
27. Endelos (Akamantis?), killed at Korinth in 394 B.C. (*IG* II² 5222; *PA* 4697)
28. Epikrates Diophantou of Peiraeus, stationed on Salamis, ca. 320 B.C. (*IG* II² 1955, line 25; *PA* 4910; related to *IG* II² 7194)
29. Eualkides Alkimachou of Kerameis, stationed on Salamis, ca. 320 B.C. (*IG* II² 1955, line 24; *PA* 5260, 5259 (*IG* II² 6338, son of our man?)
30. Euandros I Erithalionos of Euonymon, *hippeus*(?) in 404/3, so implied at scrutiny for archonship in 382/1 (Lys. 26.10; *PA* 5267; *APF* 5267, pp. 187–88.

31. Eukles Lysiou of Melite, stationed on Salamis, ca. 320 B.C. (*IG* II² 1955, line 8; *PA* 5727)
32. Euktaios II Nikiou I of Xypete, stationed on Salamis, ca. 320 B.C. (*IG* II² 1955, line 20; *PA* 5774; ephebe in 334/3, thus about 32 years old in 320; for cousin, Nikias Euktaiou, also stationed on Salamis, see *IG* II² 1955, line 5; for full stemma, see *PA* 10815)
33. Eustrophos Eugenidou of Peiraeus, stationed on Salamis, ca. 320 B.C. (*IG* II² 1955, line 6; *PA* 5996)
34. Glaukippos, ca. mid-4th century (K 8)
35. Gryllos II Xenophontos I of Erchia, son of the historian, who died at Mantineia, 362/1 (Paus. 8.9.5, 10; 8.11.6; 9.15.3; 1.3.4; Diog. Laert. "Xenophon" 2.54; *PA* 3096; Hölscher, *Historienbilder*, pp. 116–19; see Chapter Four)
36. Ischomachos, *hippeus* and interlocutor with Sokrates in Xenophon's *Oikonomikos*, pre-399 B.C.; wealthy member of liturgical class (Xen. *Oik.* 11.14–20; 6.17 to end; for confusion about identity, see *APF* 7826, pp. 265–68; see also F. D. Harvey, *Echos du Monde Classique: Classical Views* 28 [1984]: 68–70)
37. Kallias Kallikratou Kolonethen, stationed on Salamis, ca. 320 B.C. (*IG* II² 1955, line 21; *PA* 7870)
38. K[a]l[lias S]oinautou Kolonethen, stationed on Salamis, ca. 320 B.C. (*IG* II² 1955, line 26; *PA* 7871)
39. Kallistratos (Leontis), killed in 403 B.C. in retaliation for Aixone murders (Xen. *Hell.* 2.4.27; *PA* 8150)
40. Kallistra[tos], ca. mid-4th century (K 10)
41. Kimon II Miltiadou IV of Lakiadai, *hippeus*(?) in 480 B.C. (Plut. *Kim.* 5; *APF* 8429: hippotrophic family; his son, Lakedaimonios, was hipparch in mid-5th century [I² 400; *DAA* 146, no. 135; *AAA*, A34]
42. Kleon I Kleainetou I of Kydathenaion, *hippeus*(?) in 430s–420s (Ar. *Knights* with scholia; Tzetzes on *Clouds* 549a; Connor, *New Politicians*, p. 152 and n. 32; *APF* 8674, pp. 318–20; F. Bourriot, "La famille et le milieu social de Cléon," *Historia* 31 (1982): 404–35; and Chapter Three)
43. Kleophantos Themistokleous I of Phrearrhrioi, *hippeus*(?) in 5th century B.C. (Pl. *Meno* 93 D; Plut. *Them.* 32; *APF* 6669, VI, pp. 217–18; *AAA*, P99)
44. Kleophantos Kleophontos of Eleusis, stationed on Salamis, ca. 320 B.C. (*IG* II² 1955, line 19; *PA* 8633)
45. Komaios (Antiochis?), ca. mid-4th century (K 13; linked with no. 12, p. 109 and possibly *APF* 8693; pp. 320–21: Komaios Komonos of Semachidai)
46. Kritoboulos Kritonos I of Alopeke, *hippeus*, major liturgist, and interlocutor with Socrates, pre-399 B.C. (Xen. *Oik.* 2.6, 3.9; *APF* 8823, pp. 336–37)
47. Lakrates, Olympic victor and recently identified as an Athenian *hippeus* who died in 403 B.C. and buried in Kerameikos (formerly identified as a

Spartan) (Xen. *Hell.* 2.4.33; *IG* II² 11678; *SEG* XXVII.23; Willemsen, *AM* 92 [1977]: 135–40; Pritchett, *GSW* 4, pp. 133–34)

48. Lykios (Polystratou I of Deiradiotai), hipparch of 40–man cavalry contingent during "Anabasis," 401/0 and possibly *hippeus* against Spartans at Dekeleia, ca. 410 B.C. (Xen. *Anab.* 3.3.20; 4.3.22, 25; 4. 7.24; [Lys.] 20.4, 28; *APF* 12076, pp. 467–68: but Lykios, hipparch in 401/0, even if the son of Polystratos, could be the *hippeus* who served in the Sicilian Expedition; in any case, a family with two sons who served in the cavalry)
49. Lysitheos (Akamantis?), killed at Korinth in 394 B.C. (*IG* II² 5222; *GHI* 104, pp. 18–20)
50. Makartatos, one of two *hippeis* killed at Tanagra (458) or in Peloponnesian War (Paus. 1.29.6; Meritt, *Hesperia* 16 [1947]: 147–48, no. 36; *SEG* X.426; Pritchett, *GSW* 4, p. 148, n. 161)
51. Mantitheos, young (ca. 30) aristocrat at scrutiny of Council of Five Hundred, ca. 392–390 B.C. (Lys. 16; Wevers, *Isaeus*, pp. 77–80; *APF* 9667, pp. 364–68, though Davies is disinclined to identify Mantitheos with father of Mantias of Thorikos, subject of Demosthenic speeches, nos. 39 and 40; see Chapter Four)
52. Meixis[---], ca. mid-4th century B.C. (K 23)
53. Melanopos, one of two *hippeis* killed at Tanagra (458) or in Peloponnesian War (Paus. 1.29.6; Meritt, *Hesperia* 16 (1947): 147–48 no. 36; *SEG* X.426; Pritchett, *GSW* 4, p. 148, n. 161)
54. Melesias (Akamantis?), killed at Korinth in 394 B.C. (*IG* II² 5222; *GHI* 104, pp. 18–20)
55. Moschos Antiphanous of Anaphlystos, mover of decree of cavalry of Antiochis, ca. 350–300 B.C. (*SEG* III.115, lines 3–4; *APF* 1227)
56. Nausikles Nausigenou of Anagyrous, stationed on Salamis, ca. 320 B.C. (*IG* II² 1955, line 10; *PA* 10550)
57. Neokleides (Akamantis?), killed at Koroneia in 394 B.C. (*IG* II² 5222; *GHI* 104, pp. 18–20)
58. Nikandros, ca. mid-4th century (K 11)
59. Nikias II Euktaiou II of Xypete, stationed on Salamis, ca. 320 B.C. (*IG* II² 1955, line 5; for cousin, see II² 1955, line 20; for family, see *APF* 10814, pp. 407–8; Nikias must be about 30 years old: ephebe in 334/3 [*IG* II² 1156, line 21]; Reinmuth, *Ephebic Inscriptions*, no. 2; of interest is fact that family has property on Salamis: *IG* II² 6930 and *APF*, p. 407)
60. Nikomachos (Akamantis?), killed at Korinth in 394 B.C. (*IG* II² 5222; *GHI* 104, pp. 18–20)
61. Nikostratos ("the Fair"), killed near Phyle, 404/3 B.C. (Xen. *Hell.* 2.4.6; *PA* 11005)
62. Nikostratos II Nikostratou of Cholargos, stationed on Salamis, ca. 320 B.C. (*IG* II² 1955, line 7; *PA* 11058)
63. Onetorides (Akamantis?), killed at Korinth in 394 B.C. (*IG* II² 5222; *GHI* 104; *APF* 11473, (G), p. 422, which identifies this man as possibly Onetorides II of Melite of Kekropis, thus rejecting casualty list as limited to Akamantis)

64. Pandios (Akamantis?), killed at Korinth in 394 B.C. (*IG* II² 5222; *GHI* 104, pp. 18–20)
65. Paralos Perikleous I of Cholargos, *hippeus*, son of the famous Perikles, who also died in the plague, ca. 430 B.C. (Pl. *Meno* 94 B; Plut. *Per.* 36.8; *APF* 11811, pp. 457–59; *AAA*, P106; his brother, Xanthippos, also identified as *hippeus* and victim of the plague)
66. Perikles I Xanthippou I of Cholargos, possible *hippeus*, instructed sons to be such, died in plague, ca. 430 B.C. (Pl. *Meno* 94 B; *APF* 11811)
67. Polystratos II Polystratos I of Deiradiotai, perhaps the son who served as a cavalryman in the Sicilian Expedition 415–413, s.v. Lykios ([Lys.] 20.24–25; *APF* 12076)
68. Prokles Erxigenou of Kollytos, stationed on Salamis, ca. 320 B.C. (*IG* II² 1955, line 23; *PA* 12230)
69. Prokles Gniphonos (of Thorai), *hippeus* (or cavalry officer?) associated with cavalry relief, ca. mid-4th century (*IG* II² 12523; Meritt, *Hesperia* 15 (1946): 217, and n. 30)
70. Pronapes of Aixone, litigant in lawsuit over property of Apollodoros II Thrasyllou II of Leukonoion; aspires to offices appropriate to Solonian census class of *hippeus*, but no evidence of cavalry service, pre-350s (Is. 7.39; Ste. Croix, *C&M* 14 (1953): 44; Wevers, *Isaeus*, p. 63; *APF* 1395, pp. 45–46)
71. Phainippos Kallippou, young, fit, and rich *hippeus* who retired from cavalry service in his twenties, ca. 330 ([Dem.] 42.24; Ste. Croix, "The Estate of Phaenippus (Ps.-Dem., xlii)," *Ancient Society and Institutions: Studies in Honour of Victor Ehrenberg* (Oxford, 1966), 109–14; *APF* 14734, pp. 552–54)
72. Phanes (Akamantis?), killed at Korinth in 394 B.C. (*IG* II² 5222; *GHI* 104, pp. 18–20)
73. Pheidippides, son of Strepsiades, character in Aristophanic play of 423 B.C. with hippic interests, long hair, but actual cavalry service is problematic inasmuch as he sells his horses and turns to philosophy apparently without state restriction. But this is a comic play and a fictional character (Ar. *Clouds*)
74. Phileas Diokleous of Elaious, stationed on Salamis, ca. 320 B.C. (*IG* II² 1955, line 17; *APF* 4021 on father or brother as trierarch in 330s)
75. Philokles, ca. mid-4th century B.C. (K 20)
76. Philok[---], ca. mid-4th century B.C. (K 25)
77. Philoktemon I Euktemonos of Kephisia, cavalryman and trierarch many times and died while serving in the latter capacity off Chios in the 370s (Is. 6.5; Diod. 15.32.2; Lacey, *The Family in Classical Greece*; Wevers, *Isaeus*, p. 61; *APF* 15164, pp. 562–66 with stemma)
78. Philomelos II Philippidou I of Paiania, a *hippotrophos* who loaned a horse to Meidias in 349/48, but not attested as a *hippeus* (Dem. 21.174; *APF* 14670, pp. 548–50 and linked by marriage with the family of Xenophon Euripidou, see *APF* 5951 with stemma, p. 203)
79. Philon Aristokleos of Melite, *hippeus* depicted in cavalry-infantry combat

on a funerary loutrophoros, ca. 370 B.C. (Nikopoulou, *AAA* 2 [1969]: 329–33; Clairmont, *GRBS* 13 [1972]: 53, n. 4, i)

80. Philostratos Teisandrou of Peiraeus, stationed on Salamis, ca. 320 B.C. (*IG* II² 1955, line 11; *PA* 14743)
81. Smikythos Sosippou of Aixone, stationed on Salamis, ca. 320 B.C. (*IG* II² 1955, line 18; *PA* 12784; *SEG* XVIII.97)
82. Smoios, described as "wearing a hippic garment" in Aristophanic play of 390s (Ar. *Ekklesiazusai* 846 with scholia)
83. Sokles Aleximachou of Euonymon, stationed on Salamis, ca. 320 B.C. (*IG* II² 1955, line 22; cousin of Charmantides Sokleou also stationed on Salamis [II² 1955, line 9]; *PA* 13069, 13070 with stemma)
84. S[o]k[r]ates Eudramonos of Aphidna, stationed on Salamis, ca. 320 B.C. (*IG* II² 1955, line 12; *PA* 13105)
85. Strat[--], ca. mid-4th century B.C. (K 21a; reinscribed tablet with other name as Chairephon)
86. Theangelos (Akamantis?), killed at Korinth in 394 B.C. (*IG* II² 5222; *GHI* 104, pp. 18–20)
87. Themistokles I Neokleous I of Phrearrhioi, implied as *hippeus* inasmuch as he taught his son Kleophantos to be one, 6th–5th centuries B.C. (Pl. *Meno* 93 D; Plut. *Them.* 32; some evidence to associate him with *hippotrophia*, see *APF* 6669, p. 215 and *AAA*, P91)
88. Theogenes, ca. mid-4th century B.C. (K 4)
89. Thoudes, ca. mid-4th century B.C. (K 2)
90. Thrasippos, ca. mid-4th century B.C. (K 9)
91. Timokrates Antiphontos of Krioa, *hippotrophos* and winner of equestrian event at Olympic Games, mid-4th century, but nowhere attested as *hippeus* (*IG* II² 3127; L. Moretti, *Olympionikai* [Rome, 1957], 124, no. 440; *APF* 13772, pp. 513–14; *AAA*, A65)
92. Xanthippos Perikleous I of Cholargos, *hippeus*, son of the famous Perikles, who also died in the plague, ca. 430 B.C. (Pl. *Meno* 94 B; *APF* 11811, pp. 457, 459; *AAA*, P105; his brother Paralos, also *hippeus* and died in same plague, see above)
93. Xenophon of Melite, ca. mid-4th century; his paternal grandfather is probably the hipparch Xenophon Euripidou (K 24, pp. 111–12; *APF* 11313 and 5951 with stemma, now requiring updating; *AAA*, A78)
94. Xenophon I Gryllou I of Erchia, the historian, probably *hippeus* in 404/3, certainly a *hippotrophos* on basis of equestrian treatises (*Hipparchikos*, *Peri Hippikes*; *PA* 11307; Anderson, *Xenophon*; his son Gryllos died as a *hippeus* at Mantineia, 362 B.C.)

B. Hellenistic Period, Third Century B.C.

1. Aischylides, ca. mid-3d century B.C. (B 8)
2. Alkibiades (Leontis?), ca. mid-3d century, possibly linked with famous Kleinias-Alkibiades family of Classical period (B 9; K 74, p. 130)
3. Alkmeon of Aphidna, ca. mid-3d century (B 13, 14)

4. Ameinokles, ca. mid-3d century (B 15–19)
5. Andro[---], ca. mid-3d century (B 20–21)
6. Antiandros, ca. mid-3d century (B 22–23)
7. Antikrates of Hermos, ca. mid-3d century; probably descendant of Antikrates Sokratous Hermeios, hipparch in 4th century (B 24–25, p. 243)
8. Antimachos of Rhamnous, ca. mid-3d century (B 27–28, p. 245; K, p. 104 [d]; and now Petrakos, *AE* [1979]: 27–29, no. 8; *SEG* XXX.207, 208: Antimachos Thrasymachou Rhamnousios, 4th century)
9. Antimedes, ca. mid-3d century (B 31)
10. Antiphon, ca. mid-3d century (B 32–34)
11. Antichares, ca. mid-3d century (B 35)
12. Apollodoros, ca. mid-3d century (B 36)
13. Apollophanes Neoptolemou, ca. mid-3d century (B 37, 39)
14. Apoll[---], ca. mid-3d century (B 38)
15. Archeboulos, ca. mid-3d century (B 65–66)
16. Archedemos, ca. mid-3d century (B 67)
17. Archestra[tos], ca. mid-3d century (B 68)
18. Archippos, ca. mid-3d century (B 69)
19. Aristaios, ca. mid-3d century (B 40–42)
20. Aristeas, ca. mid-3d century (B 43)
21. Aristeides, ca. mid-3d century (B 501, 1)
22. Aristion (Leontis?), ca. mid-3d century (B 44–45; K 75, 81b, 82)
23. Aristogenes, ca. mid-3d century (B 46)
24. Aristodikos (Hippothontis?) ca. mid-3d century (B 47–49; K, p. 133, n. 69 links B 47 with Hippothontis in Series H [nos. 85–88])
25. Aristokrates, ca. mid-3d century (B 50; K 104)
26. Aristomachos of Myrrhinous, ca. mid-3d century (B 52)
27. Ariston, ca. mid-3d century (B 56)
28. Aristonymos of Thorai, ca. mid-3d century (B 57)
29. Aristonymos, ca. mid-3d century (B 58–59)
30. Ariston[---], ca. mid-3d century (K 103)
31. Aristo[---], ca. mid-3d century (B 55)
32. Aristophanes Amo[---], ca. mid-3d century (B 54)
33. Aristophon, ca. mid-3d century (K 92)
34. Aristoteles, ca. mid-3d century (B 53)
35. Arkesas of Euonymon, ca. mid-3d century, son of Nikogenes, hipparch in 282/1 (K 27, 36; B 60, 61; TV, p. 104, lines 25–26)
36. Arkesilas, ca. mid-3d century (B 62–63; K 85, 89 and p. 133, n. 69 for affiliation with Hippothontis tribe)
37. Arketo[s], ca. mid-3d century (B 64)
38. Asklepiades Zenonos of Phyle, ca. mid-3d century (B 70 with p. 245; K, p. 106, x: *strategos* in 225/4 [*IG* II² 2978])
39. Asklepiades tou Xenonos, of Phyle ca. mid-3d century (B 71 with p. 245; K, p. 106[w])

40. [As]klepiades, ca. mid-3d century (B 72–74; K, p. 133, n. 69 links B 73 with Hippothontis)
41. Astynomos of Oion, ca. mid-3d century (B 75–76); identified with Astynomos Lysiadou, member of well-recorded and politically important family of the 4th–3d centuries B.C.: *SEG* XXXII.559.
42. Astynomos, ca. mid-3d century (B 77–80); probably same as no. 41.
43. Athenodaios, ca. mid-3d century (B 6)
44. Athenodoros, ca. mid-3d century (B 7)
45. Autokles, ca. mid-3d century (B 81–83)
46. Automenes, ca. mid-3d century (B 84)
47. Boiotos, ca. mid-3d century (B 86–88)
48. Chaireas, ca. mid-3d century (B 546)
49. Chairephon of Alopeke, ca. mid-3d century (K 55, 61, 63a, 72; p. 122: eponymous archon in 219/8 (*IG* II² 1706; *SEG* XXVI.143)
50. Chairestratos, ca. mid-3d century (B 547–549)
51. Chairion of Lamptrai, ca. mid-3d century (B 550–53; K 47)
52. Chares of Aphidna, ca. mid-3d century (B 27[2], 554–55; fairly active political family in 3d century: B, p. 249 and K, p. 104[e]; see also Habicht, *Untersuchungen*, p. 136)
53. Chariklei[des], ca. mid-3d century (B 567)
54. Charinos of Ikaria, ca. mid-3d century (B 568; K 111)
55. Charmiades, ca. mid-3d century (B 569)
56. Charopinos, ca. mid-3d century (B 570–72)
57. Chniphonides of Prospalta, ca. mid-3d century (B 573)
58. Deinias, ca. mid-3d century (B 90–92)
59. Deinon, ca. mid-3d century (B 94)
60. Demeas, ca. mid-3d century (B 95)
61. Demetrios, ca. mid-3d century (B 96–98)
62. Demochares, ca. mid-3d century (B 99–100)
63. Demokl[---], ca. mid-3d century (K 93; may be same man as K 105 and restored as Demokles, p. 138)
64. Dexandros of Anaphystos, ca. mid-3d century (K 49, 57; for prosopography see K, p. 105[v] and p. 119: a Dexandros, son of Philinos of Anaphlystos, was winner of horse race in 161/0 B.C. at Theseia [*IG* II² 956.II.89–90])
65. Diogenes of Paiania, ca. mid-3d century (B 101–3; K, pp. 92–93, n. 26)
66. Diokles, ca. mid-3d century (B 104–5)
67. Dionysodoros, ca. mid-3d century (B 106; K 86 (Hippothontis?)
68. Diopeithes, ca. mid-3d century (B 108–112)
69. Dioskourides of Euonymon (B 113–116 with p. 245; K 37 with pp. 104[g] and 116; Traill, *Hesperia* 51 [1982]: 199 = Dioskourides II, son of Theophanes, proxenos in Oropos; father agonothete, 270/69)
70. Diphilides of Paiania, ca. mid-3d century (B 120; *PA* 4460)
71. Diphilos, ca. mid-3d century (B 121)
72. Diotimos, ca. mid-3d century (B 117–19)

73. Dromokles of Kephisia, ca. mid-3d century (B 122–123; K 38 and possibly K 35 with restoration)
74. Ekphantos, ca. mid-3d century (B 124–29)
75. Endios, ca. mid-3d century (B 130)
76. Epainetos of Euonymon, ca. mid-3d century (B 132; K 28, 39)
77. Epameinon, ca. mid-3d century (B 133–34)
78. Epichares, ca. mid-3d century (B 144–48)
79. Epigonos, ca. mid-3d century (B 138)
80. Erastos, ca. mid-3d century (B 149–50)
81. Eraton, ca. mid-3d century (B 151)
82. Eubios, ca. mid-3d century (B 152)
83. Euboulos, ca. mid-3d century (B 153–55, 338[1], 465[1])
84. Eucharistos, ca. mid-3d century (B 182–83)
85. Eudemos of Oinoe, ca. mid-3d century (B 156[1]; K, p. 92, n. 26 for ascription to Aiantis)
86. Eudromos, ca. mid-3d century (B 157)
87. [Eu]klinos, ca. mid-3d century (B 174)
88. Euktimenos (Leontis?), ca. mid-3d century (B 175–77; K 81[a], 83, and p. 132 for prosopography and linkage with Polyktemon Euktimenou Eupyrides; K, *contra* B, lists the *nomen* as Euktimenes, not Euktimenos)
89. Euphanes, ca. mid-3d century (B 179–80)
90. Euphranor, ca. mid-3d century (B 181)
91. Eupolemos, ca. mid-3d century (B 178)
92. Euthetos, ca. mid-3d century (B 158)
93. Euthias, ca. mid-3d century (B 159)
94. Euthippos, ca. mid-3d century (B 160)
95. Euthoinides, ca. mid-3d century (B 161–63)
96. Euthydemos, ca. mid-3d century (B 164–65)
97. Euthydomos of Paiania, ca. mid-3d century (B 167–69)
98. Euthykles, ca. mid-3d century (B 170)
99. Euthykrates, ca. mid-3d century (B 171–72)
100. Euthymenes (Leontis?), ca. mid-3d century (B 173; K 76)
101. Exekias, ca. mid-3d century (B 131)
102. Gniphonides, ca. mid-3d century (B 89)
103. Habron, ca. mid-3d century (B 1–2)
104. Hagnias, ca. mid-3d century (B 3)
105. Hagnostratos (Leontis?), ca. mid-3d century (B 4, 5; K 73)
106. Haphron, ca. mid-3d century (B 85)
107. Hegesandros, ca. mid-3d century (B 184)
108. Hiero[---], ca. mid-3d century (B 237)
109. Hierokleid[es], ca. mid-3d century (B 238–40)
110. Hieromnemon, ca. mid-3d century (B 241)
111. Hieronymos, ca. mid-3d century (B 242–43)
112. Hippokrates (Antiochis?), ca. mid-3d century (K 106; may be father or son of Konon Hippokratous (K 69; p. 138)

113. Ion, ca. mid-3d century (B 248)
114. Kalaides, ca. mid-3d century (B 249)
115. Kallenikos, ca. mid-3d century (B 104, 250–52)
116. Kalliades, ca. mid-3d century (B 253–55)
117. [Kallias of Euo]nymon, ca. mid-3d century (K 48, p. 119 for this likely restoration)
118. Kallias of Thorikos, ca. mid-3d century (B 257 and p. 247; K, p. 104[c] identifies this man with grandson of Kallias II, trierarch in 322, see *APF* 7865)
119. Kallias [of Euonymon], ca. mid-3d century (B 261, with p. 247; K 31 with p. 114 for prosopography)
120. Kallias, ca. mid-3d century (K 107)
121. Kallido[mos], ca. mid-3d century (B 262)
122. Kallikrates of Phaleron, ca. mid-3d century (B 263 with p. 247)
123. Kallippos, ca. mid-3d century (B 264–68)
124. Kallisthenes Hegesippou, ca. mid-3d century (B 269–75)
125. Kallistratos, ca. mid-3d century (B 276–79)
126. Kalliteles, ca. mid-3d century (K 94)
127. Kephision of Hagnous, ca. mid-3d century (B 280 with p. 247)
128. Kephisodoros, ca. mid-3d century (B 289)
129. Kephisodotos, ca. mid-3d century (B 281–87)
130. Kephis[---], ca. mid-3d century (B 288)
131. Kineas of Euonymon, ca. mid-3d century (B 290–93; K, p. 90)
132. Kleidemos, ca. mid-3d century (B 294–95)
133. [K]leippid[es], ca. mid-3d century (B 200)
134. Kleippos, ca. mid-3d century (B 296)
135. Kleoch[ares] of Kephisia, ca. mid-3d century (B 302–5 and p. 247: grandfather was trierarch in 340–320s, *APF* 2954; K 32 and pp. 104[h] and 114 [family member was delegate to Pythais between 330 and 324]; see also Habicht, *Untersuchungen*, p. 149)
136. [Kleomedo]n (Hippothontis?), ca. mid-3d century (B 298–301; K 90 with prosopography, p. 135 and ascription to Hippothontis, p. 133, n. 69)
137. Konon of Alo(peke), ca. mid-3d century (K 59[b], 71, see also K 69 and 106; for problems of identification and linkage with either Alkmeonidai or famous Konons of Classical period, see K, pp. 105[t] and 124; K 69: Konon Hippokratous)
138. Konon, ca. mid-3d century (B 306)
139. Korragos, ca. mid-3d century (B 307)
140. K[r]atin[o]s, ca. mid-3d century (K 59[a]; Antiochis?)
141. Ktesikles, ca. mid-3d century (B 308–9)
142. [Ky]bernis, ca. mid-3d century (B 310)
143. Kydias (Leontis?), ca. mid-3d century (B 311–14; K 77, 84 with p. 130 for prosopography)
144. Laches, ca. mid-3d century (B 315–17)

145. Leontichos (Hippothontis?), ca. mid-3d century (B 318–20; K 87 and p. 133, n. 69)
146. Leochares (Hippothontis?), ca. mid-3d century (B 327–29; K 91 and p. 133, n. 69)
147. Leosthenes, ca. mid-3d century (B 324–26)
148. Lyandros (Antiochis?), ca. mid-3d century (K 108 with p. 138)
149. Lysandrides, ca. mid-3d century (B 330)
150. Lysiades of Xypete, ca. mid-3d century (B 331–33)
151. Lysikles, ca. mid-3d century (B 338–41)
152. Lysimachos, ca. mid-3d century (B 342–43)
153. Lysippos, ca. mid-3d century (B 344–45)
154. Lysistratos (Leontis?), ca. mid-3d century (B 346–49; K 78)
155. Lysitheides, ca. mid-3d century (B 335–37)
156. Megakleides, ca. mid-3d century (B 350–51)
157. Melesias, ca. mid-3d century (B 352–54)
158. Melesippos, ca. mid-3d century (B 356)
159. Menandros, ca. mid-3d century (K 95)
160. Menedemos, ca. mid-3d century (B 358–66)
161. Menestra[tos], ca. mid-3d century (B 367)
162. Menipos, ca. mid-3d century (B 368–69)
163. Menon, ca. mid-3d century (B 370–72)
164. Mikion, ca. mid-3d century (B 373–79)
164. Mnesides, ca. mid-3d century (K 96 with p. 136)
165. Mnesitheos, ca. mid-3d century (B 380–85)
166. Moiriades, ca. mid-3d century (B 386–87)
167. Moschos, ca. mid-3d century (B 388–93)
168. Myrmidon, ca. mid-3d century (B 394)
169. Nausistratos, ca. mid-3d century (396–97)
170. Nauthikos, ca. mid-3d century (B 395)
171. Nau[---], ca. mid-3d century (B 398)
172. Nearchos, ca. mid-3d century (B 399)
173. Neokles, ca. mid-3d century (B 400)
174. Nikenor, ca. mid-3d century (B 401–2)
175. Nikeratos, ca. mid-3d century (B 403–9)
176. Nikesias of Phaleron, ca. mid-3d century (B 410; for possible ancestor as *tamias tou theou* in 377/76, see *PA* 10748: Nikesias Nikesidikou Phalereos)
177. Niketes of Alopeke, ca. mid-3d century (K 51 with p. 119)
178. Nikiades, ca. mid-3d century (B 411)
179. Nikias, ca. mid-3d century (B 88, 412–13)
180. Nikogenes, ca. mid-3d century (B 414)
181. [Nik]omachos of Lamptrai, ca. mid-3d century (B 415–19 with p. 249; K 48a; notable family, see K, pp. 104[1] and pp. 119; *APF* 12883, pp. 491–93; Kineas [s.v.], hipparch, ca. 350–270 B.C. [Vanderpool, *Hesperia*

43 {1974}: 312–13] is suggested as father or great-grandfather of this Nikomachos: K, p. 119)

182. Nikostratos, ca. mid-3d century (B 420–21)
183. Noemon, ca. mid-3d century (B 422)
184. Olympiodoros (Leontis?), ca. mid-3d century (B 423; K 79)
185. Olynpiodoros of Anaphlystos, ca. mid-3d century (K 52, 62a; well-known family, see K, p. 121; *APF* 13905, pp. 515–16)
186. Onetor, ca. mid-3d century (B 424–27)
187. Pamphilos, ca. mid-3d century (B 428)
188. Pantak[les], ca. mid-3d century (B 429)
189. Pant[a]leon, ca. mid-3d century (K 109)
190. Pausanias of Oinoe (Aiantis), ca. mid-3d century (B 156[2], 430–32; K, p. 104[f] for prosopography and dating importance)
191. Pausias, ca. mid-3d century (B 433)
192. Petraios, ca. mid-3d century (B 434)
193. Phaidros, ca. mid-3d century (B 502)
194. Phanokles of Marathon, ca. mid-3d century (B 451[2], 503)
195. Phanostratos, ca. mid-3d century (B 504–5)
196. Pheidon, ca. mid-3d century (B 507)
197. Pheidostratos, ca. mid-3d century (B 506)
198. Philagros (Leontis?), ca. mid-3d century (B 508–13; K 80 and p. 132)
199. Philaios, ca. mid-3d century (B 514–16)
200. Philias, ca. mid-3d century (K 110)
201. Philion, ca. mid-3d century (B 521)
202. Philistides of Anaphlystos, ca. mid-3d century (B 518–20; K 53, 64b, 66?, 70 with discussion on the dissimilarities between the Agora and Kerameikos tablets, p. 121)
203. Philokedes, ca. mid-3d century (B 427[1], 523–25)
204. Philokrates Batrach(ou) of Euonymon, ca. mid-3d century (B 526–28 and possibly 529 [cf. K 33] with p. 249; K 45 with p. 118)
205. Ph[ilo]krates of Kephisia, ca. mid-3d century (B 530; K 33, 46 with p. 114 [and possibly B 529])
206. Philokydes, ca. mid-3d century (B 531–32)
207. Philon, ca. mid-3d century (B 534)
208. Philonides of Angele, ca. mid-3d century (B 535)
209. Philotheos, ca. mid-3d century (B 522)
210. Philoxenos, ca. mid-3d century (B 181[1])
211. Phormion, ca. mid-3d century (B 341[2], 541–42)
212. Phoryskides of Alopeke, ca. mid-3d century (K 54, 60, 65b; see K, pp. 105[r] and 121–22 for importance of this man for questions of the Athenian calendar in 3d century; Walbank [*Hesperia* 52 {1983}: 133] suggests he was granted Athenian citizenship)
213. Phoxias, ca. mid-3d century (B 536–40)
214. Phrynichos, ca. mid-3d century (B 543)
215. Phyromachos, ca. mid-3d century (B 544–45)

216. Platon, ca. mid-3d century (B 435)
217. Polemon, ca. mid-3d century (B 436)
218. Polyaratos, ca. mid-3d century (B 437)
219. Polyeuktos, ca. mid-3d century (B 438–39)
220. Polykleides, ca. mid-3d century (B 440–41)
221. Polykrates, ca. mid-3d century (B 442–43)
222. Polymedes, ca. mid-3d century (B 444)
223. Polyoktos, ca. mid-3d century (B 446)
224. Polyxenos of Erchia, ca. mid-3d century (B 445)
225. Poseidippos [An]ticha[rou or -rmou], ca. mid-3d century (B 445[2], 447–48; cf. *PA* 1317–20 for ending -armou)
226. Prokleides of Aphidna, ca. mid-3d century (B 449–51[1] with p. 249 and K, p. 104[b]; see *APF* 1916 for grandfather as trierarch in 320s)
227. Prokles, ca. mid-3d century (B 452)
228. Protarchos (Hippothontis?), ca. mid-3d century (B 458–61; K 88 with p. 133, n. 69)
229. Proxenos (Hippothontis?), ca. mid-3d century (B 453–57; K, p. 133, n. 69)
230. Pythodorides, ca. mid-3d century (B 462–63)
231. Pythodoros Home[rou], ca. mid-3d century (B 464–65)
232. Pythokles (Hippothontis?), ca. mid-3d century (B 329[2], 466–67; K, p. 133, n. 69)
233. Python, ca. mid-3d century (B 468)
234. Satyros, ca. mid-3d century (B 469)
235. Semon, ca. mid-3d century (B 470)
236. Simias, ca. mid-3d century (B 471; K 98)
237. Smikythos, ca. mid-3d century (B 472–74)
238. Sogenes, ca. mid-3d century (B 481–83)
239. Sokleides, ca. mid-3d century (B 484)
240. Sokrates, ca. mid-3d century (B 485)
241. Sosikrates, ca. mid-3d century (B 486)
242. Sosistratos, ca. mid-3d century (B 487)
243. Sostratos of Lamptrai, ca. mid-3d century (B 352[1], 488; possibly related to Sostratos Deinippou Lamptreus, holder of multiple trierarchies in 320s [*PA* 13361, 13362])
244. Spoudias, ca. mid-3d century (B 475–79)
245. Sthennides (Hippothontis?), ca. mid-3d century (B 467[2]; K, p. 133, n. 69)
246. Strepheneos, ca. mid-3d century (B 480)
247. Teisandros, ca. mid-3d century (B 489–91)
248. Theaios (Antiochis), ca. mid-3d century (K 67 with pp. 105[q] and 128)
249. Theaitetos, ca. mid-3d century (B 185–89)
250. Theocharides, ca. mid-3d century (B 213)
251. Theodoros of Alopeke, ca. mid-3d century (K 50, 58, and 65a with pp. 105[s] and 119 (thesmothete in 229/8)

252. Theodoros of Euonymon, ca. mid-3d century (B 198 and possibly 196–97; K 29; K, p. 112, suggests connection with Evander, implied in Lysianic speech of having served in the cavalry in 404/3 at his scrutiny in 382/1 (see *APF* 5267, pp. 187–88)

253. Theodoros of Rhamnous, ca. mid-3d century (B 13[1], 196–98?; cf. K 29 for different linkage with 196–98; for prosopography, see B, p. 245; K, pp. 103–04[a] with reference to Pouilloux, *La forteresse de Rhamnonte*, p. 163, no. 61)

254. Theodotos (Hippothontis?), ca. mid-3d century (B 193–94; K, p. 133, n. 69)

255. Theogeiton, ca. mid-3d century (B 190–91)

256. Theogenes (Antiochis?), ca. mid-3d century (K 68 with pp. 105[p] and 128: possibly Theogenes Poseidoniou Amphitropethen)

257. Theognis, ca. mid-3d century (B 192)

258. Theokles, ca. mid-3d century (B 199–200)

259. Theomedes of Anagyrous, ca. mid-3d century (B 201–3; K 42)

260. Theomnestos, ca. mid-3d century (B 204)

261. Theophan[es], ca. mid-3d century (B 209)

262. Theophilos, ca. mid-3d century (B 210–11)

263. Theophrastos of Lamptrai, ca. mid-3d century (B 151[1], 212)

264. Theoxenos (Hippothontis?), ca. mid-3d century (B 205–8; K, p. 133, n. 69)

265. Therippides, ca. mid-3d century (B 216)

266. Theudoros, ca. mid-3d century (B 215)

267. Theudotos, ca. mid-3d century (B 214)

268. Thoukritos, ca. mid-3d century (B 217–18; perhaps the same as Thoukritos Alkimachou Myrrhinousios, hipparch in 252/1 [new dating] and son of Alkimachos Kleouboulou, hipparch in 282/1, see earlier listings; on the career of Thoukritos as critical for the dating of the mid-3d century archons, see M. J. Osborne, "The Archonship of Nikias Hysteros," *ZPE* 58 [1985]: 289–93)

269. Thoumorios (Erechtheis), ca. mid-3d century (B 220; K, p. 105[o] with link to Thoumorios of Euonymon, contributor in 247/6)

270. Thrasippos, ca. mid-3d century (B 221–23)

271. Thrasyboulos, ca. mid-3d century (B 224–25)

272. Thrasykles of Lamptrai, ca. mid-3d century (B 226–28)

273. Thrasykles (Antiochis), ca. mid-3d century (K 63b, 64a)

274. Thrasymedes (Hippothontis?), ca. mid-3d century (B 229–31; K, p. 133, n. 69)

275. Thymochares of Sphettos, ca. mid-3d century (B 232–33 with pp. 245–49 and K, p. 103: full prosopography for this important family [cf. *APF* 13964, pp. 527–28 with Meritt, *Hesperia* 28 {1969}: 433] with considerations for dating of tablets)

276. Timokleides [--]ypono[--], ca. mid-3d century (B 496)

277. Timokrates, ca. mid-3d century (B 497)

278. Timonides of Euonymon, ca. mid-3d century (B 499–500; K 44 with p. 118 for prosopography)
279. Timostratos, ca. mid-3d century (B 498)
280. Timotheos, ca. mid-3d century (B 492–95)
281. Tmoleides, ca. mid-3d century (B 501)
282. Xenon, ca. mid-3d century (K 97)
283. [nomen, L?]amp[---] (Erechtheis), ca. mid-3d century (K 34; K, p. 114, suggests that it may have been inscribed for Nikomachos or Chairion, only known members of deme Lamptrai from Erechtheis)
284. [---]doros, ca. mid-3d century (K 99)
285. [---k]les, ca. mid-3d century (K 100)
286. [---]patros, ca. mid-3d century (K 101)

C. Hellenistic Period, Second to First Centuries B.C.

1. Achaios Kekropidos, Pythais of 106/5 (*FD* III.2, no. 28.II.35; probably the son of Dios Achaiou Kekropidos, winner of four prestigious equestrian events at Theseia of 157/6 [*IG* II² 957.II.86–90]; but also identified as ephebe in 117/6 [*IG* II² 1009.III.81]: Achaios Bou[---Me]liteus, by Raubitschek, *Hesperia* 11 [1942]: 307; but he cannot be both if the patronymic for the ephebe is correct)
2. Agathippos Kekropidos, Pythais of 106/5 (*FD* III.2, no. 28.II.31; *SIG*³. 711, p. 355; perhaps related to Hermias Aga[---] [no. 9, line 2] and no. 12.I.16; cf. Philemon Hermiou [no. 17.II.24, 98/7 B.C.]; for possible mint magistrate, 131/0 (Thompson, *New Style Silver Coinage*, p. 547) or 99/8 (Mattingly, *JHS* 91 [1971]: 91); it also seems very likely that [Xan]thippos Hermiou, winner of equestrian event in Pythais of 106/5 [no. 44.I], should be restored to read [Aga]thippos Hermiou, and to be identified with our man)
3. Alexandros Antiochidos, Pythais of 106/5 (*FD* III.2, no. 28.III.23; Colin, *FD* III.2, p. 44 and *SIG*³ 711, pp. 354–56 identify this man with Alexandros Nikagorou, winner of equestrian event, Pythais, 106/5 [*FD* III.2, no. 44.III], but Sundwall correctly disassociates them [*NPA*, pp. 8–10], as deme of Alexandros Nikagorou is Phlya, belonging to tribe Ptolemais, not Antiochis)
4. Alexandros II Nikagorou I of Phlya (Ptolemais), not attested in tribal register, but winner of equestrian event in Pythais of 106/5 (*FD* III.2, no. 44.III—see preceding entry; his cousin Polykleitos [s.v.] was a *hippeus* in Pythais of 106/5 and this cousin's sons participated as child pythaists in 106/5 and 98/7 *FD* III.2, no. 15.III.8–9; no. 17.II.9–10); brothers were adult pythaists in 98/7 (*FD* III.2, no. 17.I.20–21); for stemma of this family, see *NPA* p. 10)
5. Alkidamos (Euphanous of Euonymon) Erechtheidos, Pythais of 128/7 (*FD* III.2, no. 27.I.15; 33, line 5; his brother, Eumareides [now restored s.v.] was also a *hippeus* in same Pythais [*FD* III.2, no. 27.I.14]; both mint magistrates for 113/2 [Mattingly, *JHS* 91 {1971}: 91]; *NPA*, p. 9)

6. Amyntas Akamantidos, Pythais of 128/7 (*FD* III.2, no. 27.II.1; *NPA*, p. 13)
7. Anaxikrates (Attalidos), Pythais of 128/7 (*FD* III.2, no. 27.II.39; *NPA*, p. 14; correct tribal affiliation for this list is complicated by the fact that the stonecutter omitted the heading for Attalis, but Tracy, *BCH* 99 [1975]: 193, has shown that Menophilos [s.v.] belongs to Attalis; since this man's name appears higher in the list than Anaxikrates, the latter must also belong to Attalis)
8. Androkles Antiochidos, escort *hippeus* for Pythais of 106/5 (*FD* III.2, no. 28.IV.55; *NPA*, p. 14)
9. Andron Zoilou (Paianieus) Pandionidos, tarantine hippic winner in Theseia, ca. 140 B.C. (*IG* II[2] 961.I.34–35; *NPA*, p. 15; *Agora* XV 258. 10, 12)
10. Antipatros Oineidos, Pythais of 128/7 (*FD* III.2, no. 27.II.6; *NPA*, p. 17; perhaps related to Antipater Antipatrou honored with statue dedicated by the *koinon* of Amphictyons at Delphi, ca. 130 B.C. [*SEG* XVIII.223])
11. Apollodoros Kekropidos, escort *hippeus* for Pythais of 106/5 (*FD* III.2, no. 28.IV.49; *NPA*, p. 19)
12. Ap[o......o]s Eub[---], Pythais of 106/5 (*FD* III.2, no. 28.III.35; probably same man as equestrian winner, A[---] Eu[---] [*FD* III.2, no. 44.II]; Tracy, *BCH* 99 [1975]: 193 reads *-es* for ending of *nomen*, thereby making the likely restoration Ap[ollonid]es)
13. Apollonides, one of 30 *systratiôtai* for Pythais of 106/5 (*FD* III.2, no. 28.IV.28; *SIG*[3] 711, pp. 354–55; suggested as *hippeis* by Tracy, *HSCP* 83 [1979]: 233, but now questioned)
14. Apollonides II Kratonos, pythaist of the *hippeis* for Pythais of 106/5 and winner of equestrian event *ek pantôn* (*FD* III.2, no. 28.I.5–6; 43.III; son of Kraton Apollonidou, hipparch for Pythais of 106/5 [*FD* III.2, no. 28.I.4–5]; *NPA*, p. 22)
15. Apollonios Aigeidos, Pythais of 106/5 (*FD* III.2, no. 28.I.22; *NPA*, p. 23, suggests that Apollonios Apolloniou of Ankyle, teenager in Theseia of 161/0 [*IG* II[2] 956.II.55–56] may be same man, but this seems very unlikely; perhaps his son?
16. Apollonios Antiochidos or Attalidos, Pythais of 128/27 (*FD* III.2, no. 27.II.34; *NPA*, p. 23)
17. Apollonios (Attalidos), Pythais of 128/7 (*FD* III.2, no. 27.II.40; *NPA*, p. 23; for tribal affiliation, see comments s.v. Anaxikrates)
18. Apollonios Dionysiou Kekropidos, one of 30 *systratiôtai* for Pythais of 106/5 (*FD* III.2, no. 28.IV.13; *NPA* p. 23 for possible identification with ephebe of 128/7 [*IG* II[2] 1960, line 45; *Hesperia* 24 {1955}: 232, line 216]; for identification as a *hippeus*, see remarks s.v. Apollonides)
19. Archestratos Oineidos, Pythais of 128/7 (*FD* III.2, no. 27.II.5; *NPA*, p. 35; possibly to be identified with Archestratos II Archestratou I Acharneus, high religious official on Delos, 115/4: Roussel, *BCH* [1908]: 103)
20. Archytas Kekropidos, Pythais of 128/7 (*FD* III.2, no. 27.II.16; *NPA*, p. 37)

21. Arist[ok]es, one of 30 *systratiôtai* for Pythais of 106/5 (*FD* III.2, no. 28.IV.32; for identification as *hippeus*, see remarks s.v. Apollonides)
22. Ariston, one of 30 *systratiôtai* for Pythais of 106/5 (*FD* III.2, no. 28.IV.34; for identification as *hippeus*, see remarks s.v. Apollonides)
23. Ariston Aiantidos, Pythais of 106/5 (*FD* III.2, no. 28.III.18; *NPA*, p. 31)
24. A[rtem]idoros, one of 30 *systratiôtai* for Pythais of 106/5 (*FD* III.2, no. 28.III.44; for identification as *hippeus*, see remarks s.v. Apollonides)
25. Artemon Antiochidos, Pythais of 106/5 (*FD* III.2, no. 28.III.20; *NPA*, p. 33)
26. Artemon Ptolemaidos, Pythais of 128/7 (*FD* III.2, no. 27.I.35; *NPA*, p. 33)
27. [As]klepiades Akamantidos, Pythais of 128/7 (*FD* III.2, no. 27.I.43; *NPA*, p. 38)
28. Asklepiades Aristoboulou Aiantidos, equestrian winner in Theseia, 157/6 and teenage winner of running event, 161/0 (*IG* II^{2} 957.II.78; 956.I.81; *PA* 2591)
29. Asklepiogenes, one of 30 *systratiôtai* for Pythais of 106/5 (*FD* III.2, no. 28.IV.37; for identification as *hippeus*, see remarks s.v. Apollonides)
30. Athenobios (Dionysiou I of Eupyridai) Leontidos, Pythais of 128/7 (*FD* III.2, no. 27.I.26; his two sons, Dionysios and Niketes, were child pythaists in 128/7 [*FD* III.2, no. 12.II.5–6] and Niketes became a *hippeus* in Pythais of 106/5 [*FD* III.2, no. 28.IV.46]; both were mint magistrates in 98/97 [Mattingly, *JHS* 91 {1971}: 92]; Athenobios had been a teenage winner in Theseia in 149/8 [*IG* II^{2} 958.I.60–62]; for stemma of family, see *NPA*, pp. 6–7)
31. [Ath]enodoros Erechtheidos, Pythais of 106/5 (*FD* III.2, no. 28.I.19; *NPA*, p. 7; perhaps related to Athenodoros D[emet]riou Erechtheidos, teenage winner of running event in ca. 140 (*IG* II^{2} 961.I.46–47)
32. Attalos, one of 30 *systratiôtai* for Pythais of 106/5 (*FD* III.2, no. 28.IV.38; for identification as *hippeus*, see remarks s.v. Apollonides)
33. Autandros Antiochidos or Attalidos, Pythais of 128/7 (*FD* III.2, no. 27.II.33; *NPA*, p. 41)
34. [Bak]chios, one of 30 *systratiôtai* for Pythais of 106/5 (*FD* III.2, no. 28.IV.18; for identification as *hippeus*, see remarks s.v. Apollonides)
35. Basileides Apollophanou Ptolemaidos, Pythais of 106/5 and winner of equestrian event in same Pythais (*FD* III.2, no. 28.I.42; 44.IV; *NPA*, p. 42: his son is probably Basileides II Basileidou I, ephebe in 106/5 [*FD* III.2, no. 25.II.8]; if this is true, it would make Basileides I quite an old winner of the "war-horse" event, but perhaps he married in his early twenties)
36. Boukattes Leontidos, Pythais of 128/7 (*FD* III.2, no. 27.I.31; *NPA*, p. 42, identifies with mint magistrate [*PA* 2905; also M. Thompson, *New Style Silver Coinage*, p. 557 (156/55]; but issue is complicated by [B]oukatte[s L]ykinou, thesmothete in Pythais of 128/7 (no. 3.9): Mattingly [*Historia* 20 {1971}: 38 and n. 78] identifies him as the mint magistrate without linkage to the *hippeus*. It is unlikely that Boukattes would be both thes-

mothete and *hippeus* in the same year, unless the obligations of cavalry service had become negligible by this date; I have argued otherwise, Chapter Six)

37. Boulon Antiochidos, Pythais of 106/5 (*FD* III.2, no. 28.III.22; *NPA*, p. 43, plausibly identifies this man with Boulon Leostratou Palleneus, child pythaist in 128/7 [*FD* III.2, no. 12.IV.5] and thesmothete in 98/7 [*FD* III.2, no. 2, line 11]; see also Tracy, *First Fruits*, p. 194)
38. Buttakos III Buttakou II of Lamptrai, Pythais of 128/7 and winner of equestrian event in same (*FD* III.2, no. 27.I.17; no. 37.I with restoration, p. 290; Tracy, *First Fruits*, p. 194 and n. 2 has now disassociated this Buttakos from Buttakos IV, *hippeus* in 106/5 (*FD* III.2, no. 28.I.16), *contra NPA*, p. 43; this important family with strong Delphian and Delian connections also provides Pyrrhos [s.v.], *hippeus* in 128/7 and nephew of Buttakos III)
39. Buttakos IV Pyrrhou II of Lamptrai, Pythais of 106/5 (*FD* III.2, no. 28.I.16; nephew of Buttakos III Buttakou II; *NPA*, pp. 43–44, with corrections by Tracy, *First Fruits*, p. 194, n. 2 and prosopography)
40. Charin[os Dio]kleous Oineidos, possibly a *hippeus*, winner of *diaulon ek pantôn* at Theseia, 149/8 and ephebic winner in Theseia of 157/6 (*IG* II² 958.II.91–92; 957.II.61; Martin, *Cavaliers*, p. 279; *PA* 15442; for father, Diokles Charinou, winner in equestrian event at Panathenaia of 182/1 or 178/7, see *IG* II² 2314, lines 77–78)
41. Charinos (Lykinou) Leontidos, Pythais of 128/7 and equestrian winner in same (*FD* III.2, no. 27.I.27; no. 40.2; *NPA*, p. 171)
42. Demetrios Aiantidos, Pythais of 106/5 (*FD* III.2, no. 28.III.14; *NPA*, p. 48)
43. Demetrios Akamantidos, Pythais of 106/5 (*FD* III.2, no. 28.II.15; *NPA*, p. 48)
44. Demetrios Antiochidos(?), Pythais of 128/7 (*FD* III.2, no. 27.II.32; *NPA*, p. 48; Tracy, *BCH* 99 [1975]: 193, notes that his name probably marks the break-off point between Antiochis and Attalis [omitted], but probably still safely Antiochis)
45. [D]emetrios Hyllou (Lamptreus?) Erechtheidos, possible *hippeus*, equestrian winner at Theseia, 149/8 and teenage winner of boxing event at Theseia of 157/6 (*IG* II² 958.II.95–96; 957.II.33–34; *PA* 3361. *IG* II² 2332.I.76)
46. [D]emetrios O(a)then Pandionidos, Pythais of 128/7 (*FD* III.2, no. 27.I.24; *NPA*, p. 49)
47. Demokleides [Timok]leous (Erechtheidos), escort *hippeus* for ephebes to Pythais of 98/7 (*FD* III.2, no. 26.IV.15; Colin supplied the restoration of [Philok]leous, but now emended on basis of Timokles Demokleidou [*FD* III.2, no. 24.II.39]; see *NPA*, p. 49, 160 and Tracy, *The Lettering of an Athenian Mason* (Princeton, 1975; *Hesperia Suppl.* 15), 56–58, 7d; for tribal affiliation, see Reinmuth, *Hesperia* 24 [1955]: 231, line 149)

48. Demokles, one of 30 *systratiôtai* for Pythais of 106/5 (*FD* III.2, no. 28.IV.39; for identification as *hippeus*, see remarks s.v. Apollonides)
49. Demokrates Oineidos, Pythais of 106/5 (*FD* III.2, no. 28.II.19; *NPA*, p. 50)
50. Demophi[lo]s Antiochidos(?), Pythais of 128/7 (*FD* III.2, no. 27.II.30; *NPA*, p. 51: probably same man as *PA* 3673: Demophilos II Polykleous of Alopeke, priest of Sarapis, 123/2; see Tracy, *BCH* 99 [1975]: 193, for tribal affiliation)
51. Demoteles Akamantidos, Pythais of 128/7 (*FD* III.2, no. 27.I.41; *NPA*, p. 50)
52. Dexandros Philinou of Anaphlystos, possible *hippeus*, winner *akampion ek pantôn* at Theseia, 161/0 (*IG* II² 956.II.89–90; *PA* 3212; homonymous ancestor belonged to the cavalry in ca. mid-3d century [K 49, 57 with p. 119; probably his grandfather])
53. Dexandros Saty(rou) Aigeidos, Pythais of 106/5 (*FD* III.2, no. 28.I.30; *NPA*, p. 47)
54. Dexiphon Kalliphanous Oineidos, equestrian winner at Theseia, 149/8 and teenage winner at Theseia, 157/6 (*IG* II² 958.II.90–91; 957.I.47–48; *PA* 3240)
55. Diodoros (Attalidos), Pythais of 128/7 (*FD* III.2, no. 27.II.38; *NPA*, p. 55; for tribal affiliation, see comments s.v. Anaxikrates)
56. Dio[dotos] Diodotou Kekropidos, Pythaides of 128/7, 106/5 and winner of equestrian event in Pythais of 128/7 (*FD* III.2, no. 27.II.13; 40.3; 28.II.30; *SIG*³ 697, p. 307, n. 18; *NPA*, p. 54, differentiates between the *hippeus* of 128/7 and that of 106/5; furthermore, Tracy, *BCH* 99 [1975]: 217, n. 12, calls the restoration of Dio[dotos] on basis of no. 40.3 "arbitrary and unsupported"—there are too many other possibilities, e.g., no. 27.II.27, 29, 38, 41; finally, even if correct, the homonymous patronymic makes identification of *hippeus* of 128/7 with 106/5 very hazardous; more likely, father and son)
57. Diogenes, one of 30 *systratiôtai* for Pythais of 106/5 (*FD* III.2, no. 28.III.42; for identification as *hippeus*, see remarks s.v. Apollonides)
58. Diokles, one of 30 *systratiôtai* for Pythais of 106/5 (*FD* III.2, no. 28.IV.16; for identification as *hippeus*, see remarks s.v. Apollonides)
59. Diokles Charinou Oineidos, not attested as *hippeus*, but equestrian winner in Panathenaia, 178/7 (*IG* II² 2314, lines 77–78, and son, Charinos Diokleous, equestrian winner in Theseia, 149/8 [s.v.]; *PA* 4008)
60. Diomenes (Attalidos), Pythais of 128/7 (*FD* III.2, no. 27.II.41; *NPA*, p. 57; for tribal affiliation, see comments s.v. Anaxikrates)
61. Dionysodoros Akamantidos, Pythais of 106/5 (*FD* III.2, no. 28.II.16; *NPA*, p. 63)
62. Dionysodoros Antiochidos, Pythais of 128/7 (*FD* III.2, no. 27.II.27; *NPA*, p. 63)
63. Dionysi[os], one of 30 *systratiôtai* for Pythais of 106/5 (*FD* III.2, no. 28.IV.31; for identification as *hippeus*, see remarks s.v. Apollonides)

64. [Di]onysios Agathokleous (Marathonios) Aiantidos, equestrian winner among horsemen and also in *ek pantôn* category in Theseia of 149/8 (*IG* II² 958.II.88 and 93–94; for additional prosopography, see Raubitschek, *Hesperia* 11 [1942]: 313—*IG* II² 1028.III.116: Dionysios Agathokleous, ephebe in 102/1)
65. Dionysi[os] Antiochidos, Pythais of 128/7 (*FD* III.2, no. 27.II.29; *NPA*, p. 58)
66. Dionysios (II Athenobiou) Leontidos, Pythais of 106/5 (*FD* III.2, no. 28.I.37; *NPA*, p. 59 with stemma of family, pp. 6–7; Dionysios, along with brother Niketes [s.v.], was a child pythaist in 128/7 [*FD* III.2, no. 12.II.5–6]; whole family, brother and father Athenobios [s.v.], served in the cavalry in this period; see Chapter Six)
67. Dionys[ios Diony]siou, escort *hippeus* for ephebes of Pythais of 98/7 (*FD* III.2, no. 26.IV.16; Tracy, *Mason*, pp. 56–58, 7d; possibly to be identified with the ephebe of 106/5 to Delphi [*FD* III.2, no. 25.II.31], but compare no. 17.I.9 and no. 24.I.43; the commonness of the name makes identification hazardous, e.g., Dionysios Dionysiou Prospaltios [*Hesperia* 24 {1955}: 232, line 186], ephebe in 128/7)
68. Dionysios Hippothontis, Pythais of 106/5 (*FD* III.2, no. 28.II.41; *NPA*, p. 59)
69. Dionysios Kekropidos, Pythais of 106/5 (*FD* III.2, no. 28.II.36; *NPA*, p. 59)
70. Dionysios Xeno[---] Hippothontis, Pythais of 106/5 (*FD* III.2, no. 28.II.39; *NPA*, p. 59)
71. Diopeithes Diodorou of Halai, Pythais of 106/5 and *theôros* in 98/7 (*FD* III.2, no. 28.II.28; no. 10.II.4; cf. no. 205, lines i, 12: *NPA*, p. 56 [stemma], now updated by Geagan, *Hesperia* 52 [1983]: 155–61, no. 2 and Lazzarini, *RFIC* 112 [1984]: 330–37; very strong cavalry traditions in family, his two brothers Theo[phil]os [s.v.] and Philanthes [s.v.], also served with him in the Pythais of 106/5 [*FD* III.2, no. 28.II.26–27])
72. Dioskourides Aiantidos, Pythais of 106/5 (*FD* III.2, no. 28.II.43; *NPA* p. 64, suggests possible identification with [Dioskou]rid[es Marathonios], herald of the Areopagos (*FD* III.2, no. 4, line 1)
73. Dositheos Pandionidos, escort *hippeus* for Pythais of 106/5 (*FD* III.2, no. 28.IV.53; *NPA*, p. 67)
74. Dromeas Dromeou (Erchieus) Aigeidos, equestrian winner at Theseia, 157/6 (*IG* II² 957.II.77; *PA* 4560, 4023)
75. Echedemos IV Arketou I of Kydathenaion equestrian winner (multiple) in Theseia, 149/8 (*IG* II² 958.I.67–68; II.81, 83–84; *PA* 6165 for stemma and MacKendrick, *AthArist.*, pp. 44–46, for discussion of this powerful cavalry and hippotrophic family)
76. [Ei]renaios Erechtheidos, Pythais of 106/5 (*FD* III.2, no. 28.I.18; *NPA*, p. 67)
77. Eirenaios Leontidos, Pythais of 106/5 (*FD* III.2, no. 28.I.38; *NPA*, p. 67, identifies him with Eirenaios II Eirenaiou I Skambonides, *ho epi tas pro-*

sodous in 106/5 [*FD* III.2, no. 5, lines 21–22]; see also no. 51, line 3 (proxeny decree), with new fragment, Bommelaer, *BCH Suppl.* 4 [1977]: 139–57, esp. 140ff.)

78. Epikles Leontidos, Pythais of 106/5 (*FD* III.2, no. 28.I.35; *NPA*, p. 69; perhaps same as ephebe in 123/2, Epikles Ep[ikleo]us Kropides [*IG* II² 1006.II.111]; Reinmuth, *Hesperia* 41 [1972]: 186–91 [col. 2, line 136])
79. Ergokles Aigeidos, Pythais of 106/5 (*FD* III.2, no. 28.I.29; *NPA*, p. 71)
80. Eu[.....]n Aigeidos, Pythais of 128/7 (*FD* III.2, no. 27.I.19)
81. Euandros Kall[---] Pandionidos, Pythais of 106/5? (*FD* III.2, no. 9, lines 3–4; identified as a *hippeus* by *NPA*, p. 75, but there is no evidence for this attribution)
82. [E]ubios, one of 30 *systratiôtai* for Pythais of 106/5 (*FD* III.2, no. 28.IV.17; for identification as *hippeus*, see remarks s.v. Apollonides)
83. Eubios Oineidos, Pythais of 128/7 (*FD* III.2, no. 27.II.8; *NPA* p. 75)
84. Eudemos Antiochidos, Pythais of 106/5 (*FD* III.2, no. 28.III.21; *NPA*, p. 75; see Tracy, *First Fruits*, p. 202 for brother, Euktimenos [probably]; if his son is the ephebe in 102/1 [Tracy, *Mason*, p. 39; *IG* II² 1028, line 238], then Eudemos must be at least in his mid-to-late thirties)
85. Eudemos Kekropidos, Pythais of 128/7 and *theôros* for same? (*FD* III.2, no. 27.II.14; no. 8, line 7; *NPA*, pp. 75–76 identifies *hippeus* with *theôros*, [Eud]emos Sarapionos, biological son of Gorgippos; for stemma, see *NPA*, p. 46)
86. Eudoxos Hippothontidos, Pythaides of 128/7, 106/5 (*FD* III.2, no. 27.II.19; 28.II.38; *NPA*, p. 76, identifies as same man and perhaps *cosmêtês* of ephebes in 107/6, Eudoxos Eudoxou Acherdousios [*PA* 5435]; he is then probably the same man as winner of boys' events in Theseia festivals of 157/56 and 149/8 [*IG* II² 957.II.49–50; 958.II.75–76]; despite *NPA*, p. 76 and K, p. 103, n. 60, the repetition of Eudoxos for father and son suggests that the *hippeus* of 106/5 may be the son of the *hippeus* of 128/7 and not identical; furthermore, there is a Eudoxos Eudoxou, ephebe in 106/5 [*FD* III.2, no. 25.III.14], so we must be careful not to assume too much on basis of *nomen* and tribal affiliation)
87. Eukles, one of 30 *systratiôtai* for Pythais of 106/5 (*FD* III.2, no. 28.IV.30; for identification as *hippeus*, see remarks s.v. Apollonides)
88. E[umarei]des (Euphanous of Euonymon) Erechtheidos, Pythais of 128/7 (*FD* III.2, no. 27.I.14; no. 23.II.21, with patronymic; ephebe in Pythais of 138/7 (23.II.21; cf. *IG* II² 2980); *SIG*³ 697, p. 305; *NPA*, p. 79; Tracy, *BCH* 99 [1975]: 192; Habicht, *AM* 97 [1982]: 172–73, line 22 with discussion, p. 181; for brother, fellow *hippeus* in 128/7 and mint magistrate for 113/2, s.v. Alkidamos)
89. Euphanes Timo(n...) Ptolemaidos, Pythais of 106/5 (*FD* III.2, no. 28.I.43; *NPA*, p. 81)
90. Euphemos, one of 30 *systratiôtai* for Pythais of 106/5 (*FD* III.2, no. 28.IV.35; for identification as *hippeus*, see remarks s.v. Apollonides)
91. Eurykleides II Mikionos III of Kephisia, not attested as a *hippeus*, but win-

ner of equestrian event at Panathenaia in 182/1 and member of hippotrophic family (*IG* II² 2314, lines 38–39; *PA* 5966; *NPA*, p. 81; MacKendrick, *AthArist*, pp. 40–43; Mattingly, *Historia* 20 [1971]: 38 with stemma; s.v. cousin, Mikion IV, chariot winner at Panathenaia, 182/1 and 178/7)

92. Glaukos Hippothontidos, Pythais of 128/7 (*FD* III.2, no. 27.II.18; *NPA*, pp. 44–45 identifies with Glaukos II Lysandrou I of Peiraeus, polemarch in 128/7 (*FD* III.2, no. 3, line 5), but the duplication of military offices makes this unlikely—the *hippeus* is probably the son of the polemarch; another son, Lysandros Glaukou, was tarantinarch at Pythais of 106/5 (*FD* III.2, no. 28, line 8); for Glaukos (father or son) and Lysandros as mint magistrates in 127/6, see Mattingly, *JHS* 91 ([1971]: 90; family also has Delian connections)
93. Herakleides Ptolemaidos, Pythais of 128/7 (*FD* III.2, no. 27.I.34; *NPA*, pp. 88–89; perhaps to be identified with Herakleides Dionysiou of Phlya, in catalog of *epimelêtai*, ca. 130–120 B.C.; [*IG* II² 1939, line 66; *PA* 6487]; if so, perhaps also the father of Attinas Herakleidou, child pythaist in 128/7 and ephebe in 119/8 [*FD* III.2, no. 12.II.8; *IG* II² 1008.I.126])
94. [Her]akleitos Aigeidos, Pythais of 106/5 (*FD* III.2, no. 28.I.26; *NPA*, p. 89)
95. Hestiaios Kekropidos, Pythais of 106/5 (*FD* III.2, no. 28.II.33; *NPA*, p. 73)
96. Hieron Hieronos Aigeidos, Pythais of 106/5 (*FD* III.2, no. 28.I.28; *NPA*, p. 100, identifies with Hieron IV Hieronos III of Halai [*PA* 7532: list of contributions in 2d century B.C.; *IG* II² 2334, lines 17–18; 2452, lines 11, 25])
97. Hieronymos Oineidos, Pythais of 106/5 (*FD* III.2, no. 28.II.22; *NPA*, p. 101)
98. Hipparchos Hippothontidos, Pythais of 128/7 (*FD* III.2, no. 27.II.21; *NPA*, p. 101, identifies with Hipparchos Timokleous I Peiraeos [24, lines 4–5] *strategos epi to nautikon* ["general in charge of naval matters"; see Tracy, *First Fruits*, p. 123: concerned with maritime trade and associated with Delos]; this identification seems unlikely as it requires two "military" offices in same year; perhaps an homonymous son [cf. s.v. Glaukos]; a son Timokles Hipparchou won an equestrian event among the *hippeis* in 128/7, but does not appear in the cavalry register for that year—perhaps Hipparchos in list should read Timokles, a stonecutter's error?
99. Iason Pandionidos, Pythais of 106/5 (*FD* III.2, no. 28.I.32; *NPA*, p. 100, identifies with Iason Seleukou Paianieus, ephebe in 133/2 [*PA* 7438]; his cousin Seleukos Iasonos was also a *hippeus* in 106/5 [*FD* III.2, no. 28.I.33; *NPA*, p. 151])
100. K..sipos, one of 30 *systratiôtai* for Pythais of 106/5 (*FD* III.2, no. 28.IV.33; for identification as *hippeus*, see remarks s.v. Apollonides)
101. [Kallias?] Habronos (of Bate) Aigeidos, Pythais of 106/5 (*FD* III.2, no. 28.I.23; perhaps identical with Kallias VI Habronos IV, exegete elected by

the demos for Pythais of 106/5 [*FD* III.2, no. 5, lines 9–10; *NPA*, p. 103]; Tracy [*BCH* 99 {1975}: 206 has calculated 6–7 letters for missing *nomen*, so Kallias would fit, but so would Ophelas and Drakon, attested members of this powerful hippotrophic and cavalry family (esp. in Theseia); for other members participating in Pythais of 106/5 as pythaists of the Eupatridai [*FD* III.2, no. 13, lines 3–5]; for family, see MacKendrick, *AthArist.*, pp. 50–52)

102. Kallias Oineidos, Pythaides of 128/7, 106/5 (*FD* III.2, no. 27.II.10; 28.II.20; *NPA*, p. 102, however, lists him in two separate entries, therefore treating them as different men; if the same man, career would span at least 22 years)

103. Kallisthenes Oineidos, Pythais of 106/5 (*FD* III.2, no. 28.II.21; *NPA*, p. 107)

104. Kallistratos Antiochidos or Attalidos, Pythais of 128/7 (*FD* III.2, no. 27.II.35; *NPA*, p. 107: probably to be identified with Kallistratos Kallistratou, equestrian winner among *tarantinoi*, 128/7 [*FD* III.2, no. 37.3])

105. Kallitheos (Lysiadou of Berenikidai) Ptolemaidos, Pythais of 128/7 and equestrian winner in same (*FD* III.2, no. 27.I.39; no. 38.1 supplying patronymic; also appears as important individual in *IG* II² 1939, line 52 and II² 2452, line 20; see *Hesperia* 18 [1949]: 97 with stemma)

106. Kephisodoros Attalidos, escort *hippeus* for Pythais of 106/5 (*FD* III.2, no. 28.IV.57; *NPA*, p. 109)

107. Keph[is]odoros Kekropidos, Pythais of 106/5 (*FD* III.2, no. 28.II.25; *NPA*, p. 109; perhaps to be identified with Kephisodoros [Demetriou---], ephebe in 128/7 [*IG* II² 1960, line 40 = Reinmuth, *Hesperia* 24 {1955}: 232, line 211] and Kephisodoros Demetriou Aixoneus, *strategos* on Imbros and epimelete of the Peiraieus, ca. 101–100 B.C. [Meritt, *Hesperia* 35 {1967}: 89–91; *IG* II² 2336, lines 103–4 as restored by Dow, *HSCP* 51 {1940}: 119, lines 108–9; *SEG* XXIV.194, lines 41–44); if this is true, he is surely the child pythaist of 138/7 and ephebe at Pythais of 128/7 [*FD* III.2, no. 11, line 19; no. 24.I.20])

108. Kichesias (IV) Kekropidos, Pythais of 128/7 (*FD* III.2, no. 27.II.15; *NPA*, p. 110; a member of an important 3d- and 2d-century family, whose homonymous grandfather was orator in decree of hipparch of 188/7 [Habicht, *AM* 76 {1961}: 130–31]; for latest stemma of family, see Habicht, *Studien*, pp. 194–97, with possibility that Leon, *grammateus* of the cavalry in Pythais of 106/5, was also a member of the family [*FD* III.2, no. 28.III.38–40])

109. Kleitomachos Ptolemaidos, Pythais of 106/5 (*FD* III.2, no. 28.I.45; *NPA*, p. 111: probably same as *PA* 8542 (Kleitomachos [Diognetou] of Phlya—*IG* II² 2336, thesmothete in 95/4; Tracy, *First Fruits*, p. 208, thinks this is "perhaps" correct)

110. Kratermos Aiantidos, Pythais of 106/5 (*FD* III.2 no. 28.II.44; *NPA*, p. 114, identifies Kratermos III Athenadou II of Rhamnous, child pythaist in

128/7 [*FD* III.2, no. 12.I.12] and ephebe in 117/6 [*IG* II² 1009.IV.67]); respectable 2d-century family)

111. Krates, one of 30 *systratiôtai* for Pythais of 106/5 (*FD* III.2, no. 28.IV.36; for identification as *hippeus*, see remarks s.v. Apollonides)

112. Krates Aiantidos, Pythais of 106/5 (*FD* III.2, no. 28.III.13; *NPA*, p. 115, suggests link with Timokrates Kratetos Marathonios, *pompostolos* at end of 2d century [see *NPA*, p. 160])

113. Ktesarchos, one of 30 *systratiôtai* for Pythais of 106/5 (*FD* III.2, no. 28.IV.14; for identification as *hippeus*, see remarks s.v. Apollonides)

114. Ktesikles Akamantidos, Pythais of 106/5 (*FD* III.2, no. 28.II.14; *NPA*, p. 115)

115. L[.....]os Ptolemaidos, escort *hippeus* for Pythais of 106/5 (*FD* III.2, no. 28.IV.47)

116. Leon, *grammateus* of cavalry for Pythais of 106/5 (*FD* III.2, no. 28.III.38–40; *NPA* p. 117; perhaps linked to family of Kichesias [s.v.])

117. Leontichos Archippou Oineidos, equestrian winner at Panathenaia, ca. 166/5 (*IG* II² 2316, lines 38–39; *PA* 9038; perhaps same as *hieropoios*, Leontichos Acharneus, 148/7 B.C. [*IG* II² 1938, line 48])

118. Lyandros Nikogenous I of Philaidai, equestrian winner in Theseia of 161/0 (*IG* II² 956.II.87–88; *PA* 9188; stemma, *PA* 10850; son of Nikogenes Nikonos, agonothete of Theseia of 161/0 and hipparch at Theseia in 157/6 [*IG* II² 956, lines 2, 29, 40–42, I.62–63; 957.I.39–40]; wealthy hippotrophic/cavalry family in 2d century)

119. Lysias III Artemonos II (of Paiania), Pythais of 98/7, equestrian winner among *hippeis*, though not listed (but fragmentary) with horsemen for 98/7 (*FD* III.2, no. 45; no. 26.IV.?; no doubt to be identified with ephebe for 107/6 [*IG* II² 1011.I.106; *PA* 9379]; see also Tracy, *Mason*, p. 59, 7f.; stemma: *NPA*, p. 34)

120. Medeios Hippothontidos, escort *hippeus* for Pythais of 106/5 and child pythaist in 128/7 (*FD* III.2, no. 28.IV.51; no. 12.II.4; *NPA*, p. 127, identifies him with Medeios II Medeiou I Peiraieos; very important family with Delian connections as well—for most recent prosopography, see Tracy, *First Fruits*, p. 210)

121. Meliton, one of 30 *systratiôtai* for Pythais of 106/5 (*FD* III.2, no. 28.IV.29; for identification as *hippeus*, see remarks s.v. Apollonides)

122. [M]enandros Akamantidos, Pythais of 128/7 (*FD* III.2, no. 27.I.42; *NPA*, p. 125)

123. Menekrates (Antiochidos or Attalidos), Pythais of 128/7 (*FD* III.2, no. 27.II.36; *NPA*, p. 126)

124. Menekrates Oineidos, Pythaides of 128/7 and 106/5 (*FD* III.2, no. 27.II.9; no. 28.II.18; *NPA*, p. 126, lists as separate entries; if it is the same man, his cavalry career would have spanned at least 22 years)

125. Menippos, one of 30 *systratiôtai* for Pythais of 106/5 (*FD* III.2, no. 28.IV.27; for identification as *hippeus*, see remarks s.v. Apollonides)

126. Menophilos Lykophronos I (Sounieus) (Attalidos), Pythaides of 128/7

and 106/5; child pythaist for 138/7 (*FD* III.2, no. 27.II.36; no. 28.III.30; no. 11, lines 4–5 along with brother Lykophron; this is first proven member of Attalis tribe in 128/7, all members below this point must belong to Attalis [Tracy, *BCH* 99 {1975}: 193]; important family with Delphian and Delian connections, 150/100 B.C.; if it is the same man for 128/7 and 106/5, his cavalry career would have spanned at least 22 years, see *NPA*, pp. 128–29 with stemma, p. 119)

127. M[eno]phon Attalidos, Pythais of 106/5 (*FD* III.2, no. 28.III.31)

128. Mikion IV Eurykleidou III Erechtheidos, equestrian winner at Panathenaia of 182/1 and 178/7, though not attested as a *hippeus* (*IG* II² 2314, lines 44, 76; powerful family in 3d and 2d century, see MacKendrick, *AthArist*, pp. 40–43 and Mattingly, *Historia* 20 [1971]: 38 with different stemma; for cousin, Eurykleides III, equestrian winner [s.v.])

129. Mnaseas Ptolemaidos, Pythais of 106/5 (*FD* III.2, no. 28.I.44; *NPA* p. 129; there are several men with the name Mnaseas from Ptolemais, but they come from different demes; perhaps the same as Mnaseas Mnaseou Berenikides, hoplite general ca. 80 B.C. [*IG* II² 1039, lines 1–3])

130. Mnason Diodo(rou) Hippothontidos, Pythais of 106/5 (*FD* III.2, no. 28.II.40; *NPA*, p. 130)

131. Mnesitheos, one of 30 *systratiôtai* for Pythais of 106/5 (*FD* III.2, no. 28.IV.15; for identification as *hippeus* see remarks s.v. Apollonides)

132. Mnesitheos III Echedemou III of Kydathenaion, equestrian winner in Panathenaia, 178/7 (*IG* II² 2314, lines 68–69, 73–74; for this famous hippotrophic family, s.v. Echedemos IV)

133. Mnesitheos Ptolemaidos, Pythais of 128/7 (*FD* III.2, no. 27.I.36; *NPA*, p. 130)

134. [Monimos] Ptolemaidos, escort *hippeus* for Pythais of 106/5 (*FD* III.2, no. 28.IV.47; Tracy [*BCH* 99 {1975}: 207] supplies the restoration)

135. Nikanor (Nikanoros of Leukonoion) Leontidos, Pythais of 106/5 and probably child pythaist for 128/7 and ephebe for 117/6 (*FD* III.2, no. 28.I.36; no. 12.I.7; *IG* II² 1009.II.66; *NPA*, p. 133, suggests *PA* 10705, epimelete of Delos [ca. 80 B.C.], as his son, but more probably our man himself)

136. Niketes (Athenobiou Eupyrides) Leontidos, escort *hippeus* for Pythais of 106/5, child pythaist for 128/7, and probably ephebe for 117/6 (*FD* III.2, no. 28.IV.45; no. 12.II.6; *IG* II² 1009.II.65; *NPA*, p. 133 with stemma p. 7 of this cavalry family: Athenobios, father, *hippeus* in Pythais of 128/7 and brother, Dionysios, *hippeus* in Pythais of 106/5)

137. Nikodoros Nikesiou (Phrearrhios) Leontidos, equestrian winner in Theseia of 161/0 and Panathenaia ca. 166 B.C. (*IG* II² 956.II.84–87; II² 2316, lines 36–38; *PA* 10881; *IG* II² 2316.37–38)

138. Nikogenes Lyan(drou) (of Philaidai) Aigeidos, Pythais of 106/5 (*FD* III.2, no. 28.I.27; *NPA*, p. 134; a member of a prominent cavalry family in the 2d century B.C., s.v. Lyandros Nikogenous I, who is probably his father; the stemma at *PA* 10850 needs to be amended to include our man)

139. Nikokrat[es] Oineidos, Pythais of 106/5 (*FD* III.2, no. 28.II.23; *NPA*, p. 135)
140. N[iko]n Antiochidos, Pythais of 106/5 (*FD* III.2, no. 28.III.25; Tracy [*BCH* 99 {1975}; 206] supplies the restoration, but admits that N[eo]n is also possible)
141. Pa[l]aios Attalidos, Pythais of 106/5 (*FD* III.2, no. 28.III.33; *NPA*, p. 140 (Pa[..]aios); Tracy [*BCH* 99 {1975}: 206–7] disputes Colin's restoration—no such name in Attic prosopography)
142. [Pan]takles, one of 30 *systratiôtai* for Pythais of 106/5 (*FD* III.2, no. 28.IV.19; for identification as *hippeus*, see remarks s.v. Apollonides)
143. Phaennos Phaen(nou) Ptolemaidos, Pythais of 106/5 (*FD* III.2, no. 28.I.41; *NPA*, p. 161)
144. Phanoleos Aiantidos, Pythais of 106/5 (*FD* III.2, no. 28.III.16; *NPA*, p. 161: his son is probably Charisandros Phanoleo, child pythaist in 106/5 (*FD* III.2, no. 15.I.18; for prosopography, see Boethius, *Pythais*, p. 169)
145. Phila[nthes] (Diodorou Halaieus) Kekropidos, Pythais of 106/5 and *theôros* at Pythais of 98/7 (*FD* III.2, no. 28.II.27; no. 10, line 3 [along with brothers, Theophilos and Diopeithes, lines 2, 4]; *NPA*, p. 162 with stemma, p. 56; for most recent stemma and discussion, see Geagan, *Hesperia* 52 [1983]: 155–61 and Lazzarini, *RFIC* 112 [1984]: 330–37; all three brothers were *hippeis* in 106/5 [*FD* III.2, no. 28.II.26–28])
146. Philemon Kekropidos, Pythais of 106/5 (*FD* III.2, no. 28.II.29; *NPA*, p. 163, identifies with Philemon II Philemonos I Meliteus, child participant at Hermaia on Delos, ca. 125 B.C. [*ID* 2595, lines 17–18; for brother, Python, and discussion, see Tracy, *First Fruits*, pp. 213–14])
147. Philinos Antiochidos?, Pythais of 128/7 (*FD* III.2, no. 27.II.31; Tracy [*BCH* 99 {1975}: 193] argues for Antiochis, not yet transition to Attalis)
148. Philokrates Polykleitou I Ptolemaidos, multiple equestrian winner at Panathenaia, ca. 166/5 (*IG* II² 2316, lines 17–18, 20–22, 22–24, 24–26, 26–28, 56–58, 60–62, 62–64, 66–68; his brother, Hagnias Polykleitou, was also an equestrian winner, II² 2316, lines 58–60; though Philokrates is not attested as a *hippeus*, his son Polykleitos was in Pythais of 128/7, thus a hippotrophic *and* cavalry family, *contra* Martin, *Cavaliers*, pp. 278–79)
149. Philotades Ptolemaidos, Pythais of 106/5 (*FD* III.2, no. 28.I.46; *NPA*, p. 167)
150. Polykleitos Ptolemaidos, Pythais of 106/5 and probably child pythaist of 128/7 (*FD* III.2, no. 28.I.40; no. 12.III.4; *NPA*, p. 144, identifies him as Polykleitos I Alexandrou I Phlyeus, epimelete of Delos, 98/7; his sons, Alexandros and Nikagoras, were child pythaists in both 106/5 and 98/7 [*FD* III.2, no. 15.III.8–9; no. 17.II.9–10; for discussion, see Tracy, *First Fruits*, pp. 212–13])
151. Polykleitos (II Philokratou) Ptolemaidos, Pythais of 128/27 and child pythaist of 138/7 (*FD* III.2, no. 27.I.37; no. 11, line 10; *NPA*, p. 144; s.v.

Philokrates, his father, for multiple equestrian winner in Panathenaia, ca. 166 B.C.)

152. [P]rokles, one of 30 *systratiôai* for Pythais of 106/5 (*FD* III.2, no. 28.IV.18; for identification as *hippeus*, see remarks s.v. Apollonides)

153. Prokles Kekropidos, Pythais of 106/5 (*FD* III.2, no. 28.II.32; *NPA*, p. 146)

154. Protopha(n)es Heg[---ido]s *phylês*, equestrian winner at Theseia of 157/6 (*IG* II^2 957.II.75–76; *PA* 12330)

155. [P]yrikles Erechtheidos, Pythais of 106/5 (*FD* III.2, no. 28.I.17; *NPA*, p. 148, tentatively identifies as Pyrikles Asopokleous Agrylethen [*PA* 12492=*IG* II^2 2452, line 52] in catalog of noble men, ca. 125 B.C.; his sons, names lost, were pythaists in 98/7 [*FD* III.2, no. 17.II.4–5]; new readings by Tracy, *Mason*, p. 54, lines 43–45, i.e., P[.....]es may reveal one of the sons' names as P[yrikl]es, like his father; for other family members, as pythaists of 98/7 [*FD* III.2, no. 17.II.6–7; *NPA*, p. 40; cf. *IG* II^2 1939, line 18])

156. Pyrrhos III Pyrrhou II of Lamptrai Erechtheidos, Pythais of 128/7 and equestrian winner (*FD* III.2, 27.I.16; no. 38.2; *NPA*, pp. 44–45; his uncle, Buttakos III[s.v.] was *hippeus* in 128/7 and brother, Buttakos IV [s.v.], was *hippeus* in 106/5; one of his sons was child pythaist in 98/7 [*FD* III.2, no. 17.II.19—restored as Buttakos by Tracy, *Mason*, p. 54]; for most recent and full prosopography, see Tracy, *First Fruits*, p. 214)

157. [R]hadamanthos Attinou Antiochidos, Pythais of 128/7, *theôros* in 128/7, and child pythaist in 138/7 (*FD* III.2, no. 27.II.28; no. 11, line 15; no. 8.I.12; no. 47, line 25; *NPA*, p. 148; for father, Attinas, see Meritt, *Hesperia* 17 [1948]: 19, lines 105–6)

158. Rho[dipp]os Attalidos, Pythais of 106/5 (*FD* III.2, no. 18.III.34; *SIG*^{3} 711, p. 355; *NPA*, p. 149)

159. Sarapion II Sarapionos I Meleteus Kekropidos, escort *hippeus* for Pythais of 106/5 (*FD* III.2, no. 28.IV.48; *NPA*, p. 149, identifies with hoplite general of 98/7 [*FD* III.2, no. 6, line 7]; his homonymous son was child pythaist in 106/5 and ephebe in 98/7 [*FD* III.2, no. 15.I.11; no. 26.II.12]; for full prosopography of this man, hoplite general in 102/1 as well as epimelete of Delos in 100/99, see Tracy, *First Fruits*, pp. 215–16)

160. Seleukos Pandionidos, Pythais of 106/5 (*FD* III.2, no. 28.I.33; *NPA*, p. 151 identifies with Seleukos Iasonos Paianieus, child participant at Hermaia on Delos, ca. 125 [*ID* 2595, line 43; *PA* 12623]; his cousin Iason [s.v.] was also *hippeus* in 106/5 [*FD* III.2, no. 28.I.33])

161. Sosibios (of Themakos) Ptolemaidos, Pythais of 128/7 (*FD* III.2, no. 27.I.33; *NPA*, p. 155; his two sons, Apollodoros and Leontichos, were child pythaists in 128/7 [*FD* III.2, no. 12.I.5–6]; and possibly a third son, [Ploutar]chos Sosibiou, was ephebe in 119/8 [*IG* II^2 1008.I.124 as restored by Raubitschek, *Hesperia* 18 (1949): 97, n. 8])

162. Sosigenes Leontidos, Pythais of 128/7 (*FD* III.2, no. 27.I.30; *NPA*, p.

155; perhaps the same as Sosigenes Sosikratous Leontidos, winner of boys' pankration, first level, at Theseia [*IG* II² 957.II.37–38; *PA* 13202])

163. So[st]ratos Attalidos, Pythais of 106/5 (*FD* III.2, no. 28.III.32; *NPA*, p. 156)

164. Sostratos Ptolemaidos, Pythais of 128/7 (*FD* III.2, no. 27.I.38; *NPA*, p. 156)

165. Sotas Antiochidos?, Pythais of 128/7 (*FD* III.2, no. 27.II.26; *NPA*, p. 157; Tracy [*BCH* 99 {1975}: 193] argues for tribal affiliation; perhaps father of Antigonos and Dionysios, child(?) pythaists in 106/5 [*FD* III.2, no. 15.II.20–21])

166. Sotion (Nikanoros) Leontidos, Pythais of 128/7 and equestrian winner at same (*FD* III.2, no. 27.I.29; no. 40.1; *NPA*, p. 157; perhaps son, Sotion II Sotionos I Phrearrhios, ephebe in 107/6 [*IG* II² 1011.II.90] and daughter, Sostrate Sotionos [*FD* III.2, no. 29, line 3])

167. Sthenelos Sthe[nelou] (Meliteus) Kekropidos, *hippeus* in both Pythais of 106/5 and of 98/7 (*FD* III.2, no. 28.II.34; no. 26.IV.14; *NPA*, p. 151; Tracy, *Mason*, pp. 56–58, 7d)

168. Straton Boulonos, equestrian winner *ex pantôn* in Pythais of 128/7, but not in official register with tribal members (*FD* III.2, no. 42; however, if the E[.....]n in no. 27.I.13 could be emended to read S[.....]n, then Straton could be restored with confidence: S[trato]n; his brother Hagias Boulonos was phylarch at Pythais in 128/7 and equestrian winner [s.v. Hagias])

169. Theodotos Aiantidos, Pythais of 106/5 (*FD* III.2, no. 28.III.17; *NPA*, p. 92; perhaps the same as the ephebe, Theodot[os--] Aiantidos, in 123/2 [Reinmuth, *Hesperia* 41 {1972}: 190, line 200])

170. Theopeithes Theophilou (of Besa), equestrian winner at Pythais of 128/7, though not listed with tribal members, and child pythaist in 138/7 (*FD* III.2, no. 41.1; no. 11, line 18; *NPA*, pp. 94–95 with stemma; probably same as in list of noble men, *IG* II² 2452, line 40 [ca. 125 B.C.]; the case of Theopeithes is cited by Colin [*FD* III.2, p. 43] as one in which an Athenian could participate in the equestrian contests of the Pythais, yet not belong to the "classe des cavaliers," i.e., cavalry, as I argue)

171. Th[eophi]los Attalidos, Pythais of 106/5 (*FD* III.2, no. 28.III.37; *NPA*, p. 96)

172. Theo[phil]os Kekropidos, Pythais of 106/5 and *theôros* for 98/7 (*FD* III.2, no. 28.II.26; no. 10.II.2; *NPA*, p. 96, plausibly identifies as Theophilos II Diodorou I of Halai; for stemma of family, see *NPA*, p. 56; for his two brothers, Diopeithes and Philanthes, also *hippeis* for 106/5 and *theôroi* for 98/7, see earlier listing; for discussion and recent prosopography, see Habicht, *AM* 97 [1982]: 172, lines 13–14; Geagan, *Hesperia* 52 [1983]: 155–61; and Lazzarini, *RFIC* 112 [1984]: 330–37)

173. The[ophra]stos Attalidos, Pythais of 106/5 (*FD* III.2, no. 28.III.36; Tracy, *BCH* 99 [1975]: 207)

174. Theophrastos Oineidos, Pythais of 128/7 (*FD* III.2, no. 27.II.11; *NPA*, p. 97, identifies him as Theophrastos Herakleitou of Acharnai, hoplite

general at Pythais of 128/7 [*FD* III.2, no. 24, lines 3–4]; epimelete of Delos in 120s; and winner of ephebic event in Theseia of 149/8 [*IG* II^2 958.I.93–94; II.77–78—men's category]; but I doubt that the *hippeus* and hoplite-general are the same man; T. Sarikakis, *The Hoplite General in Athens, A Prosopography* [Chicago, 1976 = Ph.D. diss., Princeton, 1951], p. 86, apparently concurs)

175. Theoros Hippothontidos, Pythais of 128/7 (*FD* III.2, no. 27.II.20; *NPA*, p. 97; his son is probably Theoros The[oro]u, ephebe in 98/7 [*FD* III.2, no. 26.II.7]; for confirmation that Theoros is a name, not an office, see Tracy, *BCH* 99 [1975]: 192–93)

176. Thrasykles Archikleous (Lakiades) Oineidos, equestrian winner at Panathenaia, ca. 166/5 (*IG* II^2 2316, lines 39–41; *PA* 7324; for stemma of family, see *PA* 2501)

177. Timokles Hipparchou, equestrian winner in Pythais of 128/7, but not recorded in the official register (*FD* III.2, no. 41.3; *NPA*, p. 160, identifies with Timokles II Hipparchou Peiraieos Hippothontidos; if correct, then he was never recorded in the tribal listing of Hippothontis, as there are no lacunae there; however, Sundwall has also identified him with [Timokl]es Hipparchou, child pythaist [no. 12.I.15]—but calls him simply a pythaist, not acknowledging the accepted restoration for this list—which is impossible; he cannot be both, unless Ferguson [*Klio* 9 {1909}: 50, n. 2] is correct that as a boy he rode his father's horse, his father being the "*strategos* for naval affairs" [*FD* III.2, no. 24, lines 4–5]; Daux [*Delphes*, p. 716, n. 2] is probably correct to reject the restoration [Timokl]es Hipparchou in favor of some other *nomen* and posit another son of Hipparchos Timokleous, *strategos*; however, Daux errs in positing a third son, Phontides Hipparchou [*FD* III.2, no. 24.II.44], ephebe in 128/7, because in the ephebic inscription, *IG* II^2 1960.II.49, Phontides Hipparchou belongs to the tribe Kekropis, not Hippothontis [see Reinmuth, *Hesperia* 24 {1955}: 232, line 220], unless the stonemason has himself made an error, as the name appears directly above the tribal heading, Hippothontidos? In any event, all that can be said with certainty is that Timokles Hipparchou belonged to the Athenian cavalry in 128/7—as evidenced by the category *ek tôn hippeôn* in his hippic victory—yet does not appear on the official list; for the problem of Hipparchos, *hippeus* and Hipparchos Timokleous, *strategos*, see earlier listing)

178. Timokrates Akamantidos, Pythais of 106/5 (*FD* III.2, no. 28.II.13; *NPA*, p. 160; perhaps to be identified with the ephebe in 128/7, Timokrates Theodorou Cholargeus [Reinmuth, *Hesperia* 24 {1955}: 232, line 188])

179. [Xan]thippos Hermiou, see different restoration, s.v. [Aga]thippos, equestrian winner in 106/5.

180. Xe[no]n Deme[triou], escort *hippeus* for Pythais of 98/7 (*FD* III.2, no. 26.IV.17; for this restoration in conjunction with *FD* III.2, no. 257 [p. 176], see Tracy, *Mason*, p. 58, lines 83–84)

181. Zenion Aiantidos, Pythais of 106/5 (*FD* III.2, no. 28.III.15; *NPA*, p. 82;

this man probably same as Zenion Marathonios [*PA* 6187]; for brother, Diotimos, see Tracy, *First Fruits*, p. 200 and son, Diodotos, ephebe in 102/1 [*IG* II² 1028.III.112; Tracy, *Mason*, p. 38])

182. (Z)oilos Klean(drou) Leontidos, escort *hippeus* for Pythais of 106/5 (*FD* III.2, no. 28.IV.44; *NPA*, p. 86)

183. Zoilos Timokratou Leontidos, Pythais of 128/7 and equestrian winner in same (*FD* III.2, no. 27.I.28; no. 37.2; probably same man as Zoilos Timokratou Hybades (*PA* 6248 = *IG* II² 3863; *NPA*, p. 87)

184. Z[opy]ros, one of 30 *systratiôtai* for Pythais of 106/5 (*FD* III.2, no. 28.III.43; for identification as *hippeus*, see remarks s.v. Apollonides)

185. Zopyros (Onesandrou Oethen) Oineidos, Pythais of 128/7 and ephebe in Pythais of 138/7 (*FD* III.2, no. 27.II.7; no. 23.I.7; *NPA*, p. 87; probably same man as in list of noble men, ca. 125 B.C. [*IG* II² 2452, line 48; see line 34 for his brother, [--8-]s Onesandrou Oethen])

FIGURE 1. "Dokimasia Painter"

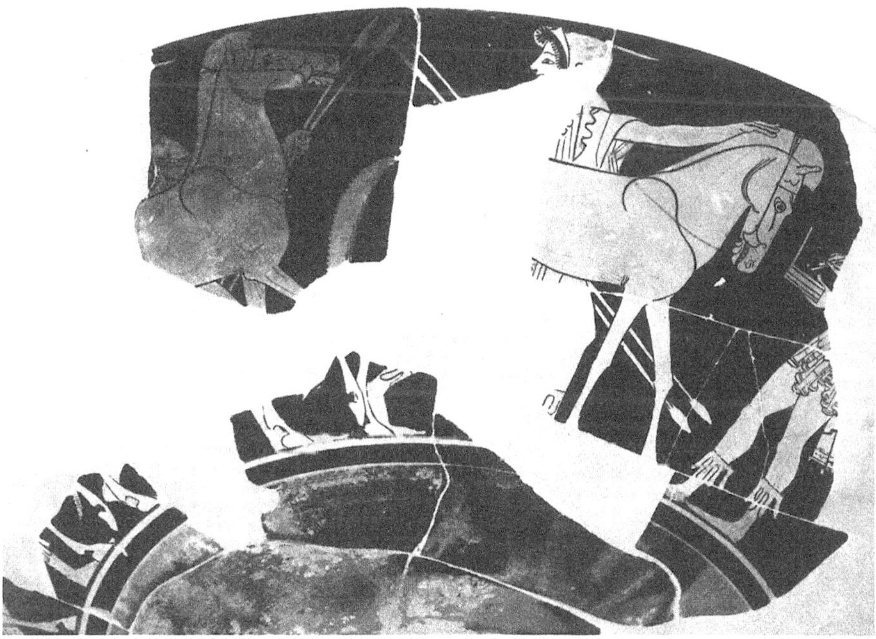

FIGURE 2. "Thalia Painter"

FIGURE 3. Parthenon frieze, W. Slab XII.22–24

a

b

FIGURE 4. Parthenon frieze, (a) W. Slab IX.16–17; (b) W. Slab VI.11–12

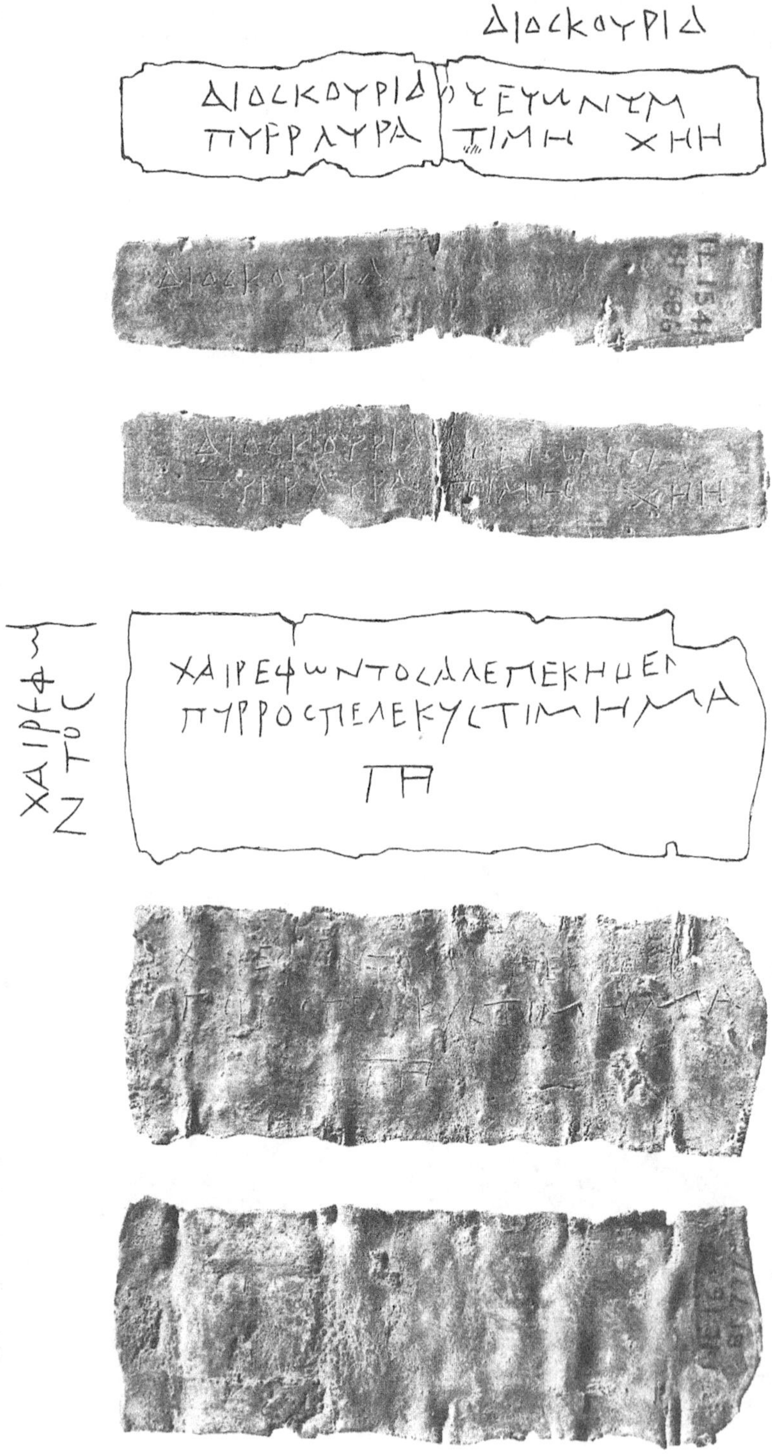

FIGURE 5. Cavalry archives, sample tablets

FIGURE 6. Bryaxis base, one of three victorious phylarchs

FIGURE 7. Leontis *anthippasia* relief

FIGURE 8. Pheidon, Hipparch to Lemnos, clay tokens

FIGURE 10. Eleusis relief, Pythodoros, hipparch

FIGURE 9. Parthenon frieze, W. Slab IV.7–8

b

FIGURE 11. Athenian casualties at Korinth and Koroneia, (a) cavalry; (b) all military forces

FIGURE 12. Dexileos relief

INDEX

GPSR Authorized Representative: Easy Access System Europe - Mustamäe tee 50, 10621 Tallinn, Estonia, gpsr.requests@easproject.com

www.ingramcontent.com/pod-product-compliance
Lightning Source LLC
LaVergne TN
LVHW011117300926
849160LV00008B/420
* 9 7 8 0 6 9 1 6 0 5 7 4 6 *